Checklist of the Hemiptera of Canada and Alaska

Cover illustration and frontispiece: *Parthenicus pallidicollis* Van Duzee

A Publication of the
National Research Council of Canada
Monograph Publishing Program

Checklist of the Hemiptera of Canada and Alaska

H.E.L. Maw, R.G. Foottit, K.G.A. Hamilton

Agriculture and Agri-Food Canada
Eastern Cereal and Oilseed Research Centre
Ottawa, Ontario, Canada
K1A 0C6

and G.G.E. Scudder

Department of Zoology
University of British Columbia
Vancouver, British Columbia, Canada
V6T 1Z4

NRC·CNRC
NRC Research Press
Ottawa 2000

Printed in Canada on acid-free paper. ♾

ISBN 0-660-18165-7
NRC No. 42850

Canadian Cataloguing in Publication Data

Maw, H.E.L. (H. Eric L.)

Checklist of the Hemiptera of Canada and Alaska

Includes bibliographical references and an index.
Issued by the National Research Council of Canada.
ISBN 0-660-18165-7

1. Hemiptera — Canada — Classification. 2. Hemiptera — Alaska — Classification.
3. Homoptera — Canada — Classification. 4. Homoptera — Alaska — Classification.

I. Maw, H.E.L. (H. Eric L.) II. Foottit, R.G. III. Hamilton, K.G.A. IV. Scudder, G.G.E.
V. National Research Council of Canada. VI. Title.

QL522.A1C43 2000 595.7'54'0971 C00-980221-5

Inquiries: Monograph Publishing Program, NRC Research Press, National Research Council of Canada, Ottawa, Ontario K1A 0R6, Canada

Correct citation for this publication: Maw, H.E.L., Foottit, R.G., Hamilton, K.G.A., and Scudder, G.G.E. 2000. *Checklist of the Hemiptera of Canada and Alaska*. NRC Research Press, Ottawa, Ontario, Canada. 220 p.

CONTENTS

Abstract / Résumé

Over 3900 species of Hemiptera (including 'Homoptera') known to occur in Canada and Alaska are listed and their distribution by major political division is given. Arrangement is by taxonomic hierarchy with cross-references from alphabetic lists of genera and species. Annotations identify introduced species and indicate pest status. A brief descriptive synopsis is included for principle taxonomic groups and references to available identification guides (keys, revisions and handbooks), catalogues and bibliographies are provided for genera and higher taxa.

Plus de 3900 les espèces d'hémiptères (y compris les homoptères) signalées au Canada et en Alaska sont recensées dans cette publication. L'information comprend aussi des données de répartition par territoire politique. La liste est présentée selon la hiérarchie taxonomique et ponctuée de renvois à de listes alphabétiques des genres et des espèces. Des annotations rendent compte des espèces étrangères et nuisibles. On trouvera aussi dans cet ouvrage de brefs résumés, dans le cas des principaux groupes taxonomique, et des renvois à des guides d'identification (clefs, révisions, manuels), à des catalogues et à des bibliographies, dans le cas des genres et des taxons supérieurs

Checklist of the Hemiptera of Canada and Alaska

Introduction

This checklist attempts to summarize all records of species of Hemiptera in the broad sense (= 'Rhynchota', *i.e.* including 'Homoptera') found in Canada and Alaska. No new combinations of genus and species names are presented here, even though in the opinion of the authors a number of species are wrongly placed.

Occurrence records are given for the primary political subdivisions of Canada (provinces and territories) and the state of Alaska, except that Labrador is listed separately from insular Newfoundland. This scheme is adequate for most users, but has, for those interested in making biogeographic inferences, the unfortunate consequence that small, biologically and climatically homogenous areas, such as Prince Edward Island, are given the same status as huge heterogenous regions. However, a more equitable treatment, such as subdividing political units based on ecoregions, would have required a much more elaborate format.

The majority of the records are supported by specimens in the Canadian National Collection of Insects, Arachnids and Nematodes (Agriculture and Agri-Food Canada, Research Branch, Ottawa), the Spencer Entomological Collection (University of British Columbia, Vancouver), the Lyman Museum (MacDonald College [MacGill University], Ste-Anne-de-Bellevue, Québec), and the M.E. MacGillivray Collection of Aphids (Agriculture and Agri-Food Canada, Research Branch, Frederiction). Additional records from the literature have been included without differentiation when there is no evidence that they are in error. Doubtful records (as for collections markedly disjunct from the known broader distribution, or older records from the Northwest Territories without specific locality, which may in fact be from the prairie provinces) are marked with a query. Records in the literature known to be in error are generally not addressed, although some identification errors with a significant history in the literature are listed in the species index with cross reference to the corresponding correct identification.

Arrangement of Taxa

Overall arrangement is by taxonomic hierarchy, *i.e.* species are placed under the suborder, infraorder, superfamily, family, subfamily, tribe (if appropriate) and genus to which they belong. Higher taxa are arranged in approximate phylogenetic order. Families, subfamilies and genera are arranged alphabetically within the appropriate higher taxon, subgenera (when defined) alphabetically within genus, and species alphabetically within (sub)genus. An index of species names ordered alphabetically by specific epithet, and an index of generic and subgeneric names, are also provided.

The arrangement and naming of suborders and infraorders follows Sorensen *et al.* (1995). This differs from traditional classifications of the hemipterous groups in that the 'Homoptera' are divided into three separate suborders: Sternorrhyncha (with its traditional scope), Clypeorrhyncha (cicadomorphans) and Archaeorrhyncha (fulgoromorphans). These suborders are coordinate with a fourth suborder, Prosorrhyncha, containing the true bugs, or Heteroptera, plus a small south-temperate group, the Coleorrhyncha.

A brief discussion and summary description is provided for higher taxa. The detail given at a particular taxonomic level varies with the suborder being considered. This is a reflection of the differences in ease of recognition and stability of the groups to the non-specialist, and corresponds in part to previous family-level classifications. Thus, in Sternorrhyncha, Clypeorrhyncha and Archaeorrhyncha, most information is provided at the superfamily level or subordinal level, since these correspond to stable entities (formerly families). On the other hand, the families of Prosorrhyncha have (with some exceptions such as Lygaeidae and related families) been used with their current meaning for some time, while the higher groupings have been subject to recent rearrangements.

Synonyms

No attempt is made to give a complete synonymy for any genus or species. However, common synonyms which have appeared in the recent literature, or for which there is an extensive older literature base (as for some economically important species), are included in the alphabetic indices with cross reference to the appropriate current name. For full synonymy, consult the catalogues cited within relevant discussions of the higher groups (families and above).

Identification Aids

References to useful identification keys, revisions and catalogues are provided for genera and higher taxa with an indication of the geographic area covered.

Format of Checklist Entries

Presence of a species in a geographic unit is indicated by a two-letter abbreviation of that unit's name, while a dash (-) indicates that there are no known collections of the species in that region. The geographic abbreviations, in the order in which they are arranged in the checklist (approximately west to east and north to south) are:

AK - Alaska
YK - Yukon Territory
NT - Northwest Territories
NU - Nunavut
BC - British Columbia
AB - Alberta
SK - Saskatchewan
MB - Manitoba
ON - Ontario
QC - Québec
NB - New Brunswick
PE - Prince Edward Island
NS - Nova Scotia
NF - insular Newfoundland
LB - Labrador

These geographic abbreviations may be accompanied or replaced by modifying symbols:

* - species is believed to be introduced and has become established (this includes species native to regions within the area covered by this checklist, and subsequently introduced to another region).
[ ] - species found in Canada only in greenhouses, conservatories or other indoor situations (sometimes escaping and found temporarily outdoors adjacent to such situations).
{ } - species is believed to have been extirpated from a region, or has become extinct.
? - indicates uncertainty about literature records or the identification of the relevant specimens.
@ - transient occurrences, known only from specimens found on imported produce or horticultural material or packing material, or captured at ports of entry, and apparently not persisting.
$ - specimens known only as interceptions by regulatory agencies at ports of entry.

Additional coded information is given in the right margin if relevant:

A - agricultural field pest
F - forest pest
G - pest in greenhouses (both horticultural and vegetable)
O - pest of ornamental plants, including nuisance pests of urban trees
r - rare or very local in occurrence
b - biological control agent

A summary of these codes in provided as a running footer.

Common names are in accord with the the insect common names published by the Entomological Society of Canada (Belton & Eidt 1993), with a few minor changes (such as replacement of "woolly adelgid" with "adelgid") and some additional names, principally among the Cicadellidae.

Acknowledgments

Dr Yves Bousquet and Dr Henri Goulet (Eastern Cereal and Oilseed Research Centre (ECORC), Agriculture and Agri-Food Canada, Ottawa), Dr Gary M. Stonedahl (Bellingham, Washington), Dr David Voegtlin (Illinois Natural History Survey (INHS), Champaign) reviewed the manuscript as a whole and provided more detailed comments and corrections on the sections falling within their respective areas of expertise. Dr Michael Schwartz (ECORC) gave us valuable insights into the status of various Miridae. However, any errors or omissions are entirely the responsiblity of the authors.

The following persons permitted unfettered access to collections and data under their care: Dr Yvan Pelletier and Dr Gilles Boiteau (M.E. MacGillivray Collection; Potato Research Centre, Agriculture and Agri-Food Canada, Fredericton), Ms Georgette Smith (FIDS collection; Atlantic Forestry Centre, Canadian Forest Service, Fredericton), and Dr Terry Wheeler (Lyman Museum, MacDonald College, Ste-Anne-de-Bellevue).

The cover illustration was produced by Mr Klaus Bolte (Canadian Forest Service, Natural Resources Canada, Ottawa). We wish also to thank Ms Karen McLachlan-Hamilton and Ms Diana Barnes for their assistance in the tedious task of proof-reading.

Development of the checklist was aided by research grants to G.G.E. Scudder from the Natural Sciences and Engineering Research Council of Canada and Forest Renewal British Columbia.

Suborder Sternorrhyncha

Aphids, psyllids, whiteflies and scales. Rostrum arising at the back margin of the head adjacent to, or between, the bases of the forelegs; antennal flagellum with few to many segments, but not bristle-like; tarsi with 1 or 2 segments.

Infraorder Aleyrododea

Superfamily Aleyrodoidea

Key to genera and pest species: Martin 1987 [world].

Catalogue: Mound & Halsey 1978 [world].

Whiteflies. Both male and female whitefly adults are winged and covered by a white wax coating. Immature whiteflies are scale-like. The last larval instar, usually called the pupa, is quiescent. Many produce a distinctive fringe of white or clear wax filaments or a surrounding wax palisade. This group is most diverse in the tropics, and very few species are found in north temperate regions. Other than the few serious greenhouse pests, whiteflies have been very poorly collected in Canada: a few more species should be expected in forest and bog habitats, and the known ranges of the others expanded. The records for the two *Bemisia* species are confounded and only one of the two may actually be present.

The superfamily is considered here to be composed of a single family.

Family ALEYRODIDAE

Subfamily ALEYRODINAE

Genus ALEUROPLATUS Quaintance & Baker

A. berbericolus Quaintance & Baker	-	-	-	-	BC	-	-	-	-	-	-	-	-	-	-
A. epigaeae Russell	-	-	-	-	-	-	-	-	-	-	NB	PE	NS	-	-
A. plumosus (Quaintance)	-	-	-	-	-	-	-	-	-	-	-	-	-	NF	-

Genus ALEYRODES Latreille

A. asumaris Shimer	-	-	-	-	-	-	-	-	ON	QC?	-	-	-	-	-
A. spiraeoides Quaintance iris whitefly	-	-	-	-	BC	-	-	-	-	-	-	-	-	-	-

Genus BEMISIA Quaintance & Baker

B. argentifolii Bellows & Perring	-	-	-	-	-	-	-	-	[ON]	-	-	-	[NS]	-	-	[G]
B. tabaci (Gennadius) sweet potato whitefly	-	-	-	-	-	-	-	-	[ON]	-	-	-	[NS]	-	-	[G]

Genus DIALEURODES Cockerell

D. chitendeni Laing	-	-	-	-	BC*	-	-	-	-	-	-	-	-	-	-

Genus TRIALEURODES Cockerell

T. merlini (Bemis)	-	-	-	-	BC	-	-	-	-	-	-	-	-	-	-	
T. packardi	-	-	-	-	-	-	-	-	ON	QC	-	-	-	-	-	
T. vaporariorum (Westwood) greenhouse whitefly; aleurode des serres	-	-	-	-	BC*	[AB]	[SK]	[MB]	[ON]	[QC]	[NB]	-	[NS]	-	-	[G]

Infraorder Psyllodea

Superfamily Psylloidea

Keys to genera and species: Crawford 1914, Tuthill 1943 [North America].

Catalogue: Hodkinson 1988 [North America; synonymies, distribution and host summaries].

Psyllids, jumping plant lice. Adult psyllids are always winged with well developed venation, and have the hind legs adapted for jumping. Immature psyllids are dorsoventrally flattened; some secrete abundant wax. Females have a well developed valved ovipositor. Parthenogenetic reproduction is very rare. Almost all psyllids are restricted to a single host plant genus or species group.

The psyllid fauna of North America, and of Canada in particular, is poorly known: a disproportionate number of species are known from Alberta and Alaska due to recent collections by Hodkinson. A total of 147 resident species are listed here, compared to the 101 species known and 23 expected given by Scudder (1979).

The classification followed here is that of Burckhardt (1987). In this scheme, Aphalaridae, Liviidae and Spondyliaspididae are included in Psyllidae.

Family CALOPHYIDAE

A small family similar to Psyllidae, but with the break in the anal margin of the forewing distant from the apex of the proximal branch of the cubital vein. The species found in Canada is associated with *Rhus*.

Genus CALOPHYA Löw

C. triozomima Schwarz - - - - - AB SK - ON - - - - - -

Family PSYLLIDAE

Basal vein of forewing bifurcate, with radius diverging from the common basal vein before the separation of media and cubitus; thickening of anal margin of hind wing interrupted adjacent to apex of proximal branch of cubitus (Cu1b). Wingpads of larvae distinctly separated from thoracic terga. Basal tarsal segment of hind leg usually with one or two heavy black saltorial spines.

Generally free living, but feeding by some species causes folding or curling of the leaf of the host plant, and a few (Pachypsyllinae) form closed galls. A few serious pests of apple and pear are included in this family, but most species have little economic impact.

Subfamily ACIZZINAE

Genus ACCIZIA Heslop-Harrison

A. acaciaebaileyanae group - - - - - - - - - $ - - - - -

Subfamily APHALARINAE

Tribe Aphalarini

Genus APHALARA Förster

A. calthae (Linnaeus) complex - - - - - AB SK MB ON QC - - - - -
A. confusa Caldwell - - - - - AB - - - - - - - - -
A. curta Caldwell - - - - - AB - - - - - - - - -
A. dentata Caldwell - - - - - AB - - - - - - - - -
A. loca Caldwell - - - - - AB - MB - - - - - - -

* introduced; [] indoors; { } extirpated; ? uncertain record; @ unestablished adventive; $ port-of-entry interception

Aphalara manitobaensis Caldwell	AK	-	-	-	-	AB	-	MB	-	-	-	-	-	-	-
A. monticola Hodkinson	-	-	-	-	-	AB	-	-	-	-	-	-	-	-	-
A. nigra Caldwell	-	-	-	-	-	-	-	MB	-	-	-	-	-	-	-
A. nubifera Patch	AK	-	-	-	-	-	-	-	-	-	-	-	-	-	-
A. persicariae Caldwell	-	-	-	-	BC	-	-	-	-	-	-	PE?	-	-	-
A. rumicis Malley	-	-	-	-	BC	AB	SK	MB	-	QC	-	-	NS	-	-
A. steironemicola Richards	-	-	-	-	-	-	-	-	ON	-	NB	-	-	-	-

Genus CRASPEDOLEPTA Enderlein

Revision, key to species: Journet & Vickery 1979 [North America].

C. angustipennis (Crawford)	AK	YK	-	-	BC	AB	SK	-	ON	QC	-	-	-	-	-
C. canadensis Journet & Vickery	-	-	-	-	BC	AB	SK	-	ON	-	-	-	-	-	-
C. flavida (Caldwell)	-	-	-	-	-	AB	SK	MB	ON	QC	-	-	-	-	-
C. fumida (Caldwell)	-	-	-	-	BC	AB	SK	-	ON	QC	NB	-	NS	-	-
C. furcata (Caldwell)	AK	-	-	-	BC	AB	SK	MB	ON	QC	NB	-	NS	-	-
C. gutierreziae (Klyver)	-	-	-	-	-	AB?	-	-	-	-	-	-	-	-	-
C. maculipilosa Journet & Vickery	-	-	-	-	-	AB	SK	-	-	-	-	-	-	-	-
C. magna Journet & Vickery	-	-	-	-	-	AB	SK	MB	ON	QC	NB	PE	NS	-	-
C. minuta (Caldwell)	-	-	-	-	-	AB	SK	-	ON	QC	-	-	-	-	-
C. minutissima (Crawford)	-	-	-	-	BC	-	-	-	-	-	-	-	-	-	-
C. nebulosa (Zetterstedt)	AK	YK	NT	-	BC	AB	SK	MB	ON	QC	-	-	-	-	LB
C. nota Journet & Vickery	-	-	-	-	-	AB	-	MB	-	-	-	-	-	-	-
C. ochracea (Provancher)	-	-	-	-	-	-	-	-	-	QC	-	-	-	-	-
C. parvula Journet & Vickery	-	-	-	-	-	AB	SK	-	-	QC	NB	-	-	-	-
C. schwarzi (Ashmead)	AK	-	-	-	BC	AB	-	-	-	-	-	-	-	-	-
C. scurra Journet & Vickery	-	-	-	-	-	-	-	-	-	QC	-	-	NS	-	-
C. sonchi (Förster)	-	-	-	-	-	AB	SK	MB	ON	-	-	-	-	-	-
C. subpunctata (Förster)	AK	YK	-	-	BC	AB	SK	MB	ON	-	-	-	-	-	-
C. vancouverensis (Klyver)	-	-	-	-	BC	AB	-	-	-	-	-	-	-	-	-
C. veaziei (Patch)	-	-	-	-	BC	AB	SK	MB	ON	QC	NB	PE	-	-	-
C. vulgaris (Journet & Vickery)	-	-	-	-	BC	AB	SK	MB	ON	QC	NB	PE	NS	-	-

Subfamily ARYTAININAE

Genus ARYTAINILLA Loginova

A. spartiophila (Förster)	-	-	-	-	BC*	-	-	-	-	-	-	-	-	-	-

Subfamily DIAPHORININAE

Tribe Psyllopseini

Genus PSYLLOPSIS Löw

P. discrepans (Flor)	-	-	-	-	-	-	-	-	ON*	-	-	-	NS*	-	-
P. fraxini (Linnaeus)	-	-	-	-	BC*	-	-	-	ON*	-	-	-	-	-	-
P. fraxinicola (Förster)	-	-	-	-	BC*	-	-	-	ON*	-	-	-	NS*	-	-

Subfamily EURYPHYLLURINAE

Genus NEOPHYLLURA Loginova

N. arbuti (Schwarz)	-	-	-	-	BC	-	-	-	-	-	-	-	-	-	-
N. arctostaphyli (Schwarz)	-	-	-	-	BC	AB	SK	MB	-	-	-	-	-	-	-

Subfamily LIVIINAE

Genus LIVIA Latreille

L. bifasciata Provancher	-	-	-	-	-	-	-	-	-	QC	-	-	-	-	-	
L. caricis Crawford	-	-	NT	-	BC	AB	-	-	-	QC	-	-	-	-	-	
L. coloradensis Crawford	-	-	-	-	-	-	-	-	ON	-	-	-	-	-	-	
L. fimbriata (Heslop-Harrison)	-	-	-	-	-	-	-	-	ON	-	-	-	-	-	-	
L. maculipennis (Fitch)	-	-	-	-	-	-	-	-	ON	QC	-	-	-	-	-	
L. marginata Patch	-	-	NT	-	-	-	-	-	ON	QC	-	-	-	-	-	
L. omissa (Heslop-Harrison)	-	-	-	-	-	-	-	-	ON	-	-	-	-	-	-	
L. salatrix Provancher	-	-	-	-	-	-	-	-	-	-	-	PE	-	-	-	
L. vernalis (Fitch)	-	-	-	-	-	-	-	-	ON	QC	-	-	-	-	-	

Subfamily PACHYPSYLLINAE

Genus PACHYPSYLLA Riley

P. celtidisgemma Riley	-	-	-	-	-	-	-	-	ON	-	-	-	-	-	-	
P. celtidismamma (Fletcher)	-	-	-	-	-	-	-	-	ON	-	-	-	-	-	-	
P. celtidisvesiculum Riley	-	-	-	-	-	-	-	-	ON	-	-	-	-	-	-	

Subfamily PSYLLINAE

Genus CACOPSYLLA Ossiannilsson

C. alaskensis (Ashmead)	AK	-	-	-	-	-	-	MB	-	-	-	-	-	-	-	
C. alba (Crawford)	-	-	-	-	-	AB	-	-	-	-	-	-	-	-	-	
C. americana (Crawford)	-	-	-	-	-	AB	-	-	-	-	-	-	NS	-	-	
C. arctica (Walker)	-	-	-	-	-	-	-	-	ON	-	-	-	-	-	-	
C. breviata (Patch)	AK	-	-	-	-	-	-	-	ON	-	-	-	-	-	-	
C. brunnea (Provancher)	-	-	-	-	-	-	-	-	-	QC	-	-	-	-	-	
C. coryli (Patch)	-	-	-	-	BC	-	-	-	-	-	-	-	-	-	-	
C. dilonchi (Caldwell)	-	-	-	-	-	AB	-	-	-	-	-	-	-	-	-	
C. fatsiae (Jensen)	-	-	-	-	[BC]	-	-	-	-	-	-	-	-	-	-	
C. fibulata (Crawford)	AK	-	-	-	-	AB	-	-	-	-	-	-	-	-	-	
C. groenlandica (Šulc)	-	-	-	-	-	-	-	-	-	-	-	-	-	-	-	
C. haliaceti (Hodkinson)	AK	-	-	-	-	-	-	-	-	-	-	-	-	-	-	
C. hamata (Tuthill)	AK	-	-	-	-	AB	-	-	-	-	-	-	-	-	-	
C. hartigii (Flor)	-	-	-	-	-	AB	-	-	-	-	-	-	NS	-	-	
C. highwoodensis (Hodkinson)	AK	-	-	-	-	AB	-	-	-	-	-	-	-	-	-	
C. jenseni (Hodkinson)	AK	-	-	-	-	-	-	-	-	-	-	-	-	-	-	
C. kananaskensis (Hodkinson)	AK	-	-	-	-	AB	-	-	-	-	-	-	-	-	-	
C. latiforceps (Tuthill)	-	-	-	-	BC?	-	-	-	-	-	-	-	-	-	-	
C. ledi (Flor)	AK	-	-	-	-	-	-	-	-	-	-	-	-	-	-	
C. longiforceps (Hodkinson)	AK	-	-	-	-	-	-	-	-	-	-	-	-	-	-	
C. macleani (Hodkinson)	AK	-	-	-	-	-	-	-	-	-	-	-	-	-	-	
C. magnicauda (Crawford)	-	-	-	-	BC	AB	SK	MB	-	-	-	-	-	-	-	
C. mali (Schmidberger) apple sucker; psylle du pommier	-	-	-	-	-	-	-	-	-	-	NB*	-	NS*	-	-	[A]
C. minor (Crawford)	AK	-	-	-	BC	AB	-	-	-	-	-	-	-	-	-	
C. myrtilli (Wagner)	AK	-	-	-	-	AB	-	-	-	-	-	-	-	-	-	
C. negundinis (Malley) boxelder psyllid; psylle de l'érable négondo	-	-	-	-	-	AB	-	MB	-	-	-	-	-	-	-	
C. palmeni (Löw)	AK	-	-	-	-	-	-	-	-	-	-	-	-	-	-	
C. parallela (Crawford)	-	-	-	-	BC?	-	-	-	-	-	-	-	-	-	-	
C. peregrina (Förster)	-	-	-	-	BC*	-	-	-	-	-	-	-	-	-	-	[O]
C. phlebophyllae (Hodkinson)	AK	-	-	-	-	-	-	-	-	-	-	-	-	-	-	
C. pyricola (Förster) pear psylla; psylle du poirier	-	-	-	-	BC*	-	-	-	ON*	-	NB*	-	NS*	-	-	[A]

* introduced; [] indoors; { } extirpated; ? uncertain record; @ unestablished adventive; $ port-of-entry interception

Cacopsylla rara (Tuthill)	AK	-	-	-	-	AB	-	-	-	-	-	-	-	-	-
C. rufipennis (Hodkinson)	AK	-	-	-	-	AB	-	-	-	-	-	-	-	-	-
C. septentrionalis (Šulc)	-	-	-	-	-	-	-	-	-	-	-	-	-	-	-
C. sinuata (Crawford)	AK	-	-	-	-	AB	-	MB	ON	-	-	-	-	-	-
C. striata (Patch)	AK	-	-	-	BC	AB	-	-	-	QC	-	-	NS	-	-
C. stricklandi (Caldwell)	AK	-	-	-	BC?	AB	-	-	-	-	-	-	-	-	-
C. subspiculata (Hodkinson)	-	-	-	-	-	AB	-	-	-	-	-	-	-	-	-
C. toolikensis (Hodkinson)	AK	-	-	-	-	-	-	-	-	-	-	-	-	-	-
C. tuthilli (Caldwell)	-	-	-	-	-	AB	-	-	-	-	-	-	-	-	-
C. zaecevi (Šulc)	AK	-	-	-	-	-	-	-	-	-	-	-	-	-	-

Genus EUGLYPTONEURA Heslop-Harrison

E. fuscipennis (Crawford)	-	-	-	-	BC	-	-	-	-	-	-	-	-	-	-
E. robusta (Crawford)	-	-	-	-	BC	-	-	-	-	-	-	-	-	-	-

Genus PSYLLA Geoffroy

P. alni (Linnaeus)	-	-	-	-	BC*	AB*	-	-	-	-	-	-	-	-	-
P. betulae (Linnaeus)	-	-	-	-	-	AB*	-	-	-	-	-	-	-	-	-
P. betulaenanae Ossiannilsson	AK	-	-	-	-	-	-	-	-	-	-	-	-	-	-
P. carpinicola Crawford	-	-	-	-	-	-	-	-	-	QC	-	-	NS	-	-
P. caudata Crawford	-	-	-	-	BC	-	-	-	-	-	-	-	-	-	-
P. floccosa Patch	AK	-	-	-	BC	AB	-	MB	ON	QC	-	-	NS	NF	-
P. sanguinea (Provancher)	-	-	-	-	BC	AB	-	MB	ON	QC	-	-	NS	NF	-
P. viridescens (Provancher)	-	-	-	-	BC	-	SK	MB	ON	QC	-	-	NS	-	-

Genus PURSHIVORA Heslop-Harrison

P. pubescens Crawford	-	-	-	-	BC	-	-	-	-	-	-	-	-	-	-

Family TRIOZIDAE

Radius, media and cubitus of forewing diverging from common basal vein at the same point (basal vein trifurcate) . Wingpads of larvae fused with thoracic terga. Basal tarsal segment of hind leg without saltorial spines.

Larvae of many species found within a depression in the host leaf; some cause extensive shoot or leaf distortion, and a few form galls. Most species are of no economic significance, but the potato psylla, *Paratrioza cockerelli* is an important vector of virus diseases of potatoes and tomatoes in the west.

Subfamily TRIOZINAE

Tribe Triozini

Genus NEOTRIOZELLA Crawford

N. pyrifolii (Forbes)	-	-	-	-	-	-	-	-	ON	-	-	-	-	-	-

Genus PARATRIOZA Crawford

P. arbolensis Crawford	-	-	-	-	-	AB	SK	-	-	-	-	-	-	-	-	
P. cockerelli (Šulc) potato (or tomato) psyllid; psylle de la tomate	-	-	-	-	BC	AB	SK	-	-	QC	-	-	-	-	-	[A]

Genus TRIOZA Förster

T. albifrons Crawford	AK	YK	-	-	BC	AB	-	MB	-	-	-	-	-	-	-
T. arctica Hodkinson	AK	-	-	-	-	-	-	-	-	-	-	-	-	-	-
T. atkasookensis Hodkinson	AK	-	-	-	-	-	-	-	-	-	-	-	-	-	-
T. aylmeriae Patch	-	-	-	-	-	-	-	-	ON	QC	-	PE	-	-	-
T. chenopodii Reuter	-	-	-	-	-	-	-	-	-	-	-	PE*	-	-	-
T. forcipula Patch	-	-	-	-	-	-	-	-	-	QC	-	PE	-	-	-

Trioza frontinalis Crawford	-	-	-	-	BC	AB	SK	MB	ON	QC	-	-	-	-	-
T. incerta Tuthill	AK	YK	NT	-	BC	-	-	-	-	-	-	-	-	-	-
T. inversa Tuthill	-	-	-	-	BC	-	-	-	-	-	-	-	-	-	-
T. longicornis Crawford	-	-	-	-	BC	-	-	-	-	-	-	-	-	-	-
T. minuta Crawford	-	-	-	-	-	AB	-	-	-	-	-	-	-	-	-
T. mira Tuthill	-	-	-	-	-	-	SK	-	-	-	-	-	-	-	-
T. obtusa Patch	-	-	-	-	-	AB	-	-	ON	QC	-	-	NS	-	-
T. occidentalis Tuthill	-	-	-	-	BC	-	-	-	-	-	-	-	-	-	-
T. pletschi Tuthill	-	-	-	-	-	AB	-	-	-	-	-	-	-	-	-
T. quadripunctata Crawford	-	-	-	-	BC	AB	-	-	-	-	-	-	-	-	-
T. salicivora Reuter	AK	YK	NT	-	BC	AB	SK	MB	ON	QC	-	-	-	-	-
T. shepherdiae Tuthill	-	-	NT	-	BC	AB	SK	MB	-	QC	-	-	-	-	-
T. stylifera Patch	AK	-	-	-	-	-	-	MB	ON	-	-	-	-	-	-
T. sulcata Crawford	-	-	-	-	-	AB	-	-	-	-	-	-	-	-	-
T. tripunctata (Fitch)	-	-	-	-	-	-	-	-	ON	QC	-	-	-	-	-
T. varians Crawford	AK	-	-	-	BC	AB	-	MB	-	QC	-	-	-	-	-

Infraorder Aphidodea

Key to genera: Foottit & Richards 1993 [Canada].

Bibliographies: Sharma 1969, 1971, 1972, 1977.

Aphids, plant lice; pucerons. The aphids are primarily northern temperate zone insects. The life cycle is characterized by alternation of sexual and asexual reproduction (although the sexual generation has been lost by some species) and alary polymorphism (both winged and wingless forms within a species). About 10% of aphid species show alternation of hosts, with sexual generations and two or more subsequent asexual generations occurring on one host (usually a woody perennial) and remaining asexual generations on other (usually herbaceous) plants. Host specificity is generally fairly high among aphids, most being restricted to a single genus or family of plants, although a few have a very broad range of acceptable food plants. Host-alternating forms generally show strong specificity in their sexual generation, but are often more eclectic in their acceptance of summer hosts.

There are several fairly well defined groups of aphids, but the interrelationships among these groups is poorly understood, and the appropriate taxonomic level to which they should be assigned subject to debate. The classification used here is essentially that of Heie (1980). The major groups are given family ranking. Superfamily Aphidoidea is equivalent to the family Aphididae, or Aphididae + Eriosomatidae of most general entomology texts, the phylloxerans and adelgids being placed in a separate superfamily, Phylloxeroidea. See Heie (1980) and Ilharco & van Harten (1987) for summaries of the history of the higher classification of aphids.

Superfamily Phylloxeroidea

Keys to species: Blackman &Eastop 1994 [world].

Parthenogenetic as well as sexual females lay eggs, and a sclerotized ovipositor may be present. The distal antennal segment does not have a distinct terminal process; primary and secondary antennal rhinaria are not distinguished and are usually absent in larvae and apterous forms; siphunculi never present; ninth tergum not forming a differentiated supra-anal structure.

* introduced; [] indoors; { } extirpated; ? uncertain record; @ unestablished adventive; $ port-of-entry interception

Family ADELGIDAE

List of species: Carter 1971 [world (as appendix to treatment of British species)].

Revision and key to species: Annand 1928 [North America].

Adelgids; conifer woolly aphids. This is a small group with only 50 species world-wide. All species feed on conifers, particularly the foliage or twigs. Many species exhibit complex heteroecious (host-alternating) life cycles, forming shoot galls on the primary host, spruce, and alternating to some other conifer. Several species are important forest pests.

Antennae of apterae with 3-4 segments, of alatae with 5 segments; each flagellar segment of alatae with a single large rhinarium; preapical setae of distal tarsal segments acuminate; pretarsal setae acuminate to weakly capitate; wax glands aggregated into facetted plates; female with well developed sclerotic ovipositor.

Subfamily ADELGINAE

Genus ADELGES Vallot

(Subgenus ADELGES Vallot)

A. laricis Vallot - - - - - - - - ON* QC* NB* - - - -
pale spruce gall adelgid; puceron à galle pâle de l'épinette

(Subgenus CHOLODKOVSKYA Börner)

A. oregonensis Annand - - - - BC - - - - - - - - - -

(Subgenus GILLETTEELLA Börner)

A. cooleyi (Gillette) - YK - - BC AB SK MB - QC* - - NS* - -
Cooley spruce gall adelgid; puceron de l'épinette de Sitka

(Subgenus APHRASTASIA Börner)

A. tsugae Annand - - - - BC* - - - - - - - - - -
hemlock adelgid; puceron lanigère de la pruche

(Subgenus SACCHIPHANTES Curtis)

A. abietis (Linnaeus) - - - - BC* - - - ON* QC* NB* - - - -
eastern spruce gall adelgid; puceron à galle conique de l'épinette

A. lariciatus (Patch) - - - - BC AB SK MB ON QC NB - NS - -
spruce gall adelgid; puceron gallicole de l'épinette

(Subgenus DREYFUSIA Börner)

A. nordmannianae (Eckstein) - - - - BC* - - - - - - - - - -

A. piceae (Ratzeburg) - - - - BC* - - - - QC* NB* PE* NS* NF* - [F]
balsam woolly aphid (or adelgid); puceron lanigère du sapin

Subfamily PINEINAE

Genus PINEUS Shimer

(Subgenus PINEUS Shimer)

P. abietinus Underwood & Balch - - - - BC - - - - - - - - - -

P. cembrae (Cholodkovsky) - - - - BC? - - - - - - - - - - [F]

P. coloradensis (Gillette) - - - - BC AB - - ON - NB - NS - - [F]

P. englemanni Annand - - - - BC - - - - - - - - - - [F]

P. floccus (Patch) - - - - - - - - - - - - NS - -
red spruce adelgid; puceron de l'épinette rouge

P. pineoides (Cholodkovsky) - - - - - - - - - - NB* - NS* - -

P. similis (Gillette) - - - - BC AB SK MB ON - NB - - - - [F]
ragged spruce gall adelgid; puceron à galle allongée de l'épinette

P. strobi (Hartig) - - - - - - - - ON QC NB - NS - -
pine bark adelgid; puceron de l'écorce du pin

P. sylvestris Annand - - - - BC - - - ON - - - - - -

(Subgenus PINEODES Börner)

P. pinifoliae (Fitch) - - - - BC AB - MB ON QC NB - NS NF - [F]
pine leaf adelgid; puceron des aiguilles du pin

Family PHYLLOXERIDAE

Phylloxerans. Members of this small group of minute insects feed on woody deciduous plants, many forming galls on the foliage. The grape phylloxeran is major pest of grape. Other species occasionally cause problems on ornamental oaks and in pecan culture.

The systematics of the family is poorly known. The genus *Phylloxera*, as it is currently used in North America, is somewhat heterogenous and assignment of many of its members to other genera is likely. Scudder (1979) indicated that 6 phylloxerid species were known from Canada based on literature records, but the exact identity and status of three of the species included in that total is uncertain.

Genus DAKTULOSPHAIRA Shimer

D. vitifoliae (Fitch) - - - - BC* - - - ON QC - - - - - - [A]
grape phylloxera; phylloxere de la vigne

Genus PHYLLOXERA Boyer de Fonscolombe

P. caryaecaulis (Fitch) - - - - - - - - ON - - - - - -
P. caryaefoliae Fitch - - - - - - - - ON? - - - - - -
P. caryaeglobuli Walsh - - - - - - - - ON? - - - - - -
P. caryaevenae Fitch - - - - - - - - ON? - - - - - -
P. corticalis Kalthenbach - - - - BC* - - - - - - - - - -
P. glabra (von Heyden) - - - - BC* - - - - - - - - - -

Genus PHYLLOXERINA Börner

P. popularia Pergande - - - - BC AB SK - - QC - - - - -

Superfamily Aphidoidea

Key to genera: Foottit & Richards 1993 [Canada].

Keys to species: Blackman & Eastop 1984 [world, species of agricultural importance], Blackman &Eastop 1994 [world, species on trees].

Regional guides: Gualtieri & McLeod 1994 [southwestern Ontario, alatae], Smith 1992 [southeastern USA, alatae] . Because many of our common aphids are introductions from Europe, the following are also useful: Heie 1980, 1982, 1986, 1992, 1994, 1995 [Scandinavia], Stroyan 1977, 1984 [Britain], Brown 1989 [northern Europe, alatae], Remaudière & Seco 1990 [southern Europe, alatae].

Regional lists: Foottit & Maw 1997 [Yukon], Robinson 1979 [Yukon, Mackenzie district, Alaska], Forbes & Chan 1989a, 1989b, 1991, 1993 [BC], Harper & Bradley 1987 [Alberta], Robinson & Lamb 1991 [Manitoba], Quednau 1966 [Québec]; Burnham (Adams) 1938, MacGillivray 1952, 1955 [Maritime provinces; note: the list of species and host records in the M.E. MacGillivray collection (Pelletier &Boiteau 1986) represents collections mainly from New Brunswick but also from elsewhere in the world], Archibald 1958 [Nova Scotia].

Bibliographies: Börner 1952, Szelegiewicz 1969, Smith 1972 with supplement in Smith & Parron 1978; Sharma 1969, 1971, 1972, 1977.

Catalogues: Eastop & Hille Ris Lambers 1976 [world; list with synonymy], Smith & Parron 1978 [North America; synonymy, type information, distributional summaries], Remaudière & Remaudière 1997 [world].

The aphid fauna of some regions of Canada is very poorly known, while that of others is comparatively well known thanks to the efforts of a few people, notably A.R. Forbes and Cho-Kai Chan in the Vancouver area, A.M. Harper in southern Alberta, A.G. Robinson in Manitoba and Ellen MacGillivray in New Brunswick. We list here 806 species established (plus 7 recorded only from Alaska and 4 unestablished interceptions), a substantial increase from the count given by Scudder (1979) of 650 confirmed and 50 estimated additional species for Canada. An unusually large proportion of these species, about 20%, represent introductions to North America.

Parthenogenetic females give birth to live young, and sclerotized ovipositors are not present. The terminal sensory setae of the antenna are generally borne on a narrow extension (sometimes very long) of the last antennal segment; antennal rhinaria (thin-walled chemosensory areas) are present at the apex of the two distal segments in all forms (including larvae), with additional rhinaria (secondary rhinaria) usually present elsewhere on the antennae of adults only (almost always present in winged forms and in males); siphunculi usually present; ninth abdominal tergum modified into a supra-anal lobe (cauda) (sometimes much reduced, but usually still recognizable).

Family ANOECIIDAE

A small family, with one Holarctic genus of 22 species, and one genus (which may not in fact be closely related) limited to southeast Asia. *Anoecia* species alternate between foliage of *Cornus* species and roots of various herbaceous plants.

Head and prothorax of apterae fused medially; eyes (except triommatidia) absent in immature apterae, present or absent in adult apterae; antennae with 6 segments; terminal process shorter than base of ultimate antennal segment; secondary rhinaria round; basal tarsal segments triangular; apical setae of distal tarsal segments acuminate; pretarsal setae acuminate; lateral thin-walled tubercles present; siphunculi with setose conical bases; anal plate entire; cauda lunate; oviparae and males apterous.

Genus ANOECIA Koch

Key: Halbert 1991 [North America].

A. corni (Fabricius)	-	-	-	-	BC*	-	-	-	ON*	-	-	PE*	-	-	-
A. cornicola (Walsh)	-	-	-	-	-	-	-	MB	-	QC	-	-	-	-	-
A. graminis Gillette & Palmer	-	-	-	-	-	-	-	MB	-	-	-	-	-	-	-
A. oenotherae Wilson	-	-	-	-	-	-	-	MB	ON	-	-	-	-	-	-
A. setariae Gillette & Palmer	-	-	-	-	-	-	-	MB	-	QC	-	PE	-	-	-

Family APHIDIDAE

This is by far the largest family of aphids, and includes most of the economically important species. Host plants are usually angiosperms (among which the Rosaceae and Compositae figure prominently), but there are species which feed on mosses, ferns, horsetails and conifers. Host-alternating species often cause distortion of leaves and new shoots of the primary host (usually a woody plant). Most species are monoecious (non-alternating) and holocyclic (include a sexual stage in the annual life cycle), but many of the pest species are exclusively or facultatively anholocyclic (continuously asexual).

The family is usually divided into two subfamilies, one (Pterocommatinae) small and restricted to Salicaceae, the other (Aphidinae) large and biologically diverse. The Aphidinae currently consist of two tribes, the Aphidini and Macrosiphini, defined by the relative position of spiracles and of lateral tubercles.

One species (*Aphis chloris*) has been introduced for biological control of the weed Saint-Johns wort.

Head and prothorax of apterae separate; eyes of apterae present; antennae of adults with 6 (rarely 4 or 5) segments; secondary rhinaria round; terminal process usually at least as long as basal part of ultimate antennal segment (usually much longer); tibiae usually without rastral strucutures; basal tarsal segments triangular; apical setae of distal tarsal segments acuminate; pretarsal setae acuminate; lateral and dorsal thin-walled tubercles present or absent; wax glands dispersed, not aggregated into discrete structures; siphunculi usually cylindrical or somewhat swollen between base and apex (rarely poriform or absent); anal plate entire; cauda various (crescentic to elongate); oviparous females apterous, males alate or apterous.

Subfamily APHIDINAE

Tribe Aphidini

Genus APHIS Linnaeus

Keys: Rojanavongse & Robinson 1977 [Manitoba], Cook 1984 [on Compositae; North America], Robinson & Chen 1969 [on *Cornus*; North America], Robinson & Rojanavongse 1976 [on Umbelliferae; North America], Robinson & Rojanavongse 1976 [on *Ribes*; North America].

(Subgenus APHIS Linnaeus)

A. armoraciae Cowen	-	-	-	-	BC	AB	-	MB	-	QC	-	-	-	-	-
A. asclepiadis Fitch	-	-	-	-	BC	-	-	MB	-	-	-	PE	-	-	-
A. astragalina H.R.L.	AK	-	-	-	-	AB	-	MB	-	-	-	-	-	-	-
A. callunae Theobald	-	-	-	-	BC*	-	-	-	-	-	-	-	-	-	-
A. ceanothi Clarke	-	-	-	-	BC	-	-	MB	-	-	-	-	-	-	-
A. chani Robinson	-	-	-	-	BC	-	-	-	-	-	-	-	-	-	-

Aphis chloris Koch	-	-	-	-	BC*	-	-	-	-	-	-	-	NS*	-	-	[b]
A. coreopsidis (Thomas)	-	-	-	-	-	-	-	-	ON	QC	NB	-	-	-	-	
A. cornifoliae Fitch	-	-	-	-	-	-	-	MB	ON	QC	NB	-	-	-	-	
A. coweni Palmer	-	-	-	-	BC	-	-	-	-	-	-	-	-	-	-	
A. craccae Linnaeus	-	-	-	-	BC*	-	-	MB*	ON*	QC*	NB*	PE*	NS*	NF*	-	
A. craccivora Koch cowpea aphid; puceron de la gourgane	-	-	-	-	BC*	AB*	SK*	-	ON*	QC*	NB*	-	-	-	-	
A. cytisorum Hartig	-	-	-	-	BC*	-	-	-	-	-	-	-	-	-	-	
A. decepta Hottes & Frison	-	-	-	-	-	-	SK	MB	ON	-	NB	-	-	-	-	
A. duckmountainensis Robinson	-	-	-	-	-	-	-	MB	-	-	-	-	-	-	-	
A. epilobii Kaltenbach	-	-	NT*	-	BC*	-	-	-	-	-	-	-	-	-	-	
A. equiseticola Ossiannilsson	-	-	-	-	BC*	-	-	-	-	-	-	-	-	-	-	
A. fabae Scopoli [sensu lato] black bean aphid; puceron noir des fèvres	-	-	-	-	BC*	AB*	SK*	MB*	ON*	QC*	NB*	PE*	NS*	-	-	[A]
A. farinosa Gmelin	AK	YK	NT	-	BC	AB	SK	MB	ON	QC	NB	-	NS	NF	-	
A. folsomii J.J. Davis	-	-	-	-	-	-	-	-	ON	-	-	-	NS	-	-	
A. forbesi Weed	-	-	-	-	BC	AB	-	MB	ON	-	NB	-	NS	-	-	
A. galiiscabri Schrank	-	-	-	-	BC*	-	-	-	-	-	-	-	-	-	-	
A. genistae Scopoli	-	-	-	-	BC*	-	-	-	-	-	-	-	-	-	-	
A. gossypii Glover melon aphid; puceron du melon	-	-	-	-	BC*	AB*	-	MB*	ON*	QC*	NB*	PE*	-	-	-	[AG]
A. hamamelidis Pepper	-	-	-	-	-	-	-	-	-	-	-	-	NS	-	-	
A. hederae Kaltenbach	-	-	-	-	BC*	-	-	-	ON*	-	-	-	-	-	-	
A. helianthi Monell	AK	YK	-	-	BC	AB	SK	MB	ON	QC	NB	-	NS	-	-	[A]
A. holodisci Robinson	-	-	-	-	BC	-	-	-	-	-	-	-	-	-	-	
A. idaei van der Goot	-	-	-	-	BC*	-	-	-	-	-	-	-	-	-	-	[r]
A. illinoisensis Shimer	-	-	-	-	-	-	-	-	ON	-	-	-	-	-	-	
A. impatientis Thomas	-	-	-	-	-	-	-	-	ON	-	-	-	-	-	-	
A. knowltoni Hottes & Frison	-	-	-	-	-	-	SK	MB	-	-	-	-	-	-	-	
A. lugentis Williams	AK	-	NT	-	BC	-	-	MB	-	-	-	-	-	-	-	
A. lupini Gillette & Palmer	-	-	-	-	BC	AB	-	-	-	-	-	-	-	-	-	
A. maculatae Oestlund spotted poplar aphid; puceron tacheté du peuplier	-	-	NT	-	BC	AB	SK	MB	ON	QC	NB	-	NS	-	-	
A. madderae Robinson	-	-	-	-	-	-	-	MB	-	-	-	-	-	-	-	
A. maidiradicis Forbes	-	-	-	-	-	-	-	-	ON	-	NB	-	-	-	-	[A]
A. manitobensis Robinson & Rojanavongse	-	-	-	-	-	-	-	MB	ON	-	-	-	-	-	-	
A. masoni Richards	AK	YK	NT	NU	-	-	-	MB	-	-	-	-	-	-	-	
A. middletoni Thomas	-	-	-	-	-	-	-	-	-	-	NB	-	-	-	-	
A. mimuli Oestlund	-	YK	-	-	-	-	-	MB	-	QC	-	-	-	-	-	
A. nasturtii Kaltenbach buckthorn aphid; puceron du nerprun	-	-	NT*	-	BC*	AB*	-	MB*	-	QC*	NB*	PE*	NS*	-	-	[A]
A. neilliae Oestlund	-	-	-	-	BC	-	-	-	ON	QC	-	-	-	-	-	
A. neogillettei Palmer	AK	-	-	-	-	-	SK	MB	-	QC	NB	-	NS	NF	-	
A. neomonardae Rojanavongse & Robinson	-	-	-	-	-	-	-	MB	ON	-	-	-	-	-	-	
A. nerii Boyer de Fonscolombe	-	-	-	-	-	-	-	MB*	ON*	QC*	-	-	-	-	-	
A. nigratibialis Robinson	-	-	-	-	BC	-	-	-	-	-	-	-	-	-	-	
A. nivalis H.R.L.	-	-	NT	NU	-	-	-	MB	-	-	-	-	-	-	-	
A. oenotherae Oestlund	AK	-	-	-	BC	AB	SK	MB	ON	QC	NB	-	NS	NF	LB	
A. oestlundi Gillette	-	-	-	-	-	-	-	MB	-	-	NB	-	-	-	-	
A. pomi DeGeer apple aphid; puceron vert du pommier	-	-	-	-	BC*	AB*	-	MB*	ON*	QC*	NB*	PE*	NS*	NF*	-	[A]
A. praeterita Walker	-	-	-	-	BC*	-	-	-	-	QC*	-	-	-	-	-	
A. ribiensis Gillette & Palmer	-	YK	-	-	BC	AB	-	MB	-	QC	-	-	-	-	-	
A. rubicola Oestlund raspberry aphid	AK	YK	-	-	-	-	-	MB	ON	QC	NB	PE	NS	-	-	
A. rubifolii (Thomas)	-	-	-	-	-	-	-	MB	ON	-	NB	PE	NS	-	-	
A. rubiradicis Robinson	-	-	-	-	-	-	SK	-	-	-	-	-	-	-	-	
A. rumicis Linnaeus	-	-	NT*	-	BC*	-	SK*	MB*	ON*	-	NB*	PE*	NS*	-	-	
A. salicariae Koch	AK	-	NT	-	BC	-	-	MB	-	QC	NB	-	-	NF	-	

* introduced; [] indoors; { } extirpated; ? uncertain record; @ unestablished adventive; $ port-of-entry interception

Aphis sambuci Linnaeus	-	-	-	-	BC*	AB*	-	-	ON*	QC*	NB*	-	-	-	-	
A. saniculae Williams	-	-	-	-	-	-	-	MB	-	-	-	-	-	-	-	
A. sedi Kaltenbach	-	-	-	-	BC*	-	-	MB*	ON*	-	NB*	-	-	-	-	
A. spiraecola Patch	-	-	-	-	BC	AB	-	MB	ON	QC	NB	-	-	-	-	[A]
A. spiraephila Patch	-	-	-	-	BC	-	-	MB	ON	QC	NB	-	NS	-	-	
A. taraxacicola (Börner)	-	-	-	-	-	-	-	-	-	QC	-	-	-	-	-	
A. thaspii Oestlund	-	-	-	-	-	-	-	MB	-	-	-	-	-	-	-	
A. valerianae Cowen	-	-	-	-	BC?	-	-	-	-	-	-	-	-	-	-	
A. varians Patch	AK	YK	NT	-	BC	AB	SK	MB	-	QC	NB	-	-	-	-	
A. viburniphila Patch	-	-	-	-	-	-	-	MB	ON	QC	NB	-	NS	-	-	
A. whiteshellensis Robinson	-	-	-	-	-	-	-	MB	-	-	-	-	-	-	-	

(Subgenus ZYXAPHIS Knowlton)

A. canae (Williams)	-	-	-	-	BC	-	-	-	-	-	-	-	-	-	-	
A. chrysothamni (Wilson)	-	-	-	-	BC	-	-	-	-	-	-	-	-	-	-	
A. crypta (Pack & Knowlton)	-	-	-	-	-	-	-	-	-	-	NB?	-	-	-	-	
A. hermistonii (Wilson)	-	-	-	-	BC?	-	-	-	-	-	-	-	-	-	-	
A. oregonensis (Wilson)	-	-	-	-	BC?	-	-	-	-	-	-	-	-	-	-	

Genus ASIPHONAPHIS Wilson

Key: Robinson 1964.

A. pruni Wilson	-	-	-	-	BC	AB	SK	MB	ON	-	-	PE	-	-	-	

Genus HYALOPTERUS Koch

H. pruni (Geoffroy) mealy plum aphid; puceron farineux du prunier	-	-	-	-	BC*	-	SK*	MB*	ON*	-	NB*	-	-	-	-	[A]

Genus HYSTERONEURA J.J. Davis

H. setariae (Thomas)	-	-	-	-	-	-	-	MB	ON	-	-	-	-	-	-	

Genus MISTURAPHIS Robinson

M. shiloensis Robinson	-	-	-	-	-	-	-	MB	-	-	-	-	-	-	-	

Genus RHOPALOSIPHUM Koch

Key: Richards 1960.

R. cerasifoliae (Fitch)	AK	-	-	-	BC	AB	SK	MB	ON	QC	NB	-	NS	-	-	
R. enigmae Hottes & Frison	-	-	-	-	BC	AB	SK	MB	ON	QC	NB	-	-	-	-	
R. insertum (Walker) apple-grain aphid; puceron des graminées	AK*	-	-	-	BC	-	SK	MB	ON	QC	NB	PE	NS	-	-	[A]
R. maidis (Fitch) corn leaf aphid; puceron du maïs	-	-	NT*	-	BC*	AB*	SK*	MB*	ON*	QC*	NB*	PE*	NS*	-	-	[A]
R. musae (Schouteden)	-	-	-	-	BC	-	-	-	-	-	-	-	-	-	-	
R. nigrum Richards	-	YK	-	-	-	-	-	MB	ON	-	NB	-	-	NF	-	
R. nymphaeae (Linnaeus)	-	-	-	-	BC*	-	-	MB*	ON*	QC*	NB*	-	-	-	-	[A]
R. padi (Linnaeus) oat-birdcherry aphid; puceron bicolore des céréales	AK*	-	-	-	BC*	AB*	-	MB*	ON*	QC*	NB*	PE*	NS*	-	-	
R. padiforme Richards	AK	-	-	-	BC	-	-	-	-	-	-	-	-	-	-	
R. rufiabdominale Sasaki	-	-	-	-	BC*	-	-	-	ON*	-	-	-	-	-	-	
R. rufulum Richards	-	-	-	-	-	-	-	MB	ON	-	-	-	-	-	-	

Genus SCHIZAPHIS Börner

Key to species: Eastop 1961.

(Subgenus EUSCHIZAPHIS Hille Ris Lambers)

S. longicornis (Richards)	-	-	-	NU	-	-	-	-	-	-	-	-	-	-	-	
S. palustris (Theobald)	-	-	-	-	BC*	-	-	-	-	QC*	-	-	-	-	-	

(Subgenus PARASCHIZAPHIS Hille Ris Lambers)

S. scirpi (Passerini)	-	-	-	-	-	-	-	-	-	-	NB?	-	-	-	-	
S. scirpicola (H.R.L.)	-	-	-	-	-	-	-	-	-	-	NB	-	NS	-	-	

(Subgenus SCHIZAPHIS Börner)

Schizaphis gracilis Richards	-	-	-	NU	-	-	-	-	-	-	-	-	-	-	-	
S. graminum (Rondani) greenbug; puceron vert des graminées	-	-	NT	-	BC	AB	-	MB	ON	QC	NB	-	NS	-	-	[A]
S. phlei Orlob	-	-	-	-	-	-	-	-	-	-	NB	-	-	-	-	
S. viridirubra (Gillette&Palmer)	-	-	-	-	-	-	-	MB	-	QC	-	-	-	-	-	

Genus TOXOPTERA Koch

T. aurantii (Boyer de Fonscolombe)	-	-	-	-	-	-	-	-	-	@	-	-	-	-	-	

Genus APHTHARGELIA Hottes

A. symphoricarpi (Thomas)	-	-	-	-	BC	AB	SK	MB	-	-	-	-	-	-	-	

Tribe Macrosiphini

Genus ABSTRUSOMYZUS Jensen & Stoetzel

Key to species: Jensen & Stoetzel 1999.

A. phloxae (Sampson)	-	-	-	-	BC	-	-	-	-	QC	-	-	-	-	-	
A. valuliae (Robinson)	-	-	-	-	-	AB	-	MB	-	-	-	-	-	-	-	

Genus ACUTICAUDA Hille Ris Lambers

A. asterensis (Gillette & Palmer)	-	-	-	-	-	-	-	-	-	QC	-	-	-	-	-	

Genus ACYRTHOSIPHON Mordvilko

Key: Eastop 1971 [world], Richards 1972d [Canada].

(Subgenus ACYRTHOSIPHON Mordvilko)

A. assiniboinensis Robinson	AK	YK	NT	-	-	-	-	MB	-	-	-	-	-	-	-	
A. borealis H.R.L.	-	-	-	NU	-	-	-	-	-	QC	-	-	-	-	-	
A. brachysiphon H.R.L.	-	-	-	NU	-	-	-	MB	-	-	-	-	-	-	-	
A. brevicornis H.R.L.	-	-	-	NU	-	-	-	MB	-	-	-	-	-	-	-	
A. caraganae (Cholodkovsky) caragana aphid; puceron du caragan	AK*	YK*	-	-	BC*	AB*	SK*	MB*	ON*	QC*	NB*	-	NS*	-	-	
A. churchillensis Robinson	-	YK	NT	-	-	-	-	MB	-	-	-	-	-	-	-	
A. lactucae (Passerini)	-	-	-	-	BC*	AB*	SK*	MB*	-	QC*	-	-	-	-	-	
A. macrosiphum (Wilson)	-	-	-	-	BC	AB	SK	-	-	-	-	-	-	-	-	
A. malvae malvae (Mosley)	-	-	-	-	BC*	-	-	-	ON*	-	-	PE*	-	-	-	
A. malvae rogersi (Theobald)	-	-	-	-	BC*	-	-	-	-	-	-	-	-	-	-	
A. pisum (Harris) pea aphid; puceron du pois	-	-	-	-	BC*	AB*	SK*	MB*	ON*	QC*	NB*	PE*	NS*	-	-	[A]
A. pseudodirhodus (Patch)	-	-	-	-	BC	-	-	MB	-	-	NB	-	NS	-	-	
A. purshiae (Palmer)	-	-	-	-	BC	-	-	-	-	-	-	-	-	-	-	

Genus ALPHITOAPHIS Hottes

A. lonicericola (Williams)	-	-	-	-	-	-	-	MB	-	-	-	-	-	-	-	

Genus AMPHICERCIDUS Oestlund

A. flocculosus (Gillette & Palmer)	-	-	-	-	-	-	-	-	-	QC	-	-	-	-	-	
A. pulverulens (Gillette)	-	-	-	-	BC	-	SK	-	-	QC	-	-	-	-	-	

Genus AMPHOROPHORA Buckton

A. agathonica Hottes	AK	YK	NT	-	BC	AB	-	MB	ON	QC	NB	PE	NS	-	-	
A. ampullata Buckton	-	-	-	-	-	-	SK	MB	ON	QC	NB	PE	-	NF	-	
A. forbesi Richards	-	-	-	-	BC	-	-	-	-	-	-	-	-	-	-	
A. geranii Gillette & Palmer	-	-	-	-	BC	-	-	-	-	-	-	-	-	-	-	
A. parviflori Hill	-	-	-	-	BC	-	-	-	-	-	-	-	-	-	-	
A. rossi Hottes & Frison	-	-	-	-	BC	-	-	-	-	QC	NB	-	-	-	-	
A. rubitoxica Knowlton	-	-	-	-	BC	-	-	-	-	-	-	-	-	-	-	
A. sensoriata Mason	-	-	-	-	-	-	-	-	-	QC	-	-	-	-	-	

* introduced; [] indoors; { } extirpated; ? uncertain record; @ unestablished adventive; $ port-of-entry interception

Amphorophora stachyophila H.R.L.	-	-	-	-	BC	-	-	-	-	-	-	-	-	-	-	
A. stolonis Robinson	-	-	-	-	-	-	-	MB	-	-	-	-	-	-	-	
A. tigwatensa Hottes	-	-	-	-	BC	-	-	-	-	-	-	-	-	-	-	
A. urtica Essig	-	-	-	-	BC	-	SK	-	-	-	-	-	-	-	-	
Genus ANTHRACOSIPHON Hille Ris Lambers																
A. crystleae C.F. Smith	-	-	-	-	-	-	-	-	-	QC	-	-	NS	-	-	
Genus ASPIDAPHIS Gillette																
A. adjuvans (Walker)	-	-	-	-	-	-	-	MB	-	QC	-	-	-	-	-	
Genus ASPIDOPHORODON Verma																
A. longicaudus (Richards)	-	-	-	-	BC	-	-	-	-	-	-	-	-	-	-	
Genus ATARSOS Gillette																
A. grindeliae Gillette	-	-	-	-	-	-	-	MB	-	-	-	-	-	-	-	
Genus AULACORTHUM Mordvilko Key: Richards 1972c.																
(Subgenus AULACORTHUM Mordvilko)																
A. capilanoense Robinson	-	-	-	-	BC	-	-	-	-	-	-	-	-	-	-	
A. dorsatum (Richards)	-	-	-	-	BC	-	-	-	-	-	-	-	-	-	-	
A. pterinigrum Richards	-	-	-	-	BC	-	-	-	-	-	-	-	-	-	-	
A. solani (Kaltenbach)	AK*	-	NT*	-	BC*	AB*	-	MB*	ON*	QC*	NB*	PE*	NS*	-	-	[AG]
(Subgenus NEOMYZUS van der Goot)																
A. circumflexum (Buckton) foxglove aphid; puceron de la digitale	-	-	-	-	[BC]	[AB]	-	[MB]	[ON]	[QC]	[NB]	-	-	[NF]	-	[G]
Genus BIPERSONA Hottes																
B. ochrocentri (Cockerell)	-	-	-	-	BC	-	SK	-	-	-	-	-	-	-	-	
Genus BRACHYCAUDUS van der Goot																
(Subgenus ACAUDUS van der Goot)																
B. cardui (Linnaeus)	-	-	-	-	BC*	AB*	-	-	ON*	QC*	NB*	-	NS*	NF*	-	
B. persicae (Passerini) black peach aphid; puceron noir du pêcher	-	-	-	-	-	-	-	-	-	-	NB*	-	-	-	-	[A]
B. rociadae (Cockerell)	-	-	-	-	-	-	-	MB	-	-	-	-	NS	-	-	
(Subgenus BRACHYCAUDUS van der Goot)																
B. helichrysi (Kaltenbach)	-	-	-	-	BC*	AB*	-	MB*	ON*	QC*	NB*	-	-	-	-	
(Subgenus THULEAPHIS Hille Ris Lambers)																
B. acaudatus H.R.L.	-	-	-	NU	-	-	-	-	-	-	-	-	-	-	-	
B. rumexicolens (Patch)	-	-	NT	-	BC	-	-	-	-	QC	NB	PE	NS	-	-	
Genus BRACHYCOLUS Buckton																
B. brachysiphon Richards	-	-	-	NU	-	-	-	-	-	-	-	-	-	-	-	
Genus BRACHYCORYNELLA Ajzenberg																
B. asparagi (Mordvilko) asparagus aphid	-	-	-	-	BC*	AB*	-	-	ON*	QC*	-	-	-	-	-	[A]
Genus BREVICORYNE van der Goot																
B. arctica Richards	-	-	-	NU	-	-	-	-	-	-	-	-	-	-	-	
B. brassicae (Linnaeus) cabbage aphid; puceron du chou	-	-	-	-	BC*	AB*	-	MB*	ON*	QC*	NB*	-	NS*	-	-	[A]

Genus CACHRYPHORA Oestlund

Key: Richards 1972b.

C. canadensis H.R.L.	-	-	-	-	-	-	-	-	ON	QC	NB	-	NS	-	-
C. serotinae (Oestlund)	-	-	-	-	-	-	SK	-	-	-	NB	-	-	-	-

Genus CAPITOPHORUS van der Goot

Key: Corpuz-Raros & Cook 1974.

C. carduinus (Walker)	-	-	-	-	BC*	-	-	-	-	-	-	-	-	-	-
C. eleagni (Del Guercio)	AK*	-	NT*	-	BC*	AB*	-	MB*	ON*	QC*	NB*	PE*	NS*	-	-
C. hippophaes (Walker)	-	-	-	-	BC*	-	-	MB*	ON*	QC*	-	-	-	-	-
C. horni Börner	-	-	-	-	BC*	-	-	-	-	-	NB*	-	-	-	-
C. hudsonicus Robinson	AK	-	NT	-	-	-	-	MB	-	-	-	-	-	-	-
C. pakansus Hottes & Frison	-	-	-	-	-	-	-	-	ON	QC	-	-	-	-	-
C. xanthii (Oestlund)	-	-	-	-	-	-	-	MB	-	-	-	-	-	-	-

Genus CAROLINAIA Wilson

Key: Remaudière & Muñoz Viveros1993.

(Subgenus CAROLINAIA Wilson)

C. rhusifoliae (Richards)	-	-	-	-	-	-	-	-	ON	QC	-	-	-	-	-

(Subgenus GLABROMYZUS Richards)

C. howardii (Wilson)	-	-	-	-	-	-	-	MB	-	QC	-	-	NS	-	-
C. quednaui Remaudière & Muñoz Viveros	-	-	-	-	-	-	-	-	-	QC	-	-	-	-	-
C. rhois (Monell)	-	-	-	-	-	-	-	MB	ON	QC	NB	-	-	-	-
C. schlingeri (H.R.L.)	-	-	-	-	BC	-	-	-	-	-	-	-	-	-	-

Genus CATAMERGUS Oestlund

C. fulvae (Oestlund)	-	-	-	-	-	-	-	-	ON	QC	-	-	-	NF	-
C. kickapoo (Hottes & Frison)	-	-	-	-	-	-	-	MB	-	-	NB	-	-	-	-

Genus CAVARIELLA Del Guercio

Key: Stroyan 1969.

(Subgenus CAVARIELLA Del Guercio)

C. aegopodii (Scopoli) carrot-willow aphid; puceron de la carotte	AK*	-	NT*	-	BC*	AB*	-	MB*	-	QC*	NB*	PE*	-	NF*	-
C. archangelicae (Scopoli)	-	-	-	-	BC*	-	-	-	-	-	NB*	-	-	-	-
C. borealis H.R.L.	-	-	-	NU	BC	-	-	-	-	-	-	-	-	-	-
C. digitata H.R.L.	-	-	-	-	BC	-	-	-	-	-	-	-	-	-	-
C. hendersoni Knowlton & C.F. Smith	-	-	-	-	-	-	-	-	ON	-	-	-	-	-	-
C. konoi Takahashi	AK	YK	NT	-	BC	-	-	MB	-	QC	NB	-	-	-	-
C. pastinacae (Linnaeus)	AK*	-	-	-	BC*	AB*	-	MB*	-	QC*	NB*	-	NS*	-	-
C. pseudopustula H.R.L.	-	-	-	-	-	-	-	-	-	QC	-	-	-	-	-
C. salicis (Monell)	-	-	-	-	-	-	SK	MB	-	-	-	-	-	-	-
C. theobaldi (Gillette & Bragg)	-	-	-	-	BC	AB	-	-	ON	QC	NB	-	NS	NF	-

(Subgenus CAVARAIELLIA Heinze)

C. aquatica (Gillette & Bragg)	AK	-	NT	-	-	-	-	MB	-	-	-	-	-	-	-

Genus CEDOAPHIS Oestlund

C. incognita Hottes & Frison	-	-	-	-	BC	-	-	-	-	-	-	-	-	-	-

Genus CERURAPHIS Börner

Key: Chakrabarti & Medda 1985.

C. eriophori (Walker)	-	-	-	-	BC	-	-	-	-	QC	NB	-	NS	NF	-
C. viburnicola (Gillette)	-	-	NT	-	BC	AB	-	MB	ON	QC	NB	-	-	-	-

* introduced; [] indoors; { } extirpated; ? uncertain record; @ unestablished adventive; $ port-of-entry interception

Genus CHAETOSIPHON Mordvilko

Key: Richards1963b.

(Subgenus CHAETOSIPHON Mordvilko)

C. scalare Richards	-	-	-	-	-	-	-	MB	ON	-	-	-	-	-	-

(Subgenus PENTATRICHOPUS Börner)

C. fragaefolii (Cockerell) strawberry aphid; puceron du fraisier	AK	YK	NT	-	BC	-	SK	MB	ON	QC	NB	-	NS	-	-	[A]
C. hottesi Stroyan	-	-	-	-	BC	-	-	-	-	-	-	-	-	-	-	
C. minus (Forbes)	AK	-	-	-	-	-	-	MB	-	QC	NB	-	NS	-	-	[A]
C. potentillae (Walker)	-	-	-	-	BC	-	-	-	-	-	NB	-	-	-	-	
C. tetrarhodum (Walker)	-	-	-	-	BC*	-	-	-	-	-	-	-	-	-	-	[AO]
C. thomasi H.R.L.	-	-	-	-	BC	-	-	-	-	-	NB	PE	NS	-	-	[A]

Genus COLORADOA Wilson

C. abrotani (Koch)	-	-	-	-	-	-	-	-	-	QC*	NB*	-	-	-	-
C. absinthii (Lichtenstein)	-	-	-	-	-	-	SK*	MB*	-	-	-	-	-	-	-
C. achilleae H.R.L.	-	-	-	-	-	-	-	-	ON*	QC*	-	-	-	-	-
C. artemisiae (Del Guercio)	-	-	-	-	BC*	-	SK*	MB*	-	-	-	-	-	-	-
C. rufomaculata (Wilson)	-	-	-	-	-	-	-	-	[ON]	[QC]	[NB]	-	-	-	-
C. tanacetina (Walker)	-	-	-	-	-	-	-	-	-	QC*	-	-	-	-	-

Genus CORYLOBIUM Mordvilko

C. avellanae (Schrank)	-	-	-	-	BC*	-	-	-	-	-	-	-	-	-	-

Genus CRYPTAPHIS Hille Ris Lambers

C. bromi Robinson	-	-	-	-	BC	-	-	MB	-	QC	-	-	-	-	-
C. poae (Hardy)	-	-	-	-	BC*	-	-	MB?	-	-	-	-	-	-	-

Genus CRYPTOMYZUS Oestlund

Key: Guldemond 1991.

C. galeopsidis (Kaltenbach)	-	-	-	-	BC*	-	-	-	-	QC*	NB*	PE*	NS*	NF*	-
C. ribis (Linnaeus) currant aphid; puceron jaune du groseillier	-	-	-	-	BC*	AB*	-	MB*	ON*	QC*	NB*	PE*	NS*	-	-

Genus DECOROSIPHON Börner

D. corynothrix Börner	-	-	-	-	-	-	-	-	-	QC*	NB*	-	-	-	-

Genus DELPHINIOBIUM Mordvilko

D. canadense (Robinson)	-	-	-	-	BC	-	-	-	-	-	-	-	-	-	-

Genus DIURAPHIS Ajzenberg

(Subgenus DIURAPHIS Ajzenberg)

D. mexicana (McVicar Baker)	-	-	-	-	BC	-	-	-	-	-	-	-	-	-	-	
D. noxia (Kurdjumov) Russian wheat aphid	-	-	-	-	BC*	AB*	SK*	-	-	-	-	-	-	-	-	[A]

(Subgenus HOLCAPHIS Hille Ris Lambers)

D. frequens (Walker)	-	-	-	-	BC*	-	-	MB*	ON*	QC*	NB*	-	-	-	-
D. holci (Hardy)	-	-	-	-	-	-	-	-	-	QC*	-	-	-	-	-
D. tritici (Gillette) western wheat aphid; puceron du blé	-	-	-	-	-	AB	SK	MB	-	-	-	-	-	-	-

Genus DYSAPHIS Börner

(Subgenus DYSAPHIS Börner)

D. crataegi (Kaltenbach)	-	-	-	-	-	-	-	-	ON*	QC*	-	-	-	-	-	
D. devecta (Walker)	-	-	-	-	-	-	-	-	-	QC*	-	-	-	-	-	
D. foeniculus (Theobald)	-	-	-	-	-	-	-	-	-	-	-	-	-	NF*	-	
D. tulipae (Boyer de Fonscolombe)	-	-	-	-	BC*	AB*	-	MB*	@	-	-	-	-	-	-	[O]

(Subgenus POMAPHIS Börner)

Dysaphis plantaginea (Passerini) rosy apple aphid; puceron rose du pommier	-	-	-	-	BC*	-	-	-	ON*	QC*	NB*	-	NS*	-	-	[A]
D. sorbi (Kaltenbach)	-	-	-	-	BC*	-	-	-	-	-	-	-	-	-	-	

Genus ELATOBIUM Mordvilko

E. abietinum (Walker) green spruce aphid; puceron de l'épinette	-	-	-	-	BC*	-	SK*	-	-	QC*	-	-	-	-	-	[F]

Genus EOMACROSIPHON Hille Ris Lambers

E. nigromaculosum (MacDougall)	-	-	-	-	BC	AB	-	-	-	-	-	-	-	-	-	[r]

Genus EPAMEIBAPHIS Oestlund

E. frigidae (Oestlund)	-	-	-	-	-	AB	-	MB	-	-	-	-	-	-	-

Genus ERICAPHIS Börner

E. fimbriata Richards	-	YK	NT	-	BC	-	-	-	-	QC	-	-	-	-	-	
E. gentneri (Mason)	-	-	-	-	BC	-	-	-	-	-	-	-	-	-	-	
E. scammelli (Mason)	-	-	-	-	BC	-	-	-	-	-	-	-	NS	-	-	[A]
E. siphunculata (Richards)	AK	-	-	-	BC	-	-	-	-	QC	-	-	-	-	-	
E. wakibae (Hottes)	-	-	-	-	BC	-	-	MB	-	-	-	-	-	-	-	

Genus GLENDENNINGIA MacGillivray

G. philadelphi MacGillivray	-	-	-	-	BC	-	-	-	-	-	-	-	-	-	-

Genus GYPSOAPHIS Oestlund

G. oestlundi Hottes	-	-	-	-	-	AB	-	MB	ON	-	-	-	-	-	-

Genus HAYHURSTIA Del Guercio

H. atriplicis (Linnaeus)	AK*	-	NT*	-	BC*	AB*	-	MB*	ON*	QC*	NB*	PE*	NS*	-	-

Genus HYADAPHIS Kirkaldy

Key: Voegtlin 1984.

H. foeniculi (Passerini)	-	-	-	-	BC*	-	-	MB*	-	QC*	NB*	-	-	-	-	
H. tataricae (Ajzenberg)	-	-	-	-	-	-	-	MB*	ON*	QC*	-	-	-	-	-	[O]

Genus HYALOMYZUS Richards

H. eriobotryae (Tissot)	-	-	-	-	-	-	-	-	ON	QC	NB	-	NS	-	-
H. scrophulariae (Thomas)	-	-	-	-	-	-	-	MB	-	-	NB	-	-	-	-
H. sensoriatus (Mason)	-	-	-	-	-	-	-	-	-	QC	-	-	-	-	-

Genus HYALOPTEROIDES Theobald

H. humilis (Walker)	-	-	-	-	BC*	-	-	-	-	QC*	NB*	-	-	-	-

Genus HYPEROMYZUS Börner

(Subgenus HYPEROMYZUS Börner)

H. lactucae (Linnaeus)	-	-	-	-	BC*	AB*	SK*	MB*	ON*	QC*	NB*	PE*	NS*	-	-
H. pallidus H.R.L.	-	-	-	-	BC*	-	-	MB*	ON*	QC*	NB*	PE*	NS*	-	-

(Subgenus NEONASONOVIA Hille Ris Lambers)

H. inflatus Richards	-	-	-	-	-	-	-	MB	-	-	-	-	-	-	-
H. nabali (Oestlund)	AK	-	NT	-	-	-	-	MB	-	QC	NB	-	NS	NF	-
H. ribiellus (Davis)	-	-	-	-	BC	AB	-	MB	-	-	-	-	-	-	-
H. sandilandicus (Robinson)	-	-	-	-	BC	-	-	MB	-	QC	-	-	-	-	-

(Subgenus HYPEROMYZELLA Hille Ris Lambers)

H. rhinanthi (Schouteden)	-	-	-	-	-	-	-	-	-	QC	NB	-	NS	NF	-

* introduced; [] indoors; { } extirpated; ? uncertain record; @ unestablished adventive; $ port-of-entry interception

Genus IDIOPTERUS J.J. Davis

I. nephrelepidis Davis	-	-	-	-	@	-	-	-	-	-	-	-	-	-	-	

Genus ILLINOIA Wilson

Revision: MacGillivray 1958 [world].

(Subgenus AMPHOROPHORINA MacGillivray)

I. crystleae bartholomewi (Essig)	-	-	-	-	BC	-	-	-	-	-	-	-	-	-	-	
I. crystleae crystleae (C.F. Smith & Knowlton)	-	-	-	-	BC	-	-	-	-	-	-	-	-	-	-	

(Subgenus ILLINOIA Wilson)

I. alni (Mason)	-	-	NT	-	-	-	-	MB	-	QC	NB	-	-	NF	-	
I. andromedae (MacGillivray)	-	-	-	-	-	-	-	-	-	QC	-	-	-	-	-	
I. azaleae azaleae (Mason)	-	-	NT	-	BC	-	-	MB	-	-	-	-	NS	-	-	
I. azaleae rhododendronia (Mason)	-	-	-	-	-	-	-	-	-	-	NB	-	-	-	-	
I. brevitarsis (Gillette & Palmer)	-	-	-	-	-	AB	-	-	-	QC	-	-	-	-	-	
I. canadensis (MacGillivray)	-	-	NT	-	-	-	-	MB	-	QC	NB	-	-	-	-	
I. goldamaryae (Knowlton)	AK	YK	NT	-	-	-	-	MB	ON	QC	NB	-	NS	NF	-	
I. grindeliae palmerae (MacGillivray)	-	-	-	-	-	-	-	MB	-	-	-	-	-	-	-	
I. liriodendri (Monell)	-	-	-	-	BC	-	-	-	ON	-	-	-	-	-	-	
I. morrisoni (Swain)	-	-	-	-	BC	-	-	-	-	-	-	-	-	-	-	
I. pallida (Mason)	-	-	-	-	-	-	-	-	-	-	NB	-	-	-	-	
I. patriciae (Robinson)	-	-	-	-	BC	-	-	-	-	-	-	-	-	-	-	
I. pepperi (MacGillivray)	-	-	-	-	-	-	-	-	-	-	NB	-	-	-	-	
I. richardsi (MacGillivray)	-	-	-	-	BC	-	-	MB	ON	QC	NB	PE	NS	NF	-	
I. simpsoni (MacGillivray)	-	-	-	-	-	-	-	MB	-	QC	-	-	-	-	-	
I. spiraeae (MacGillivray)	-	-	-	-	BC	-	-	-	-	-	-	-	-	-	-	
I. spiraecola (Patch)	-	-	-	-	BC	-	-	MB	-	QC	NB	-	-	-	-	
I. subviridis (MacDougall)	-	-	-	-	BC	-	-	-	-	-	-	-	-	-	-	
I. wahnaga (Hottes)	-	-	-	-	BC	-	SK	MB	-	-	-	-	-	-	-	

(Subgenus MASONAPHIS Hille Ris Lambers)

I. lambersi (MacGillivray)	-	-	-	-	BC	-	-	-	-	-	-	-	-	-	-	
I. magna (H.R.L.)	-	-	-	-	BC	AB	-	-	-	-	-	-	-	-	-	
I. menziesiae (Robinson)	-	-	-	-	BC	AB	-	-	-	-	-	-	-	-	-	
I. paqueti (MacGillivray)	AK	-	NT	-	-	-	-	MB	ON	QC	-	-	-	-	-	
I. pinawae (Robinson)	-	-	-	-	-	-	-	MB	ON	-	-	-	-	NF	-	
I. rhokalaza (Tissot & Pepper)	-	-	-	-	BC	-	-	-	-	-	-	-	-	NF	-	
I. wilhelminae (H.R.L.)	-	-	-	-	-	-	-	-	-	QC	-	-	-	-	-	

(Subgenus OESTLUNDIA Hille Ris Lambers)

I. davidsoni (Mason)	AK	-	-	-	BC	-	-	-	-	-	-	-	-	-	-	
I. maxima (Mason)	-	-	-	-	BC	AB	-	-	-	-	-	-	-	-	-	
I. rubicola (Oestlund)	AK	YK	-	-	BC	AB	-	MB	ON	QC	NB	-	NS	-	-	

Genus JACKSONIA Theobald

J. papillata Theobald	-	-	-	-	BC*	-	-	-	-	-	-	-	-	-	-	

Genus LIOSOMAPHIS Walker

L. berberidis (Kaltenbach)	-	-	-	-	BC*	AB*	-	-	ON*	QC*	NB*	-	NS*	-	-	

Genus LIPAPHIS Mordvilko

L. erysimi (Kaltenbach) turnip aphid; puceron du navet	AK*	-	-	-	BC*	AB*	SK*	MB*	-	QC*	NB*	-	NS*	-	-	[A]

Genus LONGICAUDUS van der Goot

L. trirhodus (Walker)	-	-	-	-	BC*	-	-	-	ON*	QC*	NB*	-	-	-	-	

Genus MACROSIPHONIELLA Del Guercio

Key: Robinson 1987.

M. abrotani (Walker)	-	-	-	-	-	-	-	-	ON*	QC*	NB*	-	NS*	-	-	
M. absinthi (Linnaeus)	-	-	-	-	BC*	-	-	MB*	-	-	NB*	-	-	-	-	
M. artemisiae (Boyer de Fonscolombe)	-	-	-	-	BC*	-	-	-	ON*	QC*	NB*	PE*	NS*	-	-	
M. frigidicola Gillette & Palmer	-	YK	-	-	-	-	-	MB	-	QC	NB	-	-	-	-	
M. leucanthemi (Ferrari)	-	-	-	-	BC*	AB?	-	-	ON*	-	NB*	-	-	-	-	
M. ludovicianae (Oestlund)	-	YK	-	-	BC	AB	-	MB	-	QC	-	-	-	-	-	
M. millefolii (DeGeer)	-	-	-	-	BC*	-	-	-	ON*	QC*	NB*	PE*	NS*	NF*	-	
M. paucisetosa Robinson	-	-	-	-	BC	-	-	MB	-	-	-	-	-	-	-	
M. pennsylvanica (Pepper)	-	-	NT	-	-	-	-	MB	-	QC	NB	-	-	-	-	
M. sanborni (Gillette) chrysanthemum aphid; puceron du chrysanthème	-	-	-	-	[BC]	-	-	-	[ON]	-	-	-	-	-	-	[O]
M. subterranea (Koch)	-	-	-	-	-	-	-	-	ON*	QC*	NB*	-	-	-	-	
M. tanacetaria (Kaltenbach)	-	-	-	-	BC*	AB*	-	MB*	ON*	QC*	NB*	PE*	NS*	-	-	
M. tapuskae (Hottes & Frison)	AK	YK	-	-	-	-	-	MB	-	QC	-	-	-	-	-	
M. usquertensis H.R.L.	-	-	-	-	-	-	-	-	-	QC?	-	-	-	-	-	

Genus MACROSIPHUM Passerini

Key: Palmer 1952 [Rocky Mountain region (with various other genera included in Macrosiphum)], Robinson 1980 [North American species on ferns], Jensen 1998 [North American species with dark siphunculi].

(Subgenus MACROSIPHUM Passerini)

M. adianti (Oestlund)	-	-	-	-	BC	-	-	-	-	QC	-	-	-	-	-	
M. aetheocornum C.F. Smith & Knowlton	-	-	-	-	BC	-	-	-	-	-	-	-	-	-	-	
M. agrimoniellum (Cockerell)	-	-	-	-	-	-	-	-	-	-	NB	-	-	-	-	
M. albifrons Essig	AK	-	NT	-	BC	AB	-	-	ON*	-	NB*	PE*	NS*	NF*	-	
M. audeni MacDougall	-	-	-	-	BC	-	-	-	-	-	-	-	-	-	-	
M. bisensoriatum MacDougall	-	-	-	-	BC	-	-	-	-	-	-	-	-	-	-	
M. californicum (Clarke)	AK	-	NT	-	BC	AB	-	MB	ON	QC	NB	PE	-	-	-	
M. constrictum Patch	AK	-	-	-	-	-	-	-	-	-	-	-	-	-	-	
M. corallorhizae (Cockerell)	-	-	-	-	BC	-	-	-	-	-	-	-	-	-	-	
M. creelii J.J. Davis	-	-	-	-	BC	-	-	-	-	-	-	-	-	-	-	[A]
M. crysopteris Robinson	-	-	-	-	-	-	-	-	-	QC	-	-	-	-	-	
M. daphnidis Börner	-	-	-	-	BC	-	-	MB	-	-	-	-	-	-	-	
M. diervillae Patch	-	-	-	-	-	-	-	MB	ON	QC	NB	-	NS	-	-	
M. equiseti (Holman)	-	-	-	-	BC	-	-	-	-	-	-	-	-	-	-	
M. euphorbiae (Thomas) potato aphid; puceron de la pomme de terre	AK	YK	NT	-	BC	AB	-	MB	ON	QC	NB	PE	NS	NF	-	[AO]
M. funestum (Macchiati)	-	-	-	-	-	-	-	-	-	-	NB?	-	-	-	-	
M. fuscicornis MacDougall	-	-	-	-	BC	-	-	-	-	-	-	-	-	-	-	
M. gaurae (Williams)	-	-	-	-	-	-	-	-	ON	QC	NB	-	-	-	-	
M. geranii (Oestlund)	-	-	-	-	-	-	-	MB	-	-	-	-	-	-	-	
M. hamiltoni Robinson	-	-	-	-	-	-	-	MB	-	-	NB	-	-	-	-	
M. insulare (Pergande)	AK	-	-	-	-	-	-	-	-	-	-	-	-	-	-	
M. jeanae Robinson	-	-	-	-	-	-	-	MB	-	-	-	-	-	-	-	
M. kiowanepus (Hottes)	-	-	-	-	BC	AB	-	MB	-	-	-	-	-	-	-	
M. lambi Robinson	-	-	-	-	-	-	-	-	ON	-	-	-	-	-	-	
M. manitobense Robinson	-	-	NT	-	BC	-	-	MB	-	-	-	-	-	-	-	
M. mentzeliae Wilson	-	-	-	-	BC	-	-	-	-	-	-	-	-	-	-	
M. olmsteadi Robinson	-	-	-	-	-	-	-	-	ON	-	-	-	-	-	-	
M. osmaroniae (Wilson)	-	-	-	-	BC	-	-	-	-	-	-	-	-	-	-	
M. parvifolii Richards	-	-	-	-	BC	-	-	-	-	-	-	-	-	-	-	
M. pseudorosae Patch	-	YK	-	-	BC	-	-	MB	-	-	NB	-	NS	NF	-	
M. ptericolens Patch	-	-	-	-	BC	-	-	MB	ON	QC	NB	-	NS	-	-	
M. pteridis Wilson	-	-	-	-	BC	-	-	-	-	-	-	-	-	-	-	
M. pyrifoliae MacDougall	-	-	-	-	BC	-	-	-	-	-	-	-	-	-	-	
M. rhamni Clarke	-	-	-	-	BC	-	-	-	-	-	-	-	-	-	-	
M. rosae (Linnaeus) rose aphid; puceron du rosier	-	-	-	-	BC*	-	-	MB*	ON*	QC*	NB*	-	NS*	-	-	[O]

* introduced; [] indoors; { } extirpated; ? uncertain record; @ unestablished adventive; $ port-of-entry interception

Macrosiphum stanleyi Wilson	AK	-	-	-	BC	-	-	-	-	-	-	-	-	-	-	
M. stellariae Theobald	-	-	-	-	BC*	-	-	-	-	-	-	-	-	-	-	
M. subarcticum Robinson	AK	YK	NT	-	-	-	SK	MB	-	-	-	-	-	-	-	
M. tenuicauda Bartholomew	-	-	-	-	BC	-	-	-	-	-	-	-	-	-	-	
M. tiliae (Monell)	-	-	-	-	-	-	-	MB	-	-	-	-	-	-	-	
M. tolmiea (Essig)	-	-	-	-	BC	-	-	-	-	-	-	-	-	-	-	
M. valerianae (Clarke)	-	-	-	-	-	AB	-	MB	-	QC	-	-	NS	-	-	
M. venafuscae J.J. Davis	-	-	-	-	-	-	-	-	-	-	NB	-	NS	-	-	
M. walkeri Robinson	-	-	-	-	BC	-	-	-	-	-	-	-	-	-	-	
M. woodsiae Robinson	-	-	NT	-	BC	-	-	MB	-	-	-	-	-	-	-	

(Subgenus NEOCORYLOBIUM MacGillivray)

M. carpinicolens Patch	-	-	-	-	-	-	-	-	ON	-	-	-	-	-	-	
M. coryli J.J. Davis	-	-	-	-	BC	-	-	MB	ON	QC	NB	-	-	-	-	
M. pseudocoryli Patch	-	-	-	-	BC	-	-	MB	ON	-	NB	-	-	-	-	

Genus MASTOPODA Oestlund

M. pteridis Oestlund	-	-	-	-	-	-	-	MB	ON	-	-	-	-	-	-	

Genus METOPOLOPHIUM Mordvilko

M. arctogenicolens (Richards)	-	-	-	NU	-	-	-	-	-	-	-	-	-	-	-	
M. caudatum (Pergande)	AK	-	-	-	-	-	-	-	-	-	-	-	-	-	-	
M. dirhodum (Walker) rose-grass aphid	-	-	-	-	BC*	AB*	-	MB*	-	QC*	NB*	PE*	NS*	-	-	
M. pedicularis Richards	-	-	-	NU	-	-	-	-	-	-	-	-	-	-	-	

Genus MICROLOPHIUM Mordvilko

M. carnosum (Buckton)	-	-	-	-	BC*	AB*	-	MB*	-	-	NB*	PE*	-	-	-	

Genus MICROPARSUS Patch

M. variabilis Patch	-	-	-	-	-	-	-	-	ON	-	-	-	-	-	-	

Genus MICROSIPHONIELLA Hille Ris Lambers

M. artemisiae (Gillette)	-	-	-	-	BC	-	-	MB	-	-	-	-	-	-	-	
M. oregonensis (Wilson)	-	-	-	-	BC	-	-	-	-	-	-	-	-	-	-	

Genus MUSCAPHIS Börner

Key: Remaudière & Muñoz Viveros 1985.

M. canadensis (H.R.L.)	-	-	-	-	-	-	-	-	ON	-	NB	-	-	-	-	
M. escherichi Börner	-	-	-	-	BC	-	-	MB	ON	-	NB	-	-	-	-	
M. musci Börner	-	-	-	-	-	-	-	-	-	QC	-	-	-	-	-	
M. utahensis C.F. Smith & Knowlton	-	-	-	-	-	-	-	MB	-	-	NB?	-	-	-	-	

Genus MYZAPHIS van der Goot

M. canadensis Richards	-	-	-	-	-	-	-	-	ON	-	-	-	-	-	-	
M. rosarum (Kaltenbach)	AK	YK	NT	-	BC	AB	-	MB	ON	-	NB	-	NS	-	-	[O]

Genus MYZODIUM Börner

Key: Smith & Robinson 1975.

M. knowltoni C.F. Smith & Robinson	-	-	-	-	BC	-	-	-	-	-	-	-	-	-	-	
M. modestum (Hottes)	-	-	-	-	BC	-	-	MB	-	QC	NB	-	-	-	-	

Genus MYZUS Passerini

Key: Blackman 1987.

(Subgenus MYZUS Passerini)

M. cerasi (Fabricius) black cherry aphid; puceron noir du cerisier	-	-	-	-	BC	AB	SK	MB	ON	QC	NB	PE	NS	NF	-	[A]
M. lythri (Schrank)	-	-	-	-	BC*	-	-	MB*	ON*	QC*	NB*	PE*	NS*	-	-	
M. ornatus Laing	-	-	-	-	BC*	-	-	MB*	-	QC*	NB*	-	-	-	-	[G]

(Subgenus NECTAROSIPHON Schouteden)

Myzus antirrhinii (Macchiati)	-	-	-	-	BC*	-	-	-	-	-	-	-	-	-	-	
M. ascalonicus (Schouteden) shallot aphid	-	-	-	-	BC*	-	-	-	-	-	NB*	-	-	-	-	[G]
M. certus (Walker)	-	-	-	-	BC*	-	-	MB*	-	-	NB*	-	-	-	-	
M. ligustri (Mosely)	-	-	-	-	BC*	-	-	-	ON*	-	-	-	-	-	-	
M. nicotianae Blackman	-	-	-	-	BC*	-	-	-	ON*	-	-	-	-	-	-	[A]
M. persicae (Sulzer) green peach aphid; puceron vert du pêcher	[AK]	-	-	-	BC*	AB*	SK*	MB*	ON*	QC*	NB*	PE*	NS*	NF*	-	[AOG]
M. polaris H.R.L.	-	-	-	NU	BC	-	-	-	-	QC	-	-	-	-	-	

Genus NASONOVIA Mordvilko

Revision: Heie 1979.

(Subgenus NASONOVIA Mordvilko)

N. compositellae nigra (H.R.L.)	-	-	-	-	BC*	-	-	-	-	-	NB*	-	-	-	-
N. ribisnigri Mosley	-	-	-	-	BC*	-	-	-	ON*	QC*	NB*	-	NS*	-	-

(Subgenus EOKAKIMIA Heie)

N. wahinkae robinsoni Richards	AK	YK	-	-	-	AB	SK	MB	-	-	-	-	-	-	-
N. wahinkae wahinkae (Hottes)	-	-	-	-	BC	AB	-	-	-	-	-	-	-	-	-

(Subgenus KAKIMIA Hottes & Frison)

N. alpina (Gillette & Palmer)	-	-	-	-	BC	AB	SK	-	-	-	-	-	-	-	-
N. borealis Heie	-	-	-	-	BC	AB	-	MB	ON	-	-	-	-	-	-
N. castelleiae (Sampson)	AK	YK	NT	-	BC	AB	-	MB	-	-	-	-	-	-	-
N. cynosbati (Oestlund)	AK	YK	NT	-	BC	AB	-	MB	-	-	-	-	-	-	-
N. essigi (Gillette & Palmer)	-	-	-	-	BC	AB	-	MB	-	-	NB	-	-	-	-
N. grossa Heie	-	-	-	-	BC	-	-	-	-	-	-	-	-	-	-
N. heucherae (Thomas)	-	-	-	-	BC	-	-	-	-	-	-	-	-	-	-
N. houghtonensis houghtonensis (Troop)	-	-	-	-	-	-	-	-	-	-	NB	-	-	-	-
N. houghtonensis similis Heie	-	-	-	-	BC	AB	SK	MB	ON	-	-	-	-	-	-
N. muesebecki (Knowlton & Allen)	-	-	-	-	BC	-	-	-	-	-	-	-	-	-	-
N. polemonii (Gillette & Palmer)	-	-	-	-	BC	-	-	-	-	-	-	-	-	-	-
N. sampsoni Heie	-	-	-	-	-	AB	-	-	-	-	-	-	-	-	-
N. takala (Hottes)	-	-	-	-	BC	-	-	-	-	-	-	-	-	-	-
N. vockerothi (Richards)	-	-	-	NU	-	-	-	MB	-	-	-	-	-	-	-
N. williamsi Heie	-	-	-	-	BC	-	-	MB	-	-	-	-	-	-	-

(Subgenus RANAKIMIA Heie)

N. acyrthosiphon (Richards)	-	-	-	-	BC	-	-	-	-	QC	-	-	-	-	-
N. purpurescens (Oestlund)	-	-	-	-	-	-	-	MB	ON	QC	NB	-	NS	NF	-

Genus NEARCTAPHIS Shaposhnikov

Key: Robinson 1984.

N. bakeri (Cowen) clover aphid; puceron du trèfle	AK*	-	-	-	BC	AB	-	MB	ON	QC	NB	-	NS	-	-	[A]
N. californica H.R.L.	-	-	-	-	BC	-	-	-	-	-	-	-	-	-	-	
N. clydesmithi H.R.L.	-	-	-	-	-	-	-	MB	-	-	-	-	-	-	-	
N. crataegifoliae crataegifoliae (Fitch)	-	-	-	-	BC	-	-	MB	ON	QC	NB	PE	NS	-	-	
N. crataegifoliae occidentalis H.R.L.	-	-	-	-	BC	-	-	-	-	-	-	-	-	-	-	
N. sclerosa (Richards)	-	-	-	-	BC	-	-	-	-	-	-	-	-	-	-	
N. sensoriata (Gillette & Bragg)	-	-	-	-	BC	AB	-	MB	-	-	-	-	-	-	-	
N. yohoensis Bradley	AK	-	-	-	BC	AB	-	MB	-	-	-	-	-	-	-	

Genus NEOAMPHOROPHORA Mason

N. kalmiae Mason	-	-	-	-	-	-	-	-	-	QC	NB	-	NS	NF	-

Genus NEOTOXOPTERA Theobald

N. formosana (Takahashi)	-	-	-	-	-	-	@	-	-	@	-	-	-	-	-

* introduced; [] indoors; { } extirpated; ? uncertain record; @ unestablished adventive; $ port-of-entry interception

Genus OBTUSICAUDA Soliman

Key: Robinson & Halbert 1989.

(Subgenus ARTEMISAPHIS Knowlton & Roberts)

O. artemisicola (Williams)	-	-	-	-	BC	AB	SK	-	-	-	-	-	-	-	-

(Subgenus OBTUSICAUDA Soliman)

O. coweni (Hunter)	-	-	-	-	BC	-	-	-	-	-	-	-	-	-	-
O. filifoliae (Gillette & Palmer)	-	-	-	-	BC	-	-	-	-	-	-	-	-	-	-
O. frigidae (Oestlund)	-	-	-	-	BC	AB	-	MB	-	-	NB?	-	NS?	-	-

Genus OVATUS van der Goot

O. crataegarius (Walker)	-	-	-	-	BC*	AB*	-	-	-	QC*	NB*	-	NS*	NF?	-

Genus PADUCIA Hottes & Frison

P. antennata (Patch)	-	-	-	-	-	-	-	-	ON	-	NB	PE	-	-	-

Genus PHORODON Passerini

P. humuli (Schrank) hop aphid; puceron du houblon	-	-	-	-	BC*	-	-	MB*	ON*	QC*	NB*	-	-	-	-

Genus PLEOTRICHOPHORUS Börner

Key: Corpuz-Raros & Cook 1974.

P. amsinckii Richards	-	-	-	-	BC	-	-	-	-	-	-	-	-	-	-
P. asterifoliae (Strom)	-	-	-	-	-	-	-	MB	-	QC	-	-	-	-	-
P. chrysanthemi (Theobald)	-	-	-	-	BC	-	-	-	-	-	-	-	-	-	-
P. glandulosus (Kaltenbach)	-	-	-	-	-	AB*	-	-	-	QC*	NB*	-	-	-	-
P. gnaphalodes (Palmer)	-	-	-	-	BC	-	-	MB	-	QC	-	-	-	-	-
P. hottesi H.R.L.	-	-	-	-	-	-	-	MB	-	-	-	-	-	-	-
P. intermedius Corpuz-Raros & Cook	-	YK	-	-	-	-	-	-	-	-	-	-	-	-	-
P. knowltoni Corpuz-Raros & Cook	AK	-	-	-	-	-	-	-	-	-	-	-	-	-	-
P. longinectarius (Gillette & Palmer)	-	-	-	-	-	AB	-	MB	-	-	-	-	-	-	-
P. longipes (Gillette & Palmer)	-	-	-	-	BC	-	-	-	-	-	-	-	-	-	-
P. patonkus (Hottes & Frison)	-	-	-	-	-	-	-	MB	-	-	NB	-	-	-	-
P. pseudoglandulosus (Palmer)	-	-	-	-	-	-	SK	MB	-	-	NB	-	-	-	-
P. pseudopatokusellus Corpuz-Raros & Cook	-	YK	-	-	-	-	-	-	-	-	-	-	-	-	-
P. pseudopatonkus Corpuz-Raros & Cook	-	-	NT	-	-	AB	-	MB	-	-	NB	-	-	-	-
P. pullus (Gillette & Palmer)	-	-	-	-	-	AB	-	-	-	-	-	-	-	-	-
P. villosae Robinson	-	-	-	-	-	-	-	MB	-	-	-	-	-	-	-

Genus PSEUDACAUDELLA Börner

P. rubida (Börner)	-	-	-	-	-	-	-	MB	ON	QC	-	PE	-	-	-

Genus PSEUDOCERCIDIS Richards

P. rosae Richards	AK	-	NT	-	BC	-	-	MB	-	-	-	-	-	-	-

Genus PSEUDOEPAMEIBAPHIS Gillette & Palmer

P. glauca Gillette & Palmer	-	-	-	-	-	-	-	MB	-	-	-	-	-	-	-

Genus RHODOBIUM Hille Ris Lambers

R. porosum (Sanderson)	-	-	-	-	BC	-	-	MB	ON	QC	NB	-	NS	-	-

Genus RHOPALOMYZUS Mordvilko

(Subgenus JUDENKOA Hille Ris Lambers)

R. lonicerae (Siebold)	-	-	-	-	BC*	-	-	MB*	-	-	-	-	-	-	-

(Subgenus RHOPALOMYZUS Mordvilko)

R. grabhami (Cockerell)	-	-	-	-	BC	-	-	MB	-	-	-	-	-	-	-
R. poae (Gillette)	-	-	-	-	-	-	-	MB	-	QC	-	-	-	-	-

A agricultural, F forestry, G greenhouse, O ornamental pest; r rare; b biocontrol agent

Genus RHOPALOSIPHONINUS A.C. Baker

Key: Smith & Knowlton 1977.

(Subgenus MYZOSIPHON Hille Ris Lambers)

R. solani (Thomas)	-	-	-	-	-	-	-	MB	ON	QC	NB	-	-	-	-	[A]
R. staphyleae (Koch)	-	-	-	-	BC*	-	-	-	-	-	-	-	-	-	-	
R. tulipaellus (Theobald)	-	-	NT*	-	BC*	-	-	MB*	-	QC*	NB*	-	-	-	-	

(Subgenus RHOPALOSIPHONINUS A.C. Baker)

R. hydrangeae (Matsumura)	-	-	-	-	BC*	-	-	-	-	-	-	-	-	-	-
R. laytsiphon (Davidson)	-	-	-	-	BC	-	-	-	-	-	NB	-	-	-	-

Genus SITOBION Mordvilko

Key to species:Richards 1963a [Canada].

(Subgenus METOBION Heikenheimo)

S. beiquei (H.R.L.)	-	-	-	-	-	-	-	-	-	QC	-	-	-	-	-

(Subgenus SITOBION Mordvilko)

S. alopecuri (Takahashi)	-	-	-	-	BC	-	-	-	-	-	-	-	-	-	-	
S. avenae (Fabricius) English grain aphid; puceron des céréales	AK*	-	NT*	-	BC*	AB*	SK*	MB*	ON*	QC*	NB*	PE*	NS*	-	-	[A]
S. caricis (Glendenning)	-	-	-	-	BC	-	-	-	-	-	-	-	-	-	-	
S. fragariae (Walker)	-	-	-	-	BC*	-	-	-	-	-	-	-	-	-	-	

Genus UROLEUCON Mordvilko

Keys: Robinson 1985, 1986.

(Subgenus UROLEUCON Mordvilko)

U. achilleae (Koch)	-	-	-	-	BC*	-	-	-	-	-	-	-	-	-	-	
U. alaskense (Robinson)	AK	-	-	-	-	-	-	-	-	-	-	-	-	-	-	
U. ambrosiae (Thomas)	-	-	-	-	BC	-	-	MB	ON	QC	NB	-	NS	NF	-	
U. arnesense Robinson	-	-	-	-	-	-	-	MB	-	-	-	-	-	-	-	
U. astronomus (H.R.L.)	-	-	-	-	-	-	-	MB	ON	-	NB	-	-	-	-	
U. boreale Robinson	-	YK	NT	-	-	-	-	-	-	-	-	-	-	-	-	
U. chani Robinson	-	-	-	-	BC	-	-	-	-	-	-	-	-	-	-	
U. chrysanthemi (Oestlund)	-	-	-	-	-	-	-	MB	-	QC	NB	-	-	-	-	
U. cirsii (Linnaeus)	-	-	-	-	BC*	-	-	MB*	ON*	QC*	NB*	PE*	-	-	-	
U. deltense Robinson	-	-	-	-	-	-	-	MB	-	-	-	-	-	-	-	
U. epilobii (Pergande)	AK	-	-	-	-	-	-	-	-	-	-	-	-	-	-	
U. eupatoricolens (Patch)	-	-	-	-	-	-	-	MB	-	QC	NB	-	-	-	-	
U. gigantiphagum Moran	AK	-	-	-	BC	AB	SK	MB	ON	-	-	-	-	-	-	
U. hieracicola (H.R.L.)	-	-	-	-	BC	-	-	MB	ON	QC	NB	-	-	-	-	
U. impatiensicolens (Patch)	-	-	-	-	-	-	-	-	ON	-	NB	-	-	-	-	
U. ivae (Robinson)	-	-	-	-	-	-	-	MB	-	-	-	-	-	-	-	
U. lanceolatum (Patch)	-	-	-	-	-	-	-	-	ON	QC	NB	-	NS	-	-	
U. leonardi (Olive)	-	-	-	-	-	-	-	-	ON	-	NB	PE?	-	-	-	
U. maximilianicola (Robinson)	-	-	-	-	-	-	-	MB	-	-	-	-	-	-	-	
U. nigrotibium (Olive)	-	-	-	-	-	-	-	MB	ON	QC	NB	-	NS	-	-	
U. nigrotuberculatum (Olive)	-	-	-	-	BC	AB	-	MB	ON	QC	NB	PE	NS	-	-	
U. obscuricaudatum (Olive)	-	-	-	-	-	-	-	MB	ON	-	-	-	-	-	-	
U. olivei Moran	-	-	-	-	BC	-	-	MB	ON	-	-	-	-	-	-	
U. paucosensoriatum (H.R.L.)	-	-	-	-	BC	-	-	MB	ON	QC	NB	PE	-	-	-	
U. pepperi (Olive)	-	-	-	-	-	-	-	MB	-	-	NB	-	-	-	-	
U. pieloui (Richards)	-	-	-	-	-	-	-	MB	ON	QC	-	-	-	-	-	
U. pseudambrosiae (Olive)	-	-	-	-	BC	-	-	MB	ON	QC	NB?	-	-	-	-	
U. rudbeckiae (Fitch)	-	-	-	-	BC?	AB	SK?	MB	ON?	QC	NB?	-	NS?	-	-	[O]
U. russellae (H.R.L.)	-	-	-	-	BC	-	-	MB	ON	QC	NB	PE	NS	-	-	
U. sijpkensi (H.R.L.)	-	-	-	-	-	-	-	-	-	QC	-	-	-	-	-	
U. solirostratum (Richards)	-	-	-	-	-	-	-	MB	ON	-	-	-	-	-	-	
U. sonchellum (Monell)	-	-	-	-	-	-	-	-	ON	-	-	PE	-	-	-	

* introduced; [] indoors; { } extirpated; ? uncertain record; @ unestablished adventive; $ port-of-entry interception

Uroleucon sonchi (Linnaeus)	-	-	-	-	BC*	-	-	-	ON*	QC*	NB*	-	-	-	-
U. vancouverense Robinson	-	-	-	-	BC	-	-	-	-	-	-	-	-	-	-
(Subgenus LAMBERSIUS Olive)															
U. breviscriptum (Palmer)	-	-	-	-	BC	-	-	-	-	-	-	-	NS	-	-
U. brevitarsus (Robinson)	-	-	-	-	-	-	-	MB	-	-	-	-	-	-	-
U. caligatum (Richards)	-	-	-	-	-	-	-	MB	ON	QC	-	-	-	-	-
U. canadense (Richards)	-	-	-	-	BC	-	-	-	ON	QC	-	-	-	-	-
U. carberriense Robinson	-	-	-	-	BC	AB	-	MB	-	-	-	-	-	-	-
U. erigeronense (Thomas)	-	-	-	-	BC	-	-	MB	ON	QC	NB	-	NS	-	-
U. escalantii (Knowlton)	-	-	-	-	BC	-	-	MB	ON	-	-	-	-	-	-
U. gravicorne (Patch)	-	-	-	-	BC	-	-	MB	ON	QC	-	-	-	-	-
U. longirostre (Gillette & Palmer)	-	-	-	-	BC	-	-	-	-	-	-	-	-	-	-
U. luteolum (Williams)	-	-	-	-	-	-	-	-	ON	QC	NB?	-	-	-	-
U. macgillivrayae (Olive)	-	-	-	-	-	-	SK	MB	ON	QC	NB	-	-	-	-
U. manitobense (Robinson)	-	-	-	-	-	-	-	MB	-	-	-	-	-	-	-
U. nodulum (Richards)	-	-	-	-	-	-	-	-	ON	QC	-	-	-	-	-
U. penderum Robinson	-	-	-	-	BC	-	-	-	-	-	-	-	-	-	-
U. richardsi (Robinson)	-	-	-	-	-	-	-	MB	-	-	-	-	-	-	-
U. tenuitarsum (Gillette & Palmer)	-	-	-	-	-	-	-	MB	-	-	-	-	-	NF?	-
(Subgenus UROMELAN Mordvilko)															
U. eoessigi (Knowlton)	-	-	-	-	BC	AB	-	-	-	-	-	-	-	-	-
U. eupatorifoliae (Tissot)	-	-	-	-	-	-	-	-	ON	-	-	-	-	-	-
U. helianthicola (Olive)	-	-	-	-	-	-	-	MB	ON	-	-	-	-	-	-
U. simile (H.R.L.)	AK	YK	-	-	-	-	-	-	-	-	-	-	-	-	-
U. solidaginis (Fabricius)	-	-	-	-	BC?	-	-	-	-	-	-	-	-	-	-
U. taraxaci (Kaltenbach)	-	YK	-	-	BC	AB	SK	MB	ON	QC	NB	PE	-	NF	-

Genus UTAMPHOROPHORA Knowlton

Key: Cook 1984b [North America], Remaudière 1983.

U. crataegi (Monell)	-	-	-	-	BC	AB	-	MB	ON	QC	NB	-	-	-	-
U. humboldti (Essig)	-	-	-	-	BC	-	-	MB	ON	QC	NB	-	-	-	-

Genus WAHLGRENIELLA Hille Ris Lambers

W. arbuti (Davidson)	-	-	-	-	BC	-	-	-	-	-	-	-	-	-	-	[O]
W. empetri Richards	-	-	-	NU	-	-	-	-	-	QC	-	-	-	-	-	
W. nervata (Gillette)	-	YK	-	-	BC	-	-	MB	-	-	-	-	-	-	-	
W. vaccinii (Theobald)	-	-	-	-	BC	-	-	MB	-	-	-	-	-	-	-	

Subfamily PTEROCOMMATINAE

Genus FULLAWAYA Essig

Key: Richards 1967a.

F. bradleyi (Richards)	-	-	-	-	-	-	-	-	ON	-	-	-	-	-	-
F. braggii (Gillette & Palmer)	-	-	NT	-	-	-	-	-	-	-	-	-	-	-	-
F. bulbosa (Richards)	-	-	-	-	BC	-	-	-	-	-	-	-	-	-	-
F. ontarioensis (Richards)	-	-	-	-	-	-	-	-	ON	-	-	-	-	-	-
F. saliciradicis Essig	-	-	-	-	-	-	-	-	-	-	NB?	-	-	-	-

Genus PLOCAMAPHIS Oestlund

P. flocculosa (Weed)	-	-	NT	-	BC	-	-	-	ON	QC	NB	PE	-	-	-
P. martini (Richards)	AK	-	NT	NU	-	-	-	-	-	QC	-	-	-	-	-

Genus PTEROCOMMA Buckton

Key: Richards 1967b.

P. beulahensis (Cockerell)	AK	-	-	-	BC	-	-	-	-	-	-	-	-	-	-
P. bicolor (Oestlund)	AK	YK	NT	-	BC	AB	SK	MB	ON	QC	NB	-	-	NF	-

Pterocomma groenlandicum H.R.L.	-	-	-	NU	-	-	-	-	-	-	-	-	-	-	-
P. medium A.C. Baker	-	-	-	-	BC	-	-	-	-	-	NB	-	-	-	-
P. pilosum Buckton	-	-	-	-	BC*	-	-	-	-	-	NB*	-	NS*	-	-
P. populeum (Kaltenbach)	AK*	-	-	-	BC*	-	-	-	ON*	-	NB*	-	-	NF*	-
P. populifoliae (Fitch)	AK	-	-	-	BC	AB	SK	MB	ON	QC	NB	-	NS	-	-
P. pseudopopuleum Palmer	-	-	-	-	-	AB	-	-	-	QC	-	-	-	-	-
P. salicis (Linnaeus)	-	YK	NT	-	BC	AB	SK	MB	ON	QC	NB	-	-	-	-
P. sanguiceps Richards	-	-	-	-	BC	-	-	-	-	-	-	-	-	-	-
P. smithiae (Monell) black willow aphid; puceron du saule	-	YK	NT	-	BC	AB	SK	MB	ON	QC	NB	-	NS	-	-

Family DREPANOSIPHIDAE

The majority of species feed on dicotyledonous trees and shrubs (usually on the foliage), but species in subfamily Saltusaphidinae are sedge feeders, Siphini feed on grasses, and included among the Phyllaphidinae are legume and bamboo feeders. Drepanosiphids are distributed mainly in Holarctic and Oriental regions, but a few genera are restricted to the southern hemisphere.

This group is rather diverse structurally and biologically and several of its subgroups (subfamilies in the current classification scheme) should probably be elevated to the level here treated as families.

Only a few species, including the spotted alfalfa aphid, the grass-feeding Siphini, and a few pecan-feeding species are of economic importance. A few others frequently become a nuisance in urban areas due to copious honeydew production, particularly during extended periods of warm rainless weather.

Head and prothorax of apterae rarely fused; eyes of apterae rarely absent; antennae usually with 6 segments; terminal process long or short; secondary rhinaria usually round to transverse-oval (rarely transverse linear); apical setae of distal tarsal segments acuminate; pretarsal setae flabellate, capitate or acuminate; rastral structures usually present at apex of tibiae; basal tarsal segments often quadrate; many species with lateral and/or dorsal projections of the abdominal wall; wax glands dispersed or aggregated into more or less distinct multiporous structures; lateral thin-walled tubercles often present, sometimes in clusters; siphunculi vasiform to cylindrical, rarely poriform; anal plate bilobed or entire; cauda knobbed, ligulate or semicircular; oviparous females apterous or rarely alate; in many species, all adult viviparae are alate.

Subfamily CHAITOPHORINAE

Tribe Chaitophorini

Genus CHAITOPHORUS Koch

Key: Richards 1972a.

C. horii Takahashi	-	-	NT	-	-	-	-	-	-	-	-	-	-	-	-
C. leucomelas Koch	-	-	-	-	-	-	-	-	ON*	QC*	NB*	-	-	-	-
C. macgillivrayae Richards	-	-	-	-	-	-	-	-	-	QC	NB	-	-	-	-
C. macrostachyae (Essig)	-	-	-	-	BC	-	SK	MB	ON	-	-	-	-	-	-
C. monelli (Essig)	-	-	-	-	BC	-	-	-	-	-	-	-	-	-	-
C. neglectus Hottes & Frison	AK	YK	-	-	BC	AB	-	MB	ON	QC	NB	-	NS	NF	-
C. nigrae Oestlund	-	-	-	-	BC	AB	-	MB	ON	QC	NB	PE	NS	-	-
C. nigricentrus Richards	-	-	-	-	BC	-	-	MB	ON	QC	-	-	-	-	-
C. nudus Richards	-	-	-	-	-	-	-	MB	ON	QC	-	-	-	-	-
C. pallipes Richards	-	-	-	-	BC	-	-	-	ON	-	-	PE	-	-	-
C. populialbae (Boyer de Fonscolombe)	-	-	-	-	-	-	-	MB*	ON*	-	NB*	-	-	-	-
C. populicola Thomas smokeywinged poplar aphid; puceron noir du peuplier	AK	YK	NT	-	BC	AB	SK	MB	ON	QC	NB	PE	NS	NF	-
C. populifolii populifolii (Essig)	AK	YK	NT	-	BC	AB	-	MB	ON	QC	NB	-	NS	NF	-
C. pusillus Hottes & Frison	-	-	NT	-	BC	-	-	MB	-	QC	NB	-	-	-	-
C. pustulatus H.R.L.	-	-	-	-	BC	-	-	-	-	-	-	-	-	-	-
C. salicicola Essig	-	-	-	-	BC	-	-	-	-	-	-	PE	-	-	-

* introduced; [] indoors; { } extirpated; ? uncertain record; @ unestablished adventive; $ port-of-entry interception

Chaitophorus saliciniger (Knowlton)	-	-	-	-	BC	AB	-	MB	-	-	-	-	-	-	-
C. stevensis Sanborn	-	-	-	-	BC	AB	-	MB	ON	QC	NB	-	NS	-	-
C. viminalis Monell	-	-	-	-	BC	-	-	MB	ON	QC	NB	PE	NS	-	-
C. viminicola H.R.L.	-	-	-	-	-	-	-	-	ON*	-	-	-	-	-	-

Genus PERIPHYLLUS van der Hoeven

Key: Essig & Abernathy 1952 [world], Richards 1972a [Canada].

P. aceris (Linnaeus)	-	-	-	-	BC	-	-	-	-	-	-	-	-	-	-
P. americanus (A.C. Baker)	-	-	-	-	-	-	-	-	-	QC	NB	-	NS	-	-
P. brevispinosus Gillette & Palmer	-	-	-	-	BC	AB	-	-	-	-	-	-	-	-	-
P. californiensis (Shinji)	-	-	-	-	BC*	-	-	-	-	-	-	-	-	-	-
P. lyropictus (Kessler) Norway maple aphid; puceron de l'érable de Norvège	-	-	-	-	BC*	-	-	-	ON*	QC*	NB*	-	NS*	NF*	-
P. negundinis (Thomas) boxelder aphid; puceron de l'érable négondo	-	-	-	-	BC	AB	SK	MB	ON	QC	NB	PE	NS	-	-
P. testudinaceus (Fernie)	AK*	-	-	-	BC*	-	-	-	-	-	NB*	PE*	NS*	NF*	-

Genus PSEUDOPTEROCOMMA MacGillivray

Key: Richards 1972a.

P. canadensis Richards	-	-	-	-	-	-	SK	MB	ON	-	-	-	-	-	-
P. hughi (MacGillivray)	-	-	-	-	-	-	-	MB	-	QC	NB	-	-	-	-

Tribe Siphini

Genus ATHEROIDES Haliday

A. serrulatus Haliday	-	-	-	-	BC*	-	-	-	-	QC*	-	-	-	-	-

Genus SIPHA Passerini

Key: Richards 1972a.

(Subgenus RUNGSIA Mimeur)

S. elegans Del Guercio	-	-	-	-	BC*	AB*	SK*	MB*	ON*	QC*	NB*	PE*	NS*	NF*	-

(Subgenus SIPHA Passerini)

S. flava (Forbes)	-	-	-	-	BC*	-	-	MB*	-	-	-	-	-	-	-
S. glyceriae (Kaltenbach)	-	-	-	-	BC*	-	-	MB*	ON*	QC*	NB*	-	-	-	-

Subfamily DREPANOSIPHINAE

Tribe Drepanosiphini

Genus DREPANAPHIS Del Guercio

Key: Smith & Dillery 1968.

D. acerifoliae (Thomas) painted maple aphid; puceron de l'érable	-	-	-	-	-	-	-	MB	ON	QC	NB	-	NS	-	-
D. carolinensis C.F. Smith	-	-	-	-	-	-	-	-	-	QC	NB	-	-	-	-
D. granovskyi C.F. Smith & Knowlton	-	-	-	-	-	-	-	-	ON	-	-	-	-	-	-
D. kanzensis C.F. Smith	-	-	-	-	-	-	-	-	ON	QC	NB	-	NS	-	-
D. keshenae Granovsky	-	-	-	-	-	-	-	-	ON	QC	-	-	-	-	-
D. knowltoni C.F. Smith & Dillery	-	-	-	-	-	-	-	-	-	QC	-	-	-	-	-
D. monelli (J.J. Davis)	-	-	-	-	-	-	-	-	-	-	NB	-	-	-	-
D. nigricans C.F. Smith	-	-	-	-	-	-	-	-	-	-	NB	-	-	-	-
D. pallida Richards	-	-	-	-	-	-	-	-	ON	-	-	-	-	-	-
D. parva C.F. Smith	-	-	-	-	-	-	-	-	-	QC	NB	-	NS	-	-
D. simpsoni C.F. Smith	-	-	-	-	-	-	-	-	ON	QC	NB	-	-	-	-
D. spicata C.F. Smith	-	-	-	-	-	-	-	MB	ON	QC	NB	-	-	-	-
D. utahensis Knowlton & C.F. Smith	-	-	-	-	-	-	-	-	-	-	-	-	NS	-	-

Genus DREPANOSIPHUM Koch

D. platanoidis (Schrank)	-	-	-	-	BC*	-	-	-	ON*	-	-	-	-	NF*	-

Genus SHENAHWEUM Hottes & Frison

S. minuta (J.J. Davis)	-	-	-	-	-	-	-	-	ON	-	NB	-	-	-	-

Subfamily PHYLLAPHIDINAE

Tribe Phyllaphidini

Genus APPENDISETA Richards

A. robiniae (Gillette)	-	-	-	-	BC*	-	-	-	ON	-	-	-	-	-	-

Genus BETULAPHIS Glendenning

Key: Richards 1961 [North America].

B. brevipilosa Börner	-	-	-	-	BC*	-	-	-	ON*	-	-	-	-	-	-
B. pelei H.R.L.	AK	YK	NT	NU	-	-	-	MB	-	-	-	-	-	-	-
B. quadrituberculata (Kaltenbach)	-	-	NT*	-	BC*	-	-	MB*	-	QC*	NB*	-	-	-	-

Genus BOERNERINA Bramstedt

Key: Hille Ris Lambers & Hottes 1962 [world].

B. alni alaskensis H.R.L. & Hottes	AK	-	NT	-	BC	-	-	MB	ON	QC	NB	-	NS	NF	-
B. occidentalis H.R.L. & Hottes	AK	YK?	NT?	-	BC	-	-	-	-	-	-	-	-	-	-
B. variabilis Richards	-	-	NT	-	BC	AB	-	-	-	-	-	-	-	-	-

Genus CALAPHIS B.D. Walsh

C. alni A.C. Baker	-	-	-	-	-	-	-	-	-	QC	NB	-	-	-	-
C. betulaecolens (Fitch)	-	-	-	-	BC	-	-	MB	ON	QC	NB	-	NS	-	-
C. betulicola (Kaltenbach)	-	-	-	-	BC*	-	-	-	-	QC*	NB*	-	NS*	-	-
C. flava Mordvilko	-	-	-	-	BC	-	-	MB	-	QC	NB	-	-	NF	-
C. leonardi Quednau	-	-	-	-	-	-	-	-	-	QC	NB	-	-	-	-
C. manitobensis Richards	-	-	-	-	-	-	-	MB	-	-	-	-	-	-	-

Genus CALLIPTERINELLA van der Goot

C. calliptera Hartig	-	-	-	-	BC*	-	-	MB*	ON*	QC*	NB*	-	NS*	-	-
C. minutissima (Stroyan)	-	-	-	-	BC*	-	-	-	-	-	-	-	-	-	-

Genus CEPEGILLETTEA Granovsky

Key: Richards 1969.

C. betulaefoliae Granovsky	AK	-	-	-	BC	-	-	MB	ON	QC	NB	-	-	-	-
C. myricae (Patch)	-	-	-	-	-	-	-	-	ON	QC	NB	-	NS	-	-
C. viridis Richards	-	-	-	-	-	-	-	-	ON	-	-	-	-	-	-

Genus CHROMAPHIS Walker

C. juglandicola (Kaltenbach)	-	-	-	-	BC*	-	-	-	ON*	-	-	-	-	-	-

Genus CTENOCALLIS Klodnitzki

C. setosus (Kaltenbach)	-	-	-	-	BC*	-	-	-	-	-	-	-	-	-	-

Genus EUCALLIPTERUS Schouteden

E. tiliae (Linnaeus)	-	-	-	-	BC*	-	-	MB*	ON*	QC*	NB*	PE*	NS*	NF*	-	[O]

Genus EUCERAPHIS Walker

Key: Blackman 1977.

E. betulae (Koch)	AK	YK	NT	-	BC	-	-	MB	ON	-	NB	PE	NS	-	-	[O]
E. gillettei Davidson	AK	YK	NT	-	BC	AB	-	MB	ON	QC	NB	-	NS	NF	-	
E. lineata A.C. Baker	-	-	-	-	-	-	-	-	ON	QC	NB	-	NS	-	-	

* introduced; [] indoors; { } extirpated; ? uncertain record; @ unestablished adventive; $ port-of-entry interception

Euceraphis punctipennis (Zetterstedt)	-	-	-	-	BC*	AB*	-	MB*	ON*	QC*	NB*	-	NS*	NF*	-	[O]
European birch aphid; puceron européen du bouleau																

Genus HANNABURA Matsumura

H. alnosa (Pepper)	-	-	-	-	-	-	-	-	-	QC	-	-	NS	-	-

Genus HOPLOCHAITOPHORUS Granovsky

H. quercicola (Monell)	-	-	-	-	-	-	-	MB	ON	-	-	-	-	-	-

Genus LACHNOCHAITOPHORUS Granovsky

L. obscurus (Tissot)	-	-	-	-	-	-	-	-	ON	-	-	-	-	-	-

Genus MONELLIA Oestlund

Key, review: Bissel 1978.

M. caryella (Fitch)	-	-	-	-	-	-	-	-	ON	QC	-	-	-	-	-
M. hispida Quednau	-	-	-	-	-	-	-	-	ON	QC	-	-	-	-	-
M. microsetosa Richards	-	-	-	-	-	-	-	-	ON	QC	-	-	-	-	-

Genus MONELLIOPSIS Richards

Key, review: Bissel 1978.

M. caryae (Monell)	-	-	-	-	-	-	-	-	ON	QC	NB	-	-	-	-
M. nigropunctata (Granovsky)	-	-	-	-	-	-	-	-	ON	QC	-	-	-	-	-

Genus MYZOCALLIS Passerini

Key to subgenera: Quednau & Remaudière 1994.

(Subgenus MYZOCALLIS Passerini)

M. carpini (Koch)	-	-	-	-	BC*	-	-	-	-	-	-	-	-	-	-
M. coryli (Goeze)	-	-	-	-	BC*	-	-	-	ON*	-	-	-	NS*	-	-

(Subgenus AGRIOAPHIS Walker)

M. castanicola A.C. Baker	-	-	-	-	BC*	-	-	-	-	-	-	-	-	-	-

(Subgenus CASTANEOMYZOCALLIS Quednau & Remaudière)

M. castaneae (Fitch)	-	-	-	-	-	-	-	-	ON	QC	NB	-	-	-	-

(Subgenus LINEOMYZOCALLIS Richards)

M. bellus (Walsh)	-	-	-	-	BC	-	-	-	ON	QC	NB	-	NS	-	-
M. ephemeratus Richards	-	-	-	-	-	-	-	-	ON	QC	NB	PE	-	-	-
M. exultans Boudreaux & Tissot	-	-	-	-	-	-	-	-	-	QC	-	-	-	-	-
M. granovskyi Boudreaux & Tissot	-	-	-	-	-	-	-	-	ON	QC	NB	-	-	-	-
M. longirostris Richards	-	-	-	-	-	-	-	-	ON	-	-	-	-	-	-
M. longiunguis Boudreaux & Tissot	-	-	-	-	-	-	-	-	ON	-	-	-	-	-	-
M. melanocera Boudreaux & Tissot	-	-	-	-	-	-	-	-	ON	QC	NB	-	-	-	-
M. multisetis Boudreaux & Tissot	-	-	-	-	BC	-	-	-	ON	-	-	-	-	-	-
M. spinosa Boudreaux & Tissot	-	-	-	-	BC	-	-	-	ON	-	-	-	-	-	-
M. walshii (Monell)	-	-	-	-	BC	-	-	-	ON	QC	NB	-	NS	-	-

(Subgenus NEOMYZOCALLIS Richards)

M. asclepiadis (Monell)	-	-	-	-	-	-	-	MB	ON	QC	NB	-	-	-	-
M. discolor (Monell)	-	-	-	-	-	-	-	MB	ON	-	-	-	-	-	-
M. punctatus (Monell)	-	-	-	-	BC	AB	SK	MB	ON	QC	-	-	-	-	-

Genus NEOSYMYDOBIUS A.C. Baker

N. albasiphus (J.J. Davis)	-	-	-	-	-	-	-	-	ON	-	-	-	-	-	-
N. mimicus Hottes	-	-	-	-	-	-	-	MB	-	-	-	-	-	-	-

Genus OESTLUNDIELLA Granovsky

O. flava (Davidson)	AK	YK	-	-	BC	-	-	MB	-	-	NB	-	NS	-	-

Genus PANAPHIS Kirkaldy

P. juglandis (Goeze) - - - - BC* - - - - - - - - - -

Genus PHYLLAPHIS Koch

Key: Richards 1973.

P. fagi (Linnaeus) - - - - BC - - - ON QC NB PE NS NF -

Genus MONAPHIS Walker

M. antennata (Kaltenbach) - - - - - - - - ON* - - - - - -

Genus PROTOPTEROCALLIS Richards

Key: Bissel 1978.

P. fumipennellus (Fitch) - - - - - - - - ON QC - - - - -

Genus PTEROCALLIS Passerini

P. alni (DeGeer) - - - - BC* - - - ON* QC* - - - - -
P. alnifoliae (Fitch) - - - - BC - - MB ON QC NB PE NS NF -
P. rhombifoliae (Granovsky) - - - - - - - - - QC - - - - -

Genus STEGOPHYLLA Oestlund

Key: Hille Ris Lambers 1966 [world].

S. essigi H.R.L. - - - - BC - - - - - - - - - -
S. querci (Fitch) - - - - - - - MB ON QC - - - - -
S. quercicola (Monell) - - - - - - - - - QC NB - NS - -

Genus SYMYDOBIUS Mordvilko

S. americanus A.C. Baker - - - - - AB - MB ON QC NB PE NS - -
dark birch aphid; puceron noir du bouleau
S. intermedius Gillette & Palmer - - - - BC - - - ON QC? - - - - -

Genus TAKECALLIS Matsumura

T. arundicolens (Clarke) - - - - BC* - - - - - - - - - -
T. arundinariae (Essig) - - - - BC* - - - - - - - - - -

Genus THERIOAPHIS Walker

T. ononidis (Kaltenbach) - - - - BC* - - - - - NB* - - - -
T. riehmi (Börner) - - - - BC* - - MB* ON* QC* - - - - -
sweet clover aphid
T. trifolii (Monell) - - - - BC* AB* SK* MB* ON* QC* NB* - NS* - - [A]

Genus TINOCALLIS Matsumura

T. caryaefoliae (Davis) - - - - - - - - ON - - - - - -
T. platani (Kaltenbach) - - - - BC* - - - - - - - - - -
T. ulmifolii (Monell) - - - - BC AB - MB ON QC NB - - - -

Genus TUBERCULATUS Mordvilko

Revision: Richards 1968 [world].

T. tuberculatus (Richards) - - - - - - - - ON QC - - - - -

(Subgenus PACIFICALLIS Richards)

T. columbiae Richards - - - - BC - - - - - - - - - -

(Subgenus TUBERCULOIDES van der Goot)

T. annulatus (Hartig) - - - - BC* - - - ON* QC* NB* PE* - - -

* introduced; [] indoors; { } extirpated; ? uncertain record; @ unestablished adventive; $ port-of-entry interception

Subfamily SALTUSAPHIDINAE

Tribe Saltusaphidini

Genus IZIPHYA Nevsky

Key, review: Richards 1971.

I. albipes Richards	-	-	-	-	BC	-	-	-	ON	QC	-	-	-	-	-
I. americana Baker	-	-	-	-	-	-	-	-	ON	QC	-	-	-	-	-
I. flabella (Sanborn)	-	-	NT	-	BC	-	-	-	ON	QC	-	-	-	-	-
I. mackaueri Quednau	-	-	-	-	-	-	-	-	ON	-	-	-	-	-	-
I. spenceri Richards	-	-	-	-	BC	-	-	MB	-	-	-	-	-	-	-
I. vittata Richards	-	-	-	-	-	-	-	MB	ON	QC	-	-	-	-	-

Genus SALTUSAPHIS Theobald

S. elongata A.C. Baker	-	-	-	-	-	-	-	-	ON	QC	-	-	-	-	-

Tribe Thripsaphidini

Genus PELTAPHIS Frison & Ross

P. hottesi Frison & Ross	-	-	-	-	-	-	-	-	-	QC	-	-	-	-	-

Genus SUBSALTUSAPHIS Quednau

Key, review: Richards 1971.

S. canadensis Richards	-	-	-	-	-	-	-	-	ON	-	-	-	-	-	-
S. virginica (Baker)	-	-	-	-	-	-	-	-	ON	QC	-	-	-	-	-
S. wanica (Hottes & Frison)	-	-	-	-	-	-	-	-	ON	QC	-	-	-	-	-

Genus THRIPSAPHIS Gillette

Key, review: Richards 1971.

(Subgenus THRIPSAPHIS Gillette)

T. ballii ballii (Gillette)	-	-	-	-	-	-	-	-	ON	QC	-	-	-	-	-
T. ballii longisetis Richards	-	-	-	-	-	-	-	-	ON	-	-	-	-	-	-

(Subgenus TRICHOCALLIS Börner)

T. cyperi (Walker)	AK	YK	NT	-	BC	-	-	MB	ON	QC	NB	-	-	-	-
T. daviaulti (Quednau)	-	-	-	-	-	-	-	-	-	QC	-	-	-	-	-
T. producta Gillette	-	-	-	-	-	-	-	-	ON	QC	-	-	-	-	-
T. verrucosa nodulosa H.R.L.	AK	-	-	-	-	-	-	-	-	-	-	-	-	-	LB
T. verrucosa verrucosa Gillette	AK	-	-	-	BC	-	-	-	ON	QC	NB	-	-	-	LB

Subfamily TAMALIINAE

Genus TAMALIA A.C. Baker

T. coweni (Cockerell)	-	YK	-	-	BC	AB	-	MB	ON	-	-	-	-	-	-

Family HORMAPHIDIDAE

This gall-inducing group of woolly aphids is similar in many respects to the Pemphigidae. Apterous forms are flattened, sometimes with reduced appendages, and are reminiscent of scale insects or whitefly larvae. This family is primarily Oriental in distribution but the small subfamily Hormaphidinae is Holarctic. A few species of subfamily Cerataphidinae have been widely dispersed in tropical and subtropical regions by commerce, and are sometimes encountered in temperate areas on imported orchids or palms. Both native species form galls on *Hamamelis virginiana* and one alternates to the underside of the leaves of *Betula* species (or may reproduce continuously on *Betula* in the absence of *Hamamelis*).

Apterae of some generations flattened, sometimes with reduced appendages; head and pronotum of apterae fused; antennae with 3 to 5 segments; terminal process shorter than basal part of ultimate segment; secondary rhinaria of alatae annular; eyes of apterae (except triommatidium) absent; tibiae without apical rastral structures; dorsoapical setae of distal tarsal segments capitate; pretarsal setae acuminate (or pretarsus lacking in modified aptera); basal tarsal segments triangular; thin-walled tubercles absent; wax glands aggregated into facetted plates (often arranged marginally in apterae); siphunculi poriform or absent; anal plate bilobed; cauda knobbed; oviparous females and males minute.

Subfamily CERATAPHIDINAE

Genus CERATAPHIS Lichtenstein

C. fransseni (H.R.L.)	-	-	-	-	-	@	-	-	-	-	-	-	-	-	-

Subfamily HORMAPHIDINAE

Genus HAMAMELISTES Shimer

H. spinosus Shimer witch hazel spiney gall aphid	-	-	-	-	BC	-	SK	MB	ON	QC	NB	-	NS	-	-

Genus HORMAPHIS Osten-Sacken

H. hamamelidis (Fitch)	-	-	-	-	-	-	-	-	ON	-	-	-	NS	-	-

Family LACHNIDAE

Most members of this Holarctic family feed on the trunk, branches or main roots of woody plants (Lachninae on deciduous trees, Cinarinae on conifers), but *Schizolachnus* species and Eulachnini feed on conifer needles, and members of subfamily Traminae are found on roots of Compositae. Host alternation does not occur.

Head and pronotum of apterae separate; antennae with 6 segments; terminal process shorter than basal part of ultimate segment; secondary rhinaria of alatae round; eyes of apterae present; tibiae without apical rastral structures; dorsoapical setae of distal tarsal segments acuminate; pretarsal setae acuminate; basal tarsal segments usually quadrate; thin-walled tubercles absent; wax glands not aggregated into discrete structures; siphunculi poriform or very short-tubular, usually set on a setose conical base; anal plate entire; cauda semicircular; oviparous females apterous, males usually alate.

Subfamily CINARINAE

Tribe Cinarini

Genus CINARA Curtis

Key, review: Pepper & Tissot 1973 [on *Pinus*; eastern North America]. Catalogue: Voegtlin & Bridges 1988 [North America].

(Subgenus CINARA Curtis)

C. acadiana Hottes	-	-	-	-	-	-	-	-	-	-	NB	-	-	-	-
C. alaskana Hottes	AK	-	-	-	-	-	-	-	-	-	-	-	-	-	-
C. apini (Gillette & Palmer)	-	-	-	-	-	AB	-	-	-	QC	-	-	-	-	-
C. arizonica (Wilson)	-	-	-	-	BC	-	-	-	-	-	-	-	-	-	-
C. atlantica (Wilson)	-	-	-	-	-	-	-	-	ON	-	-	-	-	-	-
C. atra (Gillette & Palmer)	-	-	-	-	-	-	-	-	ON	-	-	-	-	-	-
C. banksiana Pepper & Tissot jackpine aphid; puceron du pin gris	-	-	-	-	-	AB	SK	MB	ON	QC	NB	-	-	-	-
C. bonica Hottes	AK	-	-	-	-	-	-	-	-	-	-	-	-	-	-
C. bonita Hottes	AK	-	-	-	-	-	-	-	-	-	-	-	-	-	-
C. braggii (Gillette)	AK	-	NT	-	BC	AB	SK	MB	ON	QC	NB	PE	NS	NF	-
C. brevispinosa (Gillette & Palmer)	-	-	-	-	BC	AB	-	-	-	-	NB	-	-	-	-
C. canatra Hottes & Bradley	-	-	-	-	BC	AB	SK	MB	ON	QC	NB	-	-	NF	-
C. caudelli (Wilson)	-	-	-	-	BC	-	-	-	-	-	-	-	-	-	-
C. chinookiana Hottes	-	-	-	-	-	AB	-	-	-	-	-	-	-	-	-
C. coloradensis (Gillette)	AK	YK	-	-	BC	AB	SK	MB	ON	QC	NB	PE	NS	-	-
C. confinis (Koch)	-	-	-	-	BC	AB	SK	MB	ON	QC	NB	-	NS	NF	-

* introduced; [] indoors; { } extirpated; ? uncertain record; @ unestablished adventive; $ port-of-entry interception

Cinara contortae Hottes	-	-	-	-	BC	-	-	-	-	-	-	-	-	-	-
C. costata (Zetterstedt)	-	-	-	-	-	-	-	-	ON*	QC*	-	-	-	NF*	-
C. cronartii Tissot & Pepper	-	-	-	-	-	-	-	-	-	QC	-	-	-	-	-
C. curvipes (Patch) balsam fir aphid; puceron du sapin	-	-	-	-	BC	AB	SK	MB	ON	QC	NB	-	NS	-	-
C. ferrisi (Swain)	-	-	-	-	BC	-	-	-	-	-	-	-	-	-	-
C. fornacula Hottes green spruce aphid	AK	YK	NT	-	BC	AB	SK	MB	ON	QC	NB	PE	NS	NF	-
C. glacialis Hottes	AK	-	-	-	-	-	-	-	-	-	-	-	-	-	-
C. gracilis (Wilson)	-	-	-	-	-	-	-	MB	ON	-	NB	-	-	NF	-
C. harmonia Hottes	-	-	-	-	-	-	-	MB	ON	QC	NB	-	-	-	-
C. hottesi (Gillette & Palmer)	-	-	-	-	BC	AB	SK	MB	ON	QC	NB	PE	NS	-	-
C. inscripta Hottes & Essig	-	-	-	-	BC	-	-	-	-	-	-	-	-	-	-
C. kuchea Hottes	-	-	-	-	BC	-	-	-	-	-	-	-	-	-	-
C. laricifex (Fitch) black larch aphid; puceron noir du mélèze	AK	-	NT	-	-	AB	SK	MB	ON	QC	NB	PE	NS	NF	-
C. laricifoliae (Wilson)	-	-	-	-	BC	-	-	-	-	-	NB*	-	NS*	-	-
C. laricis (Hartig) larch aphid; puceron du mélèze	-	-	-	-	-	AB?	-	-	ON*	QC*	-	-	NS*	NF*	-
C. lyallii Bradley	-	-	-	-	-	AB	-	-	-	-	-	-	-	-	-
C. manitobensis Bradley	-	-	-	-	-	-	-	MB	-	-	-	-	-	-	-
C. mariana Bradley	-	-	-	-	-	-	SK	-	-	-	-	-	-	-	-
C. medispinosa (Gillette & Palmer)	-	-	-	-	BC	AB	-	-	-	-	-	-	-	-	-
C. murrayanae (Gillette & Palmer)	-	-	-	-	BC	AB	-	-	-	-	-	-	-	-	-
C. nepticula Hottes	-	-	-	-	-	-	-	-	ON?	-	-	-	NS	-	-
C. nigra (Wilson)	-	-	-	-	BC	-	-	-	-	QC	-	-	-	-	-
C. nigripes Bradley	-	-	-	-	BC	-	-	-	ON	-	-	PE	NS	-	-
C. nimbata Hottes	AK	-	-	-	BC	-	-	-	-	-	-	-	-	-	-
C. obscura Bradley	-	-	-	-	BC	AB	SK	MB	ON	QC	NB	PE	NS	NF	-
C. occidentalis (Davidson)	-	-	-	-	BC	-	-	-	-	QC*	-	-	-	-	-
C. ontarioensis Bradley	-	-	-	-	-	AB	-	MB	ON	QC	-	-	-	-	-
C. oregonensis (Wilson)	-	-	-	-	-	AB	SK	-	-	-	-	-	-	-	-
C. oregoni Hottes & Essig	-	-	-	-	BC	-	-	-	-	-	-	-	-	-	-
C. parvicornis Hottes	-	-	-	-	BC	-	-	-	-	-	-	-	-	-	-
C. pergandei (Wilson)	-	-	NT	-	BC	AB	SK	MB	ON	QC	NB	-	-	-	-
C. petersoni Bradley	-	-	-	-	-	-	-	MB	-	-	-	-	-	-	-
C. pilicornis Hartig	-	-	-	-	BC*	AB*	-	-	ON*	QC*	-	-	NS*	-	-
C. pinea (Mordvilko) pine aphid; puceron du pin sylvestre	-	-	-	-	BC*	AB*	-	MB*	ON*	QC*	NB*	PE*	NS*	NF*	-
C. piniradicis Bradley	-	-	-	-	-	AB	SK	MB	ON	QC	-	-	-	-	-
C. pinivora (Wilson)	-	-	-	-	-	AB	SK	MB	ON	QC	NB	-	-	-	-
C. ponderosae (Williams)	-	-	-	-	BC	-	-	-	-	-	-	-	-	-	-
C. pruinosa Hartig	-	-	-	-	-	AB	SK	MB	ON	QC	-	-	NS	NF	-
C. pseudotaxifoliae Palmer	-	-	-	-	BC	AB	-	-	-	-	-	-	-	-	-
C. pseudotsugae (Wilson)	-	-	-	-	BC	AB	-	-	-	-	-	-	-	-	-
C. rubicunda (Wilson)	AK	-	-	-	-	-	-	-	ON	QC	-	-	-	-	-
C. saskensis Bradley	-	-	-	-	BC	-	SK	MB	-	-	-	-	-	-	-
C. sonata Hottes	-	-	-	-	BC	-	-	-	-	-	-	-	-	-	-
C. spiculosa Bradley	-	-	NT	-	-	AB	SK	MB	ON	QC	NB	-	NS	NF	-
C. splendens (Gillette & Palmer)	-	-	-	-	BC	-	-	-	-	-	-	-	-	-	-
C. strobi (Fitch) white pine aphid; puceron du pin blanc	-	-	-	-	-	-	-	MB	ON	QC	NB	-	NS	NF	-
C. subterranea Bradley	-	-	-	-	-	AB	-	MB	-	-	NB?	-	-	-	-
C. taedae Tissot	-	-	-	-	-	-	-	-	-	QC	-	-	-	-	-
C. thatcheri Knowlton & C.F. Smith	-	-	-	-	BC	-	-	-	-	-	-	-	-	-	-
C. tsugae Bradley	-	-	-	-	BC	-	-	-	-	-	-	-	-	-	-
C. vandykei (Wilson)	AK	-	-	-	BC	AB	-	-	ON	-	NB	-	NS	NF	-
C. villosa Gillette & Palmer	-	-	-	-	BC	AB	-	-	-	-	-	-	-	-	-
C. yukona Hottes	AK	-	-	-	-	-	-	-	-	-	-	-	-	-	-

A agricultural, F forestry, G greenhouse, O ornamental pest; r rare; b biocontrol agent

(Subgenus CUPRESSOBIUM)

Cinara cupressi (Buckton) - - - - - BC AB - - ON QC - - - - - -
C. juniperi (DeGeer) - - - - - BC AB SK MB ON QC - - NS - -

Genus SCHIZOLACHNUS Mordvilko

S. curvispinosus Hottes, Essig & Knowlton - - - - - BC - - - - - - - - - - -
S. parvus (Wilson) - - - - - - - - - ON - NB - - - -
S. pineti (Fabricius) - - - - - BC* - - - ON* QC* NB* - - - -
S. piniradiatae (Davidson) - - - - - BC - - MB ON QC NB - - - -
woolly pineneedle aphid; puceron lanigère du pin rouge

Tribe Eulachnini

Genus ESSIGELLA Del Guercio
Revision: Sorensen 1994.

(Subgenus LAMBERSELLA Sorensen)

E. fusca fusca Gillette & Palmer - - - - - BC - - - - - - - - - - -

(Subgenus ESSIGELLA Del Guercio)

E. aleyeska Sorensen AK - - - - - - - ON QC - - - - - -
E. californica (Essig) - - - - - BC AB? - - - - - - - - - -
E. knowltoni knowltoni Hottes - - - - - BC AB - - - - - - - - - -
E. pini Wilson - - - - - - - - - - QC - - - - - -
E. wilsoni Hottes - - - - - BC AB? - - - - - - - - - -

Genus EULACHNUS Del Guercio

E. agilis (Kaltenbach) - - - - - - AB* - - ON* - NB* - - - -
spotted pineneedle aphid; puceron tacheté acicole
E. brevipilosus Börner - - - - - BC* - - - - - - - - - - -
E. rileyi (Williams) - - - - - BC? AB* - - ON* QC* - - NS* NF* -

Subfamily LACHNINAE

Genus LACHNUS Burmeister

L. allegheniensis McCook - - - - - - - - MB - - - - - - -

Genus LONGISTIGMA Wilson

L. caryae (T.W. Harris) - - - - - - - - MB ON QC - - - - -

Genus MACULOLACHNUS Gaumont

M. sijpkensi H.R.L. - - - - - BC AB SK MB - - - - NS - -

Genus TUBEROLACHNUS Mordvilko

T. salignus (Gmelin) - - - - - BC* - SK* MB* ON* - NB* - - NF* -
giant willow aphid; puceron géant du saule

Subfamily TRAMINAE

Genus TRAMA von Heyden

T. rara Mordvilko - - - - - BC* AB* SK* MB* ON* QC* NB* - - - -

* introduced; [] indoors; { } extirpated; ? uncertain record; @ unestablished adventive; $ port-of-entry interception

Family MINDARIDAE

This is a Holarctic family of one genus with 6 extant and 5 extinct species described. All feed on growing tips and young cones of spruces or firs, and may become pests in seed orchards. The life cycle is abbreviated, with overwintering eggs being laid in early summer.

Head and pronotum of apterae fused; antennae with 6 segments; terminal process shorter than basal part of ultimate segment; secondary rhinaria of alatae rounded to transverse-oval; eyes of apterae varies seasonally from fully developed to reduced or absent (except triommatidia); tibiae without apical rastral structures; dorsoapical setae of distal tarsal segments acuminate; pretarsal setae acuminate; basal tarsal segments triangular; thin-walled tubercles absent; wax glands aggregated into facetted plates; siphunculi poriform; anal plate bilobed; cauda knobbed; oviparous females and males apterous, reduced in size.

Genus MINDARUS Koch

Key to species: Voegtlin 1995.

M. abietinus Koch balsam twig aphid; puceron des pousses du sapin	-	-	-	-	BC	AB	SK	MB	ON	QC	NB	-	NS	NF	-	[F]
M. obliquus (Cholodkovsky)	AK	-	-	-	BC	AB	SK	MB	ON	QC	NB	-	NS	NF	-	
M. victoria Essig	-	-	-	-	BC	-	-	-	-	-	-	-	-	-	-	

Family PEMPHIGIDAE

Woolly aphids, gall aphids. Mainly host-alternating, forming galls or pseudogalls on the primary host. Most produce copious amounts of wax filaments, resulting in woolly colonies. Many species can persist parthenogenetically for several years on the secondary hosts. Primary hosts of Eriosomatinae are Ulmaceae; those of Fordinae are Anacardiaceae; Pemphiginae Pemphigini use *Populus* species; and Pemphiginae Prociphilini occur on various deciduous trees and shrubs. Many Prociphilini use roots of conifers as secondary hosts. Both males and sexual females are minute wingless forms and lack mouthparts. The sexual female produces only one egg. Holarctic and Oriental, with a few species now widespread by introduction. This family includes some aphids of economic significance (sugarbeet root aphid, lettuce root aphid, apple woolly aphid).

Head and pronotum of apterae fused; eyes of apterae (except triommatidia) absent or greatly reduced; antennae with 6 segments; terminal process almost always shorter than basal part of ultimate segment; secondary rhinaria of alatae annular, oval, rounded or irregular; tibiae often with apical rastral setae; dorsoapical setae of distal tarsal segments often capitate in first instar larva, acuminate in adults; pretarsal setae acuminate; basal tarsal segments triangular; thin-walled tubercles absent; wax glands aggregated into facetted plates; siphunculi poriform or absent; anal plate entire; cauda semicircular; sexual females and males minute, apterous, without mouthparts; females producing a single egg.

Subfamily ERIOSOMATINAE

Tribe Eriosomatini

Genus ERIOSOMA Leach

E. americanum (Riley) woolly elm aphid; puceron lanigère de l'orme	-	-	-	-	BC	AB	SK	MB	ON	QC	NB	-	NS	-	-
E. crataegi (Oestlund) woolly hawthorn aphid; puceron lanigère de l'aubépine	-	-	-	-	BC	-	-	MB	ON	QC	NB	-	NS	-	-
E. grossulariae (Schule)	-	-	-	-	BC*	-	-	-	ON?	QC*	NB*	PE?	-	NF*	-
E. lanigerum (Hausmann) woolly apple aphid; puceron lanigère du pommier	-	-	-	-	BC	AB	-	MB	ON	QC	NB	-	NS	-	-
E. lanuginosum (T. Hartig)	-	-	-	-	-	-	-	-	ON?	-	-	-	-	-	-
E. mimicum Hottes & Frison	-	-	-	-	-	-	-	MB	ON	QC	NB	-	-	-	-
E. pyricola Baker & Davidson	-	-	-	-	BC	-	-	-	-	-	-	-	-	-	-
E. rileyi Thomas	-	-	-	-	BC	AB	-	-	-	QC	-	-	-	-	-
E. wilsoni Remaudière	-	-	-	-	-	-	-	-	ON	QC	NB	-	-	-	-

Tribe Tetraneurini

Genus GHARESIA Stroyan

G. polunini Stroyan - - - - BC - - - - - - - - - - [r]

Genus COLOPHA Monell

C. ulmicola (Fitch) - - - - BC - - MB ON QC - - NS - -
elm cockscomb gall aphid; puceron à galle de l'orme

Genus KALTENBACHIELLA Schouteden

K. ulmifusa (Walsh & Riley) - - - - - - - - ON QC - - - - -
elm pouchgall aphid; puceron gallicole de l'orme

Genus TETRANEURA Hartig

Revision: Hille Ris Lambers 1970 [world].

(Subgenus TETRANEURA Hartig)

T. ulmi (Linnaeus) - - - - BC* AB* SK* MB* ON* QC* NB* PE* - - -

(Subgenus TETRANEURELLA Hille Ris Lambers)

T. nigriabdominalis (Sasaki) - - - - - - - - ON* - - - - - -

Subfamily FORDINAE

Tribe Fordini

Genus FORDA von Heyden

F. formicaria von Heyden - - - - BC* AB* SK* MB* ON* QC* NB* - - - -
F. marginata Koch - - - - BC* AB* - MB* - - NB* - - - -

Genus GEOICA Hart

G. squamosa Hart - - - - - - - - - - NB* - - - -
G. utricularia (Passerini) - - - - BC* AB* - - - - NB* - - - -

Genus SMYNTHURODES Westwood

S. betae Westwood - - - - BC* - SK* - - QC* NB* - - - -

Tribe Melaphidini

Genus MELAPHIS Walsh

M. rhois (Fitch) - - - - - - - - ON QC NB - NS - -

Subfamily PEMPHIGINAE

Key to genera: Smith 1974.

Tribe Pemphigini

Genus MORDWILKOJA Del Guercio

M. vagabunda (Walsh) - - - - BC AB SK MB ON QC NB - NS NF -
poplar vagabond aphid; puceron vagabond du peuplier

Genus PEMPHIGUS Hartig

Key: Smith 1985, Harper 1959, Alleyne & Morrison 1977.

P. betae Doane - YK - - BC AB SK MB ON QC NB - - - - [A]
sugarbeet root aphid; puceron de la betterave à sucre
P. bursarius (Linnaeus) - - - - BC* AB* SK* MB* ON* QC* - - - - - [A]
lettuce aphid; puceron des racines maraichères
P. groenlandicus (Rübsaamen) - - - NU - - - - - - - - - - -
P. monophagus Maxson - - - - BC AB SK MB - QC NB - - - -

* introduced; [] indoors; { } extirpated; ? uncertain record; @ unestablished adventive; $ port-of-entry interception

Pemphigus nortonii Maxson	-	-	-	-	BC	AB	SK	MB	-	QC	-	-	-	-	-	
P. populicaulis Fitch	-	-	-	-	BC	AB	SK	MB	ON	QC	NB	-	-	-	-	
P. populiglobuli Fitch	AK	-	-	-	-	AB	-	MB	-	QC	NB	-	-	-	-	
P. populiramulorum Riley	-	-	-	-	BC	AB	-	MB	-	QC	-	-	-	-	-	
P. populitransversus Riley poplar petiolegall aphid; puceron des feuilles du peuplier	-	-	-	-	-	AB	-	MB	ON	QC	-	-	-	-	-	[A]
P. populivenae Fitch	-	-	-	-	BC	AB	SK	MB	ON	QC	NB	-	NS	-	-	
P. spyrothecae Passerini	-	-	-	-	BC*	-	-	-	ON*	QC*	-	PE*	-	-	-	
P. tartareus Hottes & Frison	-	-	-	-	-	AB	SK	MB	-	QC	-	-	NS	-	-	

Genus THECABIUS Koch

Key, review: Smith 1974.

(Subgenus THECABIUS Koch)

T. affinis (Kaltenbach)	AK	YK	NT	-	BC	AB	-	MB	-	QC	NB	-	-	-	-
T. populimonilis (Riley)	-	YK	NT	-	BC	AB	-	MB	ON	QC	NB	-	-	-	-

(Subgenus PARATHECABIUS Börner)

T. gravicornis (Patch)	-	-	-	-	BC	AB	-	MB	ON	QC	NB	-	-	NF	-
T. lysimachiae Börner	-	-	-	-	-	-	-	-	-	QC*	-	-	-	-	-

Tribe Prociphilini

Genus CLYDESMITHIA Danielsson

C. canadensis Danielsson	-	-	-	-	BC	AB	-	-	-	-	-	-	-	-	-

Genus GRYLLOPROCIPHILUS C.F. Smith & Pepper

G. imbricator (Fitch) beech blight aphid; puceron fugace du hêtre	-	-	-	-	-	-	-	-	ON	-	-	-	-	-	-

Genus NEOPROCIPHILUS Patch

N. aceris (Monell)	-	-	-	-	-	-	-	MB	ON	-	NB	-	-	-	-

Genus PACHYPAPPA Koch

Key, review: Smith 1974.

P. pseudobyrsa (Walsh)	-	-	-	-	-	-	-	-	ON	QC	-	-	NS	-	-
P. sacculi (Gillette)	AK	-	-	-	BC	AB	-	MB	ON	QC	NB	-	NS	-	-
P. tremulae (Linnaeus)	-	YK	-	-	BC	AB	SK	MB	ON	QC	NB	-	NS	NF	-

Genus PARAPROCIPHILUS Mordvilko

P. tessellatus (Fitch) woolly alder aphid; puceron lanigère de l'aulne	-	-	-	-	-	-	-	MB	ON	QC	NB	-	NS	-	-

Genus PROCIPHILUS Koch

Key, review: Smith 1974 [North America].

(Subgenus MELIARHIZOPHAGUS Sanborn)

P. fraxinifolii (Riley)	-	-	-	-	BC	AB	SK	MB	ON	QC	-	-	-	-	-

(Subgenus NEOPARACLETUS Strom)

P. corrugatans (Sirrine)	-	-	-	-	BC	AB	-	-	ON	-	-	-	NS	-	-

(Subgenus PROCIPHILUS Koch)

P. americanus (Walker)	-	-	-	-	BC	AB	-	-	-	QC	NB	PE	NS	-	-
P. carolinensis C.F. Smith	-	-	-	-	-	-	-	MB	ON?	QC	NB	-	-	-	-
P. caryae caryae (Fitch)	-	-	-	-	BC	-	-	-	-	-	-	-	-	-	-
P. caryae fitchii Baker & Davidson	-	-	-	-	BC	-	-	-	ON	QC	NB	PE	-	-	-
P. erigeronensis (Thomas)	-	-	-	-	-	-	-	MB	ON	QC	NB	-	-	-	-

(Subgenus PULVIUS Sanborn)

Prociphilus probosceus (Sanborn)	-	-	-	-	-	-	-	-	ON	-	-	-	-	-	-

(Subgenus STAGONA Koch)

P. crataegistrobi (C.F. Smith)	-	-	-	-	-	-	-	-	-	QC	NB	-	-	-	-
P. picearubensis (C.F. Smith)	-	-	-	-	-	-	-	-	-	QC	-	-	-	-	-
P. xylostei (DeGeer)	-	-	-	-	BC*	-	-	MB*	-	-	NB*	-	NS*	-	-

Genus TILIPHAGUS C.F. Smith

T. lycoposugus C.F. Smith	-	-	-	-	-	-	-	-	ON	-	-	-	-	-	-

Family PHLOEOMYZIDAE

One genus, probably with only a single species. Feeds on the trunk and main branches of poplar trees. In North America, known only from collections in Maine and an unspecified locality in Atlantic Canada (probably in New Brunswick). We follow Blackman & Eastop (1994) in treating these collections (originally described as *Phloeomyzus dearborni* C.F. Smith) as introduced colonies of the widespread Palearctic species.

Head and pronotum of apterae separate; antennae with 6 segments; terminal process shorter than basal part of ultimate segment; secondary always absent; eyes of apterae (except triommatidia) absent; tibiae with apical rastral structures; dorsoapical setae of distal tarsal segments acuminate; pretarsal setae acuminate; basal tarsal segments triangular; thinwalled tubercles absent; wax glands aggregated into facetted plates, restricted to abdominal segment VIII; siphunculi poriform; anal plate entire; cauda semicircular; oviparous females alate.

Genus PHLOEOMYZUS Horváth

P. passerinii (Signoret)	-	-	-	-	-	-	-	-	-	-	NB?	-	-	-	-

Family THELAXIDAE

This small family of non-host-alternating tree-feeding aphids has species in the Holarctic and Oriental regions. North American species feed on oaks and Betulaceae.

Head and pronotum of apterae fused; antennae with 5 segments; terminal process shorter than basal part of ultimate segment; secondary rhinaria of alatae oval or rounded; eyes of apterae (except triommatidia) absent; tibiae without apical rastral structures; dorsoapical setae of distal tarsal segments acuminate, ventral apical setae capitate; pretarsal setae acuminate or capitate; basal tarsal segments triangular; thin-walled tubercles absent; wax glands not aggregated into discrete structures; anal plate entire; cauda knobbed or rounded.

Genus GLYPHINA Koch

G. betulae (Linnaeus)	-	-	-	-	BC*	-	-	-	-	-	NB*	-	NS*	-	-
G. longiseta Richards	-	-	-	-	-	-	-	-	-	-	-	-	NS	-	-
G. setosa MacGillivray	-	-	-	-	-	-	-	MB	ON	-	NB	-	-	-	-

Genus THELAXES Westwood

T. californica (Davidson)	-	-	-	-	BC	-	-	-	ON?	-	-	-	-	-	-

* introduced; [] indoors; { } extirpated; ? uncertain record; @ unestablished adventive; $ port-of-entry interception

Infraorder Coccodea

Keys to families and genera: Kosztarab 1996 [eastern North America].

Regional guides: Kosztarab 1996 [eastern North America], Ferris 1937-1955 [Illustrated atlas for North America].

Regional lists: Kozár *et al.* 1989 [British Columbia].

Bibliographies: Morrison & Renk 1957, Morrison & Morrison 1965, Russel *et al.* 1974, Kosztarab & Kosztarab 1988.

World Wide Web Site (catalogue, bibliography, general information): Ben-Dov *et al.* 1996.

Scale insects and mealybugs; cochenilles. The females of all scale insects and mealybugs are wingless. Adult females of most scale insects are sessile, with reduced appendages and frequently with a covering consisting of a thickened dorsal integument or secreted wax test or 'shell'. Mealybugs generally remain motile and are usually covered with mealy or filamentous wax. The short-lived adult males have one pair of wings (some without wings) and do not feed. Although more diverse in warmer parts of the globe, a few species are found as far north as Ellesmere Island. Because of their small size, cryptic appearance and sessile habits, many species have been transported worldwide by trade in plants and plant products.

The Canadian fauna is largely unstudied, with that of British Columbia being best known. Scudder (1979) estimated there to be 56 species recorded and 4 expected in Canada. This list almost doubles the number to 116 species.

Superfamily Orthezioidea

Family MARGODIDAE

Giant scale insects; ground pearls. Adult females usually large (most more than 10 mm), motile with well developed legs, some species fossorial. The large encysted quiescent larvae of some species are known as ground pearls.

Genus ICERYA Signoret

I. purchasi Maskell - - - - - - - @ @ - - - - - -

Genus MATSUCOCCUS Cockerell

M. macrocicatrices Richards - - - - - - - - ON - NB - NS - -
white pine fungus scale; cochenille cachée du pin

M. secretus Morrison - - - - BC - - - - - - - - - -

Genus STEINGELIA Nassonov

S. gorodetskia Nassonov - - - - - - - - ON* QC* - - - - -

Genus XYLOCOCCULUS Morrison

X. betulae (Pergande) - - - - - - - - - - NB PE NS - -

Family ORTHEZIIDAE

Ensign scales. Females are active, with long well developed legs, and bear large overlapping dorsal wax plates. They feed on various woody and herbaceous plants and are often found in leaf litter.

Genus ARCTORTHEZIA Cockerell

A. cataphracta (Olafsen) AK YK NT - - AB - - - - - - - NF -

A. occidentalis (Douglas) - - - - BC - - - - - - - - - -

Genus NEWSTEADIA Green

N. americana Morrison - - - - - - - - ON - - - - - -

Genus ORTHEZIA Bosc d'Antic

O. newcomeri Morrison - - - - BC - - - - - - - - - -
O. olivacea Cockerell - - - - BC - - - - - - - - - -

Superfamily Coccoidea

Family ACLERDIDAE

Flat grass scales. This is a relatively small family of flat ovoid scale insects feeding on grasses (a few on other monocots), usually on the stem inside the leaf sheath. Females are covered by a thin transparent test or powdery wax and have antennae much reduced and legs reduced or absent.

Genus ACLERDA Signoret

A. californica Ehrhorn - - - - - - - - ON? - - - - - -

Family ASTEROLECANIIDAE

Pit scales. The body of the adult female is round to oval in outline, distinctly convex, and covered in a thin transparent to translucent test, often with a filamentous fringe. The antennae are reduced to tubercles and the legs absent or tuberculiform. Many species cause the formation of a pit in the host tissue at the feeding site. There are about 250 species worldwide, of which 10% occur in North America. Several significant pests are included in the family.

Genus ASTERODIASPIS Signoret

A. variolosa (Ratzeburg) - - - - BC - - - ON - - - - - -
golden oak scale; cochenille dorée du chêne

Genus ASTEROLECANIUM Targioni Tozzetti

A. arabidis (Signoret) - - - - - - - - ON - - - NS - -

Family COCCIDAE

Keys: Hodgson 1994 [world genera].

Catalogue: Ben-Dov 1993 [world].

Soft scale insects. Adult females sessile, but legs and antennae often well developed; body circular to oval, becoming heavily sclerotized and convex with age. Some species produce feltlike or cottony egg sacs. Most species produce copious honeydew and are frequently attended by ants.

Among the 1100 species worldwide are included a number of important pests of fruit and ornamental trees and of greenhouse plants.

Genus CHLOROPULVINARIA Borschenius

C. floccifera (Westwood) - - - - BC - - - - - - - - - -

Genus COCCUS Linnaeus

C. hesperidum Linnaeus - - - - [BC] - - - [ON][QC] - - - - - [G]
brown soft scale

Genus LECANOPSIS Targioni Tozzetti

L. formicarum Newstead - - - - - - - - ON* - - - - - -

Genus ERIOPELTIS Signoret

E. festucae (Boyer de Fonscolombe) - - - - - - - - ON - - - - - -

* introduced; [] indoors; { } extirpated; ? uncertain record; @ unestablished adventive; $ port-of-entry interception

Genus EULECANIUM Cockerell

E. caryae (Fitch)	-	-	-	-	-	-	-	-	ON	-	-	-	-	-	-	
E. cerasorum (Cockerell)	-	-	-	-	-	-	-	-	-	QC*	-	-	-	-	-	[O]
E. tiliae (Linnaeus)	-	-	-	-	BC*	-	-	-	-	-	-	-	-	-	-	

Genus MESOLECANIUM Cockerell

M. nigrofasciatum (Pergande) terrapin scale; lécanie à bandes noires	-	-	-	-	-	-	-	-	ON	QC	NB	-	NS	-	-	[A]

Genus PARTHENOLECANIUM Šulc

P. corni (Bouché) European fruit lecanium; lécanie de la vigne	-	-	-	-	BC*	AB*	-	-	ON*	-	NB*	-	NS*	-	-	[AO]
P. fletcheri (Cockerell)	-	-	-	-	-	-	-	-	ON	-	-	-	-	-	-	[F]
P. persicae (Fabricius) European peach scale; lécanie du pêcher	-	-	-	-	-	-	-	-	ON*	QC*	-	PE*	NS*	-	-	[AO]
P. pruinosum (Coquillett)	-	-	-	-	BC	-	-	-	-	-	-	-	-	-	-	
P. quercifex (Fitch) oak lecanium; lécanie du chêne	-	-	-	-	BC	-	-	-	-	-	NB	-	-	-	-	

Genus PHYSOKERMES Targioni Tozzetti

P. concolor Coleman	-	-	-	-	BC	-	-	-	-	-	-	-	-	-	-	
P. hemicryphus (Dalman)	-	-	-	-	BC	-	-	-	-	-	-	-	-	-	-	[F]
P. taxifoliae Coleman	-	-	-	-	BC	-	-	-	-	-	-	-	-	-	-	

Genus PULVINARIA Targioni Tozzetti

P. acericola (Walsh & Riley)	-	-	-	-	-	-	-	-	ON	-	-	-	-	-	-	
P. ellesmerensis Richards	-	-	-	NU	-	-	-	-	-	-	-	-	-	-	-	
P. innumerabilis (Rathvon) cottony maple scale; cochenille floconneuse de l'érable	-	-	-	-	BC	-	-	-	ON	QC	-	-	-	-	-	
P. occidentalis Cockerell	-	-	-	-	BC	-	-	-	-	-	-	-	-	-	-	
P. vitis (Linnaeus)	-	-	-	-	-	AB	SK	-	ON	-	-	-	-	-	-	

Genus TOUMEYELLA Cockerell

T. parvicornis (Cockerell) pine tortoise scale; cochenille-tortue du pin	-	-	-	-	-	-	-	MB	ON	-	-	-	-	-	-	[F]
T. pini (King)	-	-	-	-	-	-	-	-	ON	-	-	-	-	-	-	[F]

Family CRYPTOCOCCIDAE

Bark crevice scales. Members of this small family (6 species worldwide) live in fissures or under flakes on the bark of trees. The adult female is contained within a fibrous wax egg sac as in Eriococcidae; legs present or absent. *Cryptococcus fagisuga* plays an important role in the infection of beech trees with *Nectria* fungi.

Genus CRYPTOCOCCUS Douglas

C. fagisuga Lindinger beech scale; cochenille du hêtre	-	-	-	-	-	-	-	-	ON	-	NB	PE	NS	-	-	
C. williamsi Kosztarab & Hale	-	-	-	-	-	-	-	-	ON	QC	-	-	-	-	-	

Family DACTYLOPIIDAE

Review: Perez Guerra & Kosztarab 1992.

Cochineal scales. This family has a single genus with 9 species, all native to the new world. The bright red body is covered by long thin filaments of wax. Members of this family are the source of cochineal red dye and have been used as agents for the biological control of cacti.

Genus DACTYLOPIUS O.G. Costa

D. confusus (Cockerell) - - - - - - SK - - - - - - - -

Family DIASPIDIDAE

Key to genera and species: Ferris 1942 [North America].

Armoured scales. First instar larvae remain motile for only a short while before settling. The body of females and of larval males is covered by a tough covering composed of cast larval skins and wax secretions. Adult females are pyriform to elongate, with head and thorax completely fused, antennae reduced to tubercles and legs absent (or, rarely, tuberculiform). Most species feed on woody angiosperms, but some feed on herbaceous plants and a few on conifers. A number of species are important fruit and ornamental tree pests.

Genus ABGRALLASPIS Balachowsky

A. ithaeae (Ferris) - - - - - - - - - - NB - - - -
hemlock scale; cochenille de la pruche

Genus ACUTASPIS Ferris

A. morrisonorum Kosztarab - - - - - - - - - QC - - - - -

Genus AONIDIELLA Berlese & Leonardi

A. aurantii (Maskell) - - - - - - SK - - - - - - - -

Genus AULACASPIS Cockerell

A. rosae (Bouché) - - - - - - - - ON - - - - - -
rose scale; cochenille du rosier

Genus CARULASPIS MacGillivray

C. juniperi (Bouché) - - - - BC - - - ON - - - - - -
juniper scale; cochenille du genévrier

Genus CHIONASPIS Signoret

Revision: Liu *et al.* 1989 [females, North America], Bullington , Kosztarab & Jiang 1989 [males, North America].

C. corni Cooley - - - - - - - - ON - - - - - -
dogwood scale; cochenille du cornouiller

C. furfura (Fitch) - - - - - AB SK MB ON - - - - - -
scurfy scale; cochenille écailleuse

C. lintneri Comstock - - - - BC - - - ON QC NB PE NS - -
Lintner scale; cochenille de Lintner

C. ortholobis Comstock - - - - BC - - - - - - - - - -

C. pinifoliae (Fitch) - - - - BC AB SK - ON QC - - - - - [F]
pine needle scale; cochenille des aiguilles du pin

C. salicisnigrae (Walsh) - - - - - - - - ON - - - - - -
willow scurfy scale; cochenille dartreuse du saule

Genus CHRYSOMPHALUS Ashmead

C. aonidum (Linnaeus) - - - - - - - - - QC - - - - -

C. dictyospermi (Morgan) - - - - - - - - - QC - - - - -

Genus DIASPIDIOTUS Berlese & Leonardi

D. aesculi (Johnson) - - - - BC - - - ON - - - - - -

D. ancylus (Putnam) - - - - - - SK MB ON - - - - - -
Putnam scale; cochenille de Putnam

D. ehrhorni (Colman) - - - - - - SK - - - - - - - -

Genus DYNASPIDIOTUS Thiem & Gerneck

D. britannicus (Newstead) - - - - BC - - - - - - - - - -

* introduced; [] indoors; { } extirpated; ? uncertain record; @ unestablished adventive; $ port-of-entry interception

Genus HEMIBERLESIA Cockerell

H. popularum (Marlatt) - - - - BC AB - - ON - - - - - -

Genus LEPIDOSAPHES Shimer

L. ulmi (Linnaeus) - - - - BC AB - - ON - - - - - - [A]
oystershell scale; cochenille virgule du pommier

Genus MELANASPIS Cockerell

M. obscura (Comstock) - - - - - - - - ON - - PE - - -

Genus NUCULASPIS Ferris

N. californica (Coleman) - - - - BC - - - - - - - - - - [F]

Genus PINNASPIS Cockerell

P. aspidistrae (Signoret) - - - - - - - - - [QC] - - - - - [G]

Genus QUADRASPIDIOTUS MacGillivray

Q. gigas (Thiem & Gerneck) - - - - BC - - - ON - NB - NS - -
Q. juglansregiae (Comstock) - - - - BC - - - ON - - - - - -
walnut scale; cochenille du noyer
Q. ostreaeformis (Curtis) - - - - BC - SK - ON - - - NS - - [AO]
European fruit scale; cochenille ostréiforme
Q. perniciosus (Comstock) - - - - BC - - - ON QC - - {NS} - - [AO]
San José scale; cochenille de San José

Genus RHIZASPIDIOTUS MacGillivray

R. dearnessi (Cockerell) - - - - BC - - - ON - - - - - -

Genus STRAMENASPIS Ferris

S. kelloggi (Coleman) - - - - BC - - - - - - - - - -

Genus UNASPIS MacGillivray

U. euonymi (Comstock) - - - - BC - - - - - - - - - -
euonymus scale; cochenille du fusain

Family ERIOCOCCIDAE

Keys to genera and species: Miller 1992 [western North America], Miller & Miller 1993 [eastern North America].
Catalogue: Hoy 1963 [world].

Felt scales. Adult female entirely (partly in a few species) enclosed by an ovoid feltlike wax egg sac; legs and antennae usually rather well developed.

Genus ACANTHOCOCCUS Signoret

A. azaleae (Comstock) - - - - - - - - ON - - - NS - -
A. greeni (Newstead) - - - - BC - - - - - - - - - -

Genus GOSSYPARIA Signoret

G. spuria (Modeer) - - - - BC* - - - ON* QC* - - - - -
European elm scale; cochenille de l'orme

Family KERMESIDAE

Revision: Bullington & Kosztarab 1985.

Gall-like scales. Mature females of most members of this family are large, heavily sclerotized and strongly convex to spherical, often marked with bands of pigment or wax. Host plants are in Fagaceae (except for the conifer-feeding *Eriokermes* spp.). Often attended by ants due to the large amount of honeydew produced.

Genus **ALLOKERMES** Bullington & Kosztarab

A. galliformis (Riley)	-	-	-	-	-	-	-	-	ON	-	-	-	-	-	-	

Family PSEUDOCOCCIDAE

Key to genera: Ferris 1950, 1953 [North America].

Catalogue: Ben-Dov 1994 [world].

Mealybugs. Adult females of most species remain motile (a few sessile genera with reduced appendages). The body is covered by a mealy wax coat (reduced in species living in concealed habitats), and most produce a filamentous egg sac. Hosts of most species are herbaceous plants, especially grasses. Often attended by ants. Almost 2000 known species in the world.

Some authors place *Puto* and closely related genera in a separate family, Putoidae, the giant mealybugs.

Genus **ANISOCOCCUS** Ferris

A. oregonensis Ferris	-	-	-	-	BC	-	-	-	-	-	-	-	-	-	-	
A. quercus (Ehrhorn)	-	-	-	-	BC	-	-	-	-	-	-	-	-	-	-	

Genus **CHNAUROCOCCUS** Ferris

C. trifolii (Forbes)	-	-	-	-	-	-	-	-	ON	-	-	-	-	-	-	

Genus **CHORIZOCOCCUS** McKenzie

C. altoarcticus (Richards)	-	-	-	NU	-	-	-	-	-	-	-	-	-	-	-	

Genus **HETEROCOCCUS** Ferris

Review: Miller & McKenzie 1970 [world].

H. nudus (Green)	-	-	-	-	BC	-	-	-	-	-	-	-	-	-	-	

Genus **PHENACOCCUS** Cockerell

P. aceris (Signoret) apple mealybug; cochenille du pommier	-	-	-	-	BC*	-	-	-	ON*	-	-	PE*	NS*	-	-	[A]
P. capensis Ferris	-	-	-	-	BC	-	-	-	-	-	-	-	-	-	-	
P. colemani Ehrhorn	-	-	-	-	BC	-	-	-	-	-	-	-	-	-	-	
P. dearnessi King	-	-	-	-	-	-	-	-	ON	-	-	-	-	-	-	
P. minimus Tinsley	-	-	-	-	-	AB	-	-	-	-	-	-	-	-	-	
P. solani Ferris	-	-	-	-	BC	-	-	-	-	-	-	-	-	-	-	

Genus **PELIOCOCCUS** Borchsenius

P. serratus (Ferris)	-	-	-	-	-	-	-	-	ON	-	-	-	-	-	-	

Genus **PSEUDOCOCCUS** Westwood

P. affinis (Maskell)	-	-	-	-	BC*	-	-	-	-	-	NB*	-	-	-	-	[G]
P. comstocki (Kuwana) Comstock mealybug; cochenille de Comstock	-	-	-	-	BC*	-	-	-	ON*	-	-	-	-	-	-	[A]
P. maritimus (Ehrhorn) grape mealybug; cochenille de la vigne	-	-	-	-	BC	-	-	-	ON	QC	-	-	NS	-	-	

* introduced; [] indoors; { } extirpated; ? uncertain record; @ unestablished adventive; $ port-of-entry interception

Genus PUTO Signoret

P. bryanthi Ferris	-	YK	-	-	-	-	-	-	-	-	-	-	-	-	-
P. cupressi (Coleman) fir mealybug; cochenille du sapin	-	-	-	-	BC	-	-	-	-	-	-	-	-	-	-

Genus SPILOCOCCUS Ferris

S. geraniae (Rau)	-	-	-	-	BC	-	-	-	-	-	-	-	-	-	-
S. keiferi McKenzie	-	-	-	-	BC	-	-	-	-	-	-	-	-	-	-

Genus TRIDISCUS Ferris

T. oetvoesi Kozár & Foottit	-	-	-	-	BC	-	-	-	-	-	-	-	-	-	-

Genus TRIONYMUS Berg

T. americanus (Cockerell)	-	-	-	-	BC	AB	SK	-	-	-	-	-	-	-	-
T. caricis McConnel	-	-	-	-	BC	-	-	-	-	-	-	-	-	-	-
T. utahensis (Cockerell)	-	-	-	-	BC	-	-	-	-	-	-	-	-	-	-

Suborder Clypeorrhyncha

Cicadas, leafhoppers, froghoppers and treehoppers. Mouthparts arising from posterior part of head capsule; antennae short, bristle-like, arising in front of, or between eyes; lora well defined by sulcus; middle coxae short, approximate; tarsi with 3 segments; elaborate filter chamber in mesenteron.

Superfamily Cicadoidea

Keys to genera and species: Beirne 1961 [Canada].

Cicadas. Cicadas are primarily tropical insects although one species reaches the Mackenzie River, the northern world limit for cicadas. Immature cicadas live long, subterranean lives, feeding on the roots of various plants (primarily trees) and are occasionally imported to Canada on root balls of nursery stock, emerging to universal surprise in homes and hotel lobbies. The fore femur is armed with spines for digging; the hind tibia of the large adult has only 1 or 2 stout spines and usually a circlet of apical spines, but none on the tarsi. Adults are familiar to many Canadians from the buzzing sounds created by males signaling to females in midsummer, a noise that may be heard for as much as a kilometre away in the case of the common eastern "dog-day" cicada. These are the only animals that are known to be able to "turn off" their hearing by tightening a muscle attached to the eardrum.

All our species belong to the single family Cicadidae, following the classification of Moulds (1990). Subfamilies recognized here are based on monophyletic groupings: Cicadinae characterized by concealed tympana, Platypedinae by elongate male telson, and Tibicininae by the fore femur armed with three or more spines.

Few species are known from Canada, probably not many more than the 21 species listed here; 12 were estimated previously (Scudder 1979). The number may increase with analysis of mating calls, which should reveal sibling species.

Family CICADIDAE

Subfamily CICADINAE

Genus DICEROPROCTA

D. apache (Davis) citrus cicada	-	-	-	-	-	-	-	-	[ON]	-	-	[PE]	-	-	-	[A]

Genus TIBICEN Latreille

T. auletes (Germar)	-	-	-	-	-	-	-	-	ON	-	-	-	-	-	-	[r]
T. canicularis (Harris) dog-day cicada; Cigale pruinée	-	-	-	-	-	-	SK	MB	ON	QC	NB	PE	NS	-	-	
T. linnei Smith & Grossbeck	-	-	-	-	-	-	-	-	ON	-	-	-	-	-	-	
T. lyricen (DeGeer)	-	-	-	-	-	-	-	-	ON	-	-	-	-	-	-	
T. sayi Smith & Grossbeck	-	-	-	-	-	-	-	-	{ON}	-	-	-	-	-	-	

Subfamily PLATYPEDINAE

Genus OKANAGANA Davis

O. bella Davis	-	-	-	-	BC	AB	-	-	-	-	-	-	-	-	-	
O. canadensis (Provancher)	-	-	NT	-	BC	AB	SK	MB	ON	QC	NB	-	-	-	-	
O. fratercula Davis	-	-	-	-	-	AB	-	-	-	-	-	-	-	-	-	[r]
O. lurida Davis	-	-	-	-	BC	-	-	-	-	-	-	-	-	-	-	
O. luteobasalis Davis	-	-	-	-	-	AB	-	-	-	-	-	-	-	-	-	
O. noveboracensis Emmons	-	-	-	-	-	-	-	MB	ON	-	-	-	-	-	-	[r]

Okanagana occidentalis (Walker)	-	-	-	-	BC	AB	-	MB	-	-	-	-	-	-	-	
O. oregona Davis	-	-	-	-	BC	-	-	-	-	-	-	-	-	-	-	
O. ornata (Van Duzee)	-	-	-	-	BC	-	-	-	-	-	-	-	-	-	-	[r]
O. rimosa (Say)	-	-	-	-	-	AB	SK	MB	ON	QC	NB	-	NS	-	-	
O. synodica (Say)	-	-	-	-	-	AB	-	-	-	-	-	-	-	-	-	[r]
O. vanduzeei Distant	-	-	-	-	BC	-	-	-	-	-	-	-	-	-	-	

Genus PLATYPEDIA Uhler

P. areolata (Uhler) salmonfly	-	-	-	-	BC	-	-	-	-	-	-	-	-	-	-	[A]
P. putnami (Uhler)	-	-	-	-	BC	-	-	-	-	-	-	-	-	-	-	

Subfamily TIBICININAE

Genus MAGICICADA Davis

M. septendecim (Linnaeus) periodical cicada; cigale dix-sept ans	-	-	-	-	-	-	-	-	ON	{QC}	-	-	-	-	-	[A]

Superfamily Cercopoidea

Keys to genera and species: Hamilton 1982 [Canada].

Spittlebugs; froghoppers; cercopides. Froghoppers are primarily tropical insects but have a distinct fauna of common northern-adapted genera. Adults generally have long lives compared to other true bugs; one rare species in Canada (*Philaronia canadensis* Walley) overwinters and may be found in all months of the year except July. Larvae generally cover themselves with a frothy secretion.

Spittlebugs are only considered pests in confined situations such as greenhouses, although they may be vectors of xylem-borne diseases. One European species, the meadow spittlebug *Philaenus spumarius* (L.) is among the most common and ubiquitous of bugs in much of southern Canada, although it is rare on the prairies.

Classifications usually recognize two families, the world-wide, "spittle"-producing Cercopidae and the old world, tube-dwelling Machaerotidae, but many also place Aphrophoridae and Clastopteridae as separate families. The last of these makes only small, unstable frothy "spittle" masses; it appears to be related to old world genera (e.g. *Patriziana* Lallemand) that construct similar frothy masses that dry to form fragile tubes. They are probably relatives of Machaerotidae, and are here considered to merit separate family status. The boundaries between Aphrophorinae and Cercopinae are uncertain at best.

The froghopper fauna of Canada is well known with 24 species of Cercopidae and 12 species of Clastopteridae (Hamilton 1982). This compares well with the previous estimate of 38 species from Canada (Scudder 1979).

Family CERCOPIDAE

Genus APHROPHORA Germar

A. alni (Fallén) european alder spittlebug	-	-	-	-	-	-	-	-	ON*	QC*	-	-	-	-	-	
A. ampliata Hamilton	-	-	-	-	BC	-	-	-	-	-	-	-	-	-	-	[r]
A. canadensis Walley cone spittlebug	-	-	-	-	BC	-	-	-	-	-	-	-	-	-	-	
A. cribrata (Walker) pine spittlebug; aphrophore du pin	-	-	-	-	-	-	SK	MB	ON	QC	NB	PE	NS	-	-	
A. fulva Doering western pine spittlebug	-	-	-	-	BC	-	-	-	-	-	-	-	-	-	-	
A. gelida (Walker) boreal spittlebug	-	-	NT	-	BC	AB	SK	MB	ON	QC	NB	PE	NS	NF	-	
A. maculosa Doering	-	YK	-	-	BC	AB	-	-	-	-	-	-	-	-	-	

* introduced; [] indoors; { } extirpated; ? uncertain record; @ unestablished adventive; $port-of-entry interception

Aphrophora parallella (Say) spruce spittlebug; aphrophore de l'épinette	-	-	-	-	-	-	-	MB	ON	QC	NB	-	NS	-	-	
A. permutata Uhler douglas-fir spittlebug; aphrophore du Douglas	-	-	-	-	BC	-	-	-	-	-	-	-	-	-	-	
A. princeps Walley	-	-	-	-	BC	-	-	-	-	-	-	-	-	-	-	
A. quadrinotata Say four-spotted spittlebug	-	-	-	-	-	-	SK	MB	ON	QC	NB	PE	NS	-	-	
A. regina Hamilton	AK?	-	-	-	BC	-	-	-	-	-	-	-	-	-	-	
A. saratogensis (Fitch) Saratoga spittlebug; aphrophore Saratoga	-	-	-	-	-	-	-	MB	ON	QC	-	-	NS	-	-	

Genus LEPYRONIA Amyot & Serville

L. coleoptrata (Linnaeus)	-	-	-	-	-	-	-	-	ON*	QC*	-	-	-	-	-	[A]
L. gibbosa Ball hill-prairie spittlebug	-	-	-	-	-	-	SK	MB	-	-	-	-	-	-	-	
L. quadrangularis (Say) diamond-back spittlebug	-	-	-	-	-	AB	SK	MB	ON	QC	NB	-	NS	-	-	

Genus NEOPHILAENUS Haupt

N. lineatus (Linnaeus) lined spittlebug	-	-	-	-	BC*	-	-	-	ON*	QC*	NB*	PE*	NS*	NF*	-	[A]

Genus PARAPHILAENUS Vilbaste

P. parallelus (Stearns)	-	-	-	-	-	-	-	-	ON	-	-	-	-	-	-	

Genus PHILAENARCYS Hamilton

P. bilineata (Say) prairie spittlebug	AK	YK	NT	-	BC	AB	SK	MB	ON	QC	-	-	NS	-	-	
P. killa Hamilton sand-prairie spittlebug	-	-	-	-	-	-	-	-	ON	-	-	-	-	-	-	[r]
P. spartina Hamilton	-	-	-	-	-	-	-	-	-	QC	NB	PE	NS	-	-	

Genus PHILAENUS Stål

P. spumarius (Linnaeus) meadow spittlebug; cercope des prés	AK*	-	-	-	BC*	AB*	-	MB*	ON*	QC*	NB*	PE*	NS*	NF*	-	[A]

Genus PHILARONIA Ball

P. canadensis (Walley)	-	-	-	-	-	-	-	-	ON	-	-	-	-	-	-	[r]

Genus PROSAPIA Fennah

P. ignipectus (Fitch) black spittlebug	-	-	-	-	-	-	-	-	ON	-	-	-	-	-	-	

Family CLASTOPTERIDAE

Genus CLASTOPTERA Germar

C. achatina Germar pecan spittlebug	-	-	-	-	-	-	-	-	ON	-	-	-	-	-	-	
C. arborina Ball red-cedar spittlebug	-	-	-	-	-	-	-	-	ON	-	-	-	-	-	-	
C. atrapicata Hamilton	-	-	-	-	BC	-	-	-	-	-	-	-	-	-	-	
C. brunnea Ball	-	-	-	-	BC	-	-	-	-	-	-	-	-	-	-	
C. doeringae Hamilton mountain-juniper spittlebug	-	-	-	-	BC	-	-	-	-	-	-	-	-	-	-	
C. hyperici Gibson	-	-	-	-	-	-	-	-	ON	-	-	-	-	-	-	[r]
C. juniperina Ball	-	-	-	-	BC	-	-	-	-	-	-	-	-	-	-	[r]
C. obtusa (Say) alder spittlebug; cercope de l'aulne	-	-	-	-	-	AB	SK	MB	ON	QC	NB	PE	NS	-	-	
C. ovata Doering	-	-	-	-	BC	-	-	-	-	-	-	-	-	-	-	

A agricultural, F forestry, G greenhouse, O ornamental pest; r rare; b biocontrol agent

Clastoptera proteus Fitch dogwood spittlebug	-	-	-	-	BC*	-	-	MB	ON	QC	NB	PE	-	-	-	
C. saintcyri Provancher heath spittlebug	-	-	-	-	-	-	-	-	ON	QC	NB	PE	NS	-	-	
C. testacea Fitch	-	-	-	-	-	-	-	-	ON	QC	-	-	NS	-	-	

Superfamily Membracoidea

Family CICADELLIDAE

Key to genera: Oman 1949 [North America], Beirne 1956 [Canada].

Keys species: Nielson 1968 [North America (economic species)], Hamilton 1985 [Canada (species on trees)]. Because many of our common species are introductions from Europe, the following are also useful: Hamilton 1983 [North America], Ossiannilsson 1981, 1983 [northern Europe].

Leafhoppers; cicadellides. Leafhoppers are a very large family (sometimes subdivided into as many as 17 families) of world-wide distribution consisting of well over 16 000 species. The long hind tibia with one or more rows of slender spines are highly adapted for leaping, but make leafhoppers awkward at running (but excellent at dodging to the opposite side of a leaf or stem). Many "microleafhoppers" (subfamily Typhlocybinae) are wind-dispersed over enormous distances; by contrast, most "macroleafhoppers" (other subfamilies) seem to disperse at most a few kilometres a year by flight. The family includes a large proportion of the prairie endemic insect fauna and also has many species that are important vectors of plant diseases.

The generic and suprageneric classification is in a state of flux at present, with an overemphasis on male genitalic characters. Consequently, the genus groupings presented here are considered tentative.

Over 1100 species are known from Canada, with an estimated fauna of 1262 species, comparable to the previous estimate (Scudder 1979). Undescribed species are primarily in *Colladonus* and *Empoasca*.

Diplocolenus abdominalis (Fabricius) is recorded for the first time in North America.

Subfamily APHRODINAE

Genus ACINOPTERUS Van Duzee																
A. viridis Van Duzee	-	-	-	-	BC	AB	SK	-	-	-	-	-	-	-	-	
Genus AFLEXIA Oman																
A. rubranura (DeLong)	-	-	-	-	-	-	-	MB	ON	-	-	-	-	-	-	
Genus ALIGIA Ball																
A. californica Van Duzee	-	-	-	-	BC	-	-	-	-	-	-	-	-	-	-	[r]
Genus ALLYGUS Fieber																
A. atomarius (Fabricius)	-	-	-	-	-	-	-	-	ON*	QC*	-	-	-	-	-	[A]
A. mixtus (Fabricius)	-	-	-	-	BC*	-	-	-	-	-	-	-	NS*	-	-	[A]
Genus AMBLYSELLUS Sleesman																
A. curtisii (Fitch)	-	-	-	-	-	-	-	-	ON	QC	NB	-	-	-	-	
A. dorsti (Oman)	-	-	-	-	-	AB	SK	-	-	-	-	-	-	-	-	
A. grex (Oman)	-	-	-	-	BC	AB	SK	-	-	-	-	-	-	-	-	
A. punctatus (Osborn & Ball)	-	-	-	-	BC	AB	SK	MB	-	-	-	-	-	-	-	
A. valens (Beamer & Tuthill)	-	-	-	-	-	AB	SK	MB	-	-	-	-	-	-	-	
A. wyomus (Kramer)	-	-	-	-	BC	AB	SK	-	-	-	-	-	-	-	-	
Genus AMPLICEPHALUS DeLong																
A. inimicus (Say) painted leafhopper	-	-	NT	-	BC	AB	SK	MB	ON	QC	NB	PE	NS	NF	-	[A]
A. littoralis (Ball)	-	-	-	-	-	-	-	-	-	QC	-	-	NS	-	-	

* introduced; [] indoors; { } extirpated; ? uncertain record; @ unestablished adventive; $port-of-entry interception

Amplicephalus montanus (DeLong & Sleesman)	-	-	-	-	BC	-	-	-	-	-	-	-	-	-	-	
A. nebulosus (Ball)	AK	YK	NT	-	BC	-	SK	MB	-	-	-	-	-	-	-	
A. osborni (Van Duzee)	-	-	-	-	-	-	-	MB	ON	QC	NB	-	-	-	-	
A. simplarius (Osborn & Ball)	-	-	-	-	-	-	-	-	-	QC	NB	PE	NS	-	-	

Genus ANOSCOPUS Kirschbaum

A. albifrons (Linnaeus)	-	-	-	-	BC*	-	-	-	-	-	NB*	-	NS*	-	-	[A]
A. albiger (Germar)	-	-	-	-	BC*	-	-	-	ON*	QC*	NB*	-	-	-	-	[A]
A. flavostriatus (Donovan)	-	-	-	-	-	-	-	MB*	ON*	QC*	NB*	-	NS*	NF*	-	[A]
A. serratulae (Fabricius)	-	-	-	-	BC*	-	-	-	ON*	QC*	NB*	-	NS*	-	-	[A]

Genus APHRODES Curtis

A. bicinctus (Schrank) girdled leafhopper	-	-	-	-	BC*	-	-	-	ON*	QC*	NB*	PE*	NS*	NF*	-	[A]
A. costatus (Panzer)	-	-	-	-	BC*	-	-	-	ON*	QC*	-	-	-	NF*	-	[A]

Genus ATHYSANELLA Baker

A. acuticauda Baker	AK	YK	-	-	BC	AB	SK	MB	-	QC	-	-	-	-	-	
A. attenuata Baker	-	-	NT	-	BC	AB	SK	MB	-	-	-	-	-	-	-	
A. balli (Osborn)	-	-	-	-	-	-	-	-	-	QC	-	-	-	-	-	
A. bifida Ball & Beamer	-	-	-	-	-	AB	SK	MB	-	-	-	-	-	-	-	
A. incongrua Baker	-	-	-	-	-	-	-	MB	-	-	-	-	-	-	-	
A. kadokana Knull	-	-	-	-	-	AB	SK	MB	-	-	-	-	-	-	-	
A. longicauda Beirne	-	-	-	-	-	-	-	-	ON	QC	-	-	-	-	-	
A. macleani Blocker & Wesley	AK	YK	-	-	-	-	-	-	-	-	-	-	-	-	-	[r]
A. obesa Ball & Beamer	-	-	-	-	BC	AB	SK	-	-	-	-	-	-	-	-	
A. occidentalis Baker	-	-	-	-	-	AB	SK	-	-	-	-	-	-	-	-	
A. resusca Blocker & Wesley	-	YK	NT	-	-	-	-	-	-	-	-	-	-	-	-	
A. robusta Baker	-	YK	-	-	BC	AB	SK	MB	-	-	-	-	-	-	-	
A. secunda Blocker & Wesley	-	-	-	-	-	-	SK	MB	-	-	-	-	-	-	-	
A. sinuata Osborn	-	-	-	-	-	AB	SK	MB	-	-	-	-	-	-	-	
A. terebrans Gillette & Baker	-	-	-	-	-	AB	SK	MB	ON	-	-	-	-	-	-	
A. utahna (Osborn)	-	-	-	-	BC	-	-	-	-	-	-	-	-	-	-	

Genus ATHYSANUS Burmeister

A. argentarius Metcalf	-	-	-	-	-	-	SK*	MB*	ON*	QC*	NB*	PE*	NS*	-	-	[A]

Genus ATTENUIPYGA Oman

A. balli Oman	-	-	-	-	-	-	SK	-	-	-	-	-	-	-	-	
A. vanduzeei (Osborn & Ball)	-	-	-	-	-	-	-	MB?	-	-	-	-	-	-	-	[r]

Genus AURIDIUS Oman

A. auratus (Gillette & Baker)	-	-	-	-	BC	AB	SK	-	-	-	-	-	-	-	-	
A. helvus (DeLong)	-	-	-	-	BC	AB	-	MB	-	-	-	-	-	-	-	
A. ordinatus (Ball)	-	YK	-	-	BC	AB	SK	-	-	-	-	-	-	-	-	
A. sandaraca Hamilton	-	-	-	-	-	AB	-	MB	ON	-	-	-	-	-	-	

Genus BALCLUTHA Kirkaldy

B. abdominalis (Van Duzee)	-	-	-	-	-	AB	-	-	ON	QC	-	-	-	-	-	
B. confluens (Rey)	AK	YK	-	-	BC	-	-	MB	ON	QC	-	PE	NS	NF	-	
B. impicta (Van Duzee)	-	-	-	-	BC?	-	-	MB	ON	QC	NB	PE	NS	NF	-	
B. manitou (Gillette & Baker)	AK	YK	NT	-	BC	AB	SK	MB	-	-	-	-	-	NF	-	
B. neglecta (DeLong & Davidson)	-	-	-	-	BC	AB	SK	MB	ON	-	-	-	-	-	-	
B. punctata (Fabricius)	AK	YK	NT	-	BC	AB	SK	MB	ON	QC	NB	-	NS	NF	LB	

Genus BALLANA DeLong

B. atridorsum (Van Duzee)	-	-	-	-	-	AB	-	-	-	-	-	-	-	-	-	
B. calipera DeLong	-	-	-	-	BC	-	-	-	-	-	-	-	-	-	-	[r]
B. chelata DeLong	-	-	-	-	BC	-	-	-	-	-	-	-	-	-	-	[r]
B. hebea DeLong	-	-	-	-	BC	-	-	-	-	-	-	-	-	-	-	[r]
B. ortha DeLong	-	-	-	-	BC	AB	SK	-	-	-	-	-	-	-	-	
B. remissa DeLong	-	-	-	-	BC	-	-	-	-	-	-	-	-	-	-	[r]
B. titusi (Ball)	-	-	-	-	BC	-	-	-	-	-	-	-	-	-	-	[r]

Genus BANDARA Ball

B. curvata Knull	-	-	-	-	-	-	-	-	-	QC	NB	-	-	-	-	
B. johnsoni (Van Duzee)	-	-	-	-	-	-	-	-	ON	QC	-	-	-	-	-	
B. parallela Knull	-	-	-	-	-	-	-	-	ON	QC	NB	-	NS	-	-	

Genus CALADONUS Oman

C. coquilletti (Van Duzee)	-	-	-	-	BC	-	-	-	-	-	-	-	-	-	-	[r]

Genus CALANANA DeLong

C. rubralineata (Beamer)	-	-	-	-	BC	-	-	-	-	-	-	-	-	-	-	[r]

Genus CANTURA Oman

C. jucunda (Uhler)	-	-	-	-	-	-	-	-	ON	-	-	-	-	-	-	[r]

Genus CHLOROTETTIX Van Duzee

C. attenuatus Brown	-	-	-	-	-	-	-	-	ON	-	-	-	-	-	-	[r]
C. balli (Osborn)	-	-	-	-	-	-	-	MB	ON	QC	-	-	-	-	-	
C. borealis Sanders & DeLong	-	-	-	-	-	-	-	-	-	QC	-	-	-	-	-	
C. dentatus Sanders & DeLong	-	-	-	-	-	-	-	-	ON	-	-	-	-	-	-	
C. distinctus DeLong	-	-	-	-	-	-	-	-	-	QC	-	-	-	-	-	
C. fallax Sanders & DeLong	-	-	-	-	-	-	-	-	ON	-	-	-	-	-	-	[r]
C. galbanatus (Van Duzee)	-	-	-	-	-	AB?	-	-	ON	-	-	-	-	-	-	
C. lusorius (Osborn & Ball)	-	-	-	-	BC	AB	SK	MB	ON	QC	-	-	NS	-	-	
C. meriscus Cwikla	-	-	-	-	-	-	-	-	-	QC	-	-	-	-	-	
C. similis DeLong	-	-	-	-	BC	-	-	-	-	-	-	-	-	-	-	[r]
C. spatulatus Osborn & Ball	-	-	-	-	-	-	-	-	ON	QC?	-	-	-	-	-	
C. tergatus (Fitch)	-	-	-	-	-	-	-	-	ON	QC	-	-	-	-	-	
C. unicolor (Fitch)	-	-	NT	-	BC	AB	SK	MB	ON	QC	NB	PE	NS	NF	-	

Genus CICADULA Zetterstedt

C. ciliata (Osborn)	AK?	YK?	-	-	BC	AB	SK	MB	ON	QC	-	-	-	NF?	-	
C. cyperacea (Osborn)	-	-	-	-	-	-	-	-	ON	QC	NB	-	-	-	-	
C. intermedia (Boheman)	AK	YK	NT	NU	BC	AB	-	MB	-	QC	-	-	-	-	LB	
C. junea Hamilton	AK	-	-	-	-	-	SK	MB	-	-	-	-	-	-	-	
C. longiseta (Van Duzee)	-	YK	-	-	BC	AB	SK	MB	-	-	-	-	-	-	-	
C. melanogaster (Provancher)	AK	YK	NT	-	BC	AB	SK	MB	ON	QC	NB	-	NS	NF	-	
C. ornata (Melichor)	AK	YK	NT	-	BC	AB	SK	MB	-	-	NB	-	-	-	-	
C. persimilis (Edwards)	-	-	-	-	-	-	-	-	-	-	-	-	-	NF*	-	[A]
C. saliens Hamilton	-	-	-	-	-	-	-	MB	ON	QC	-	-	NS	-	-	
C. smithi (Van Duzee)	-	-	-	-	BC	AB	SK	MB	ON	QC	-	-	NS	NF	-	
C. straminea (Sanders & DeLong)	-	-	-	-	-	-	-	-	ON	QC	-	-	-	-	-	
C. subcupraea (Provancher)	AK	YK	NT	-	BC	AB	SK	MB	ON	QC	NB	PE	NS	NF	LB	
C. vittipennis (Sanders & DeLong)	-	-	-	-	BC	AB	SK	MB	ON	QC	-	-	NS	-	-	

Genus COLLADONUS Ball

C. arcanus Nielson	AK	-	-	-	BC	-	-	-	-	-	-	-	-	-	-	
C. arctostaphyli Downes	-	-	-	-	BC	-	-	-	-	-	-	-	-	-	-	[r]
C. atriflavus Downes	-	-	-	-	BC	-	-	-	-	-	-	-	-	-	-	

* introduced; [] indoors; { } extirpated; ? uncertain record; @ unestablished adventive; $port-of-entry interception

Colladonus aureolus (Van Duzee)	-	-	-	-	BC	-	-	-	-	-	-	-	-	-	-	
C. balius Hamilton	-	-	-	-	-	-	-	-	-	-	-	-	NS	-	-	[r]
C. balli Nielson	-	-	-	-	BC	-	-	-	-	-	-	-	-	-	-	
C. belli (Uhler)	-	YK	-	-	BC	AB	SK	-	-	-	-	-	-	-	-	
C. brunneus (Osborn)	-	-	-	-	-	-	-	-	ON	QC	NB	-	-	-	-	
C. clitellarius (Say) saddleback leafhopper	-	-	-	-	-	-	-	MB	ON	QC	NB	PE	NS	-	-	
C. collaris (Ball)	-	-	-	-	-	-	-	-	ON	-	-	-	-	-	-	
C. eburatus (Van Duzee)	-	-	-	-	BC	AB	SK	MB	ON	QC		PE	NS	NF	-	
C. egenus Ball	-	-	-	-	BC	-	-	-	-	-	-	-	-	-	-	
C. flavocapitatus (Van Duzee)	AK	YK	NT	-	BC	AB	-	-	-	-	-	-	-	-	-	
C. furculatus (Osborn)	-	-	-	-	-	-	-	-	ON	QC	-	-	NS	-	-	
C. geminatus (Van Duzee) geminate leafhopper	AK	-	-	-	BC	AB	-	-	-	-	-	-	-	-	-	[A]
C. holmesi Bliven	-	-	-	-	BC	-	-	-	-	-	-	-	-	-	-	
C. mendicus (Ball)	-	-	-	-	BC	-	-	-	-	-	-	-	-	-	-	[r]
C. montanus (Van Duzee) mountain leafhopper	-	-	-	-	BC	AB	SK	-	-	-	-	-	-	-	-	[A]
C. nugax (Van Duzee)	-	-	-	-	BC	-	-	-	-	-	-	-	-	-	-	
C. olus Nielson	-	-	-	-	-	-	SK	-	ON	-	-	-	-	-	-	
C. reductus (Van Duzee)	-	-	-	-	BC	-	-	-	-	-	-	-	-	-	-	
C. rupinatus (Ball)	-	-	-	-	BC	AB	-	-	-	-	-	-	-	-	-	
C. setaceus Nielson	-	-	-	-	-	-	-	-	ON	QC	NB	PE	NS	-	-	
C. tahotus Ball	AK	-	-	-	BC	-	-	-	-	-	-	-	-	-	-	
C. torneelus (Zetterstedt)	AK	YK	-	-	BC	-	-	-	-	-	-	-	-	-	-	
C. waldanus (Ball)	AK	YK	NT	-	BC	-	SK	MB	ON	-	-	-	-	-	-	
C. youngi Nielson	AK	YK	NT	-	BC	AB	SK	MB	ON	QC	NB	-	NS	NF	LB	
Genus COMMELLUS Osborn & Ball																
C. cedilla Hamilton	AK	YK	-	-	-	AB	SK	MB	ON	QC	-	-	-	-	-	
C. colon (Osborn & Ball)	-	-	-	-	-	-	-	MB	-	-	-	-	-	-	-	
C. comma (Van Duzee)	-	-	-	-	-	AB	SK	-	-	-	-	-	-	-	-	
C. hyphen Hamilton	-	-	-	-	-	-	SK	-	-	-	-	-	-	-	-	
C. sexvittatus (Van Duzee)	-	YK	-	-	BC	AB	SK	MB	-	-	-	-	-	-	-	
Genus COSMOTETTIX Ribaut																
C. acutus (Beamer)	-	-	-	-	-	-	-	-	ON	QC	-	-	-	-	-	
C. beirnei Hamilton & Ross	-	-	-	-	BC	AB	SK	MB	ON	-	-	-	-	-	-	
C. bilineatus (Gillette & Baker)	AK	YK	-	-	BC	AB	SK	MB	ON	QC	-	-	-	NF	-	
C. delector (Sanders & DeLong)	-	-	-	-	-	AB	-	-	ON	QC	-	-	-	-	-	
C. luteocephalus (Sanders & DeLong)	-	YK	-	-	BC	AB	SK	MB	ON	-	-	-	-	NF	-	
C. paludosa (Ball)	AK	-	-	-	BC	AB	-	-	ON	-	-	-	-	-	-	
C. unica Hamilton	-	-	-	-	-	-	-	-	-	-	-	PE	-	NF	-	
Genus COULINUS Beirne																
C. uladus Beirne	AK	YK	NT	NU	-	-	-	MB	-	QC	-	-	-	-	LB	
C. usnus Beirne	AK	YK	NT	NU	-	-	-	-	-	-	-	-	-	-	-	
Genus CRIBRUS Oman																
C. micmac Hamilton	-	-	-	-	-	-	-	-	-	-	-	-	NS	-	-	[r]
Genus DELTOCEPHALUS Burmester																
D. artemisiae (Gillette & Baker)	-	-	-	-	-	AB	SK	-	-	-	-	-	-	-	-	
D. balli Van Duzee	-	-	-	-	BC	AB	SK	MB	ON	QC	-	-	-	-	-	
D. castoreus Ball	-	YK	-	-	BC	AB	SK	-	-	QC?	-	-	-	-	-	
D. flavicostatus Van Duzee	-	-	-	-	-	-	-	-	-	-	NB	-	-	-	-	
D. fuscinervosus Van Duzee	-	-	-	-	BC	-	-	-	-	-	-	-	-	-	-	
D. gnarus Ball	-	-	-	-	-	AB	SK	MB	-	-	-	-	-	-	-	

Deltocephalus lineatifrons Oman	-	-	-	-	BC	AB	-	MB	-	-	-	-	-	-	-	
D. nigriventer Sanders & DeLong	AK	YK	NT	-	BC	AB	-	-	-	QC	-	-	NS	NF	LB	
D. pulicaris (Fallén)	AK	YK	-	-	-	-	-	-	ON*	QC*	NB*	PE*	NS*	NF*	-	[A]
D. serpentinus Hamilton & Ross	-	-	-	-	-	AB	SK	MB	ON	-	-	-	-	-	-	
D. sonorus Ball	-	-	-	-	-	-	-	-	-	-	NB	-	-	-	-	
D. vanduzeei Gillette & Baker	AK	-	-	-	BC	-	-	-	-	-	-	-	-	-	-	
D. vanfus Kramer	-	-	-	-	BC	-	-	-	-	-	-	-	-	-	-	
D. zephyrius Van Duzee	-	-	-	-	BC	-	-	-	-	QC	-	-	-	-	-	
Genus DESTRIA Oman																
D. crocea (Beirne)	-	-	-	-	-	-	SK	-	-	-	NB	PE	-	-	-	
D. fumida (Sanders & DeLong)	-	-	-	-	-	-	SK	MB	-	-	-	-	-	-	-	
Genus DICYPHONIA Ball																
D. ornata (Baker)	-	-	-	-	-	AB	SK	-	-	-	-	-	-	-	-	
Genus DIPLOCOLENUS Ribaut																
D. abdominalis (Fabricius)	-	-	-	-	-	-	-	-	ON*	QC*	NB*	-	-	-	-	[A]
D. aquilonius Ross & Hamilton	AK	YK	-	-	-	-	-	-	-	-	-	-	-	-	-	[r]
D. brevior Ross & Hamilton	-	YK	NT	-	BC	AB	-	-	-	-	-	-	-	-	-	
D. configuratus (Uhler)	-	-	NT	-	-	AB	SK	MB	ON	QC	-	PE	-	-	-	
D. evansi (Ashmead)	AK	YK	NT	-	BC	AB	SK	MB	ON	QC	-	-	NS	NF	LB	
Genus DOLIOTETTIX Ribaut																
D. lunulatus (Zetterstedt)	AK	YK	-	-	-	-	-	-	-	-	-	-	-	-	-	
Genus DORATURA Sahlberg																
D. stylata (Boheman)	-	-	-	-	BC*	AB*	-	MB*	ON*	QC*	NB*	PE*	NS*	-	-	[A]
Genus DORYCARA Emeljanov																
D. minor (Osborn)	-	-	-	-	-	AB	SK	MB	-	-	-	-	-	-	-	
D. omanae (Beamer)	-	-	-	-	BC	-	-	-	-	-	-	-	-	-	-	[r]
D. platyrhynchus (Osborn)	-	-	-	-	-	-	-	-	ON	-	-	-	-	-	-	
Genus DORYDIELLA Baker																
D. kansana Beamer	-	-	-	-	-	-	-	MB?	ON	-	-	-	-	-	-	
Genus DRIOTURA Osborn & Ball																
D. gammaroides (Van Duzee)	-	-	-	-	-	AB	SK	MB	ON	-	-	-	-	-	-	
D. robusta Osborn & Ball	-	-	-	-	-	-	SK	MB	-	-	-	-	-	-	-	
Genus ELYMANA DeLong																
E. circius Hamilton	-	-	-	-	BC	AB	SK	-	ON?	-	-	-	-	-	-	
E. inornata (Van Duzee)	AK	YK	NT	-	BC	AB	SK	MB	ON	QC	NB	-	NS	NF	-	
E. pacifica Hamilton	-	-	-	-	BC	-	-	-	-	-	-	-	-	-	-	
E. sulphurella (Zetterstedt)	-	-	-	-	-	-	-	-	ON*	QC*	NB*	PE*	NS*	NF*	-	[A]
Genus EUSCELIS Brullé																
E. aemulans (Kirschbaum)	AK	-	-	-	BC	-	SK	-	ON	QC	-	-	-	NF	-	
E. alpina (Ball)	AK	-	-	-	BC	AB	-	MB	-	QC	-	-	-	NF	-	
E. confinis (Reuter)	AK	YK	-	-	BC	AB	-	MB	ON	QC	-	-	-	-	-	
E. hyperborea Van Duzee	AK	YK	NT	NU	-	-	-	-	-	-	-	-	-	-	-	
E. maculipennis DeLong & Davidson	-	-	-	-	BC	AB	SK	MB	-	-	-	-	-	-	-	
E. obsoleta (Kirschbaum)	-	-	-	-	BC*	-	-	-	ON?	-	-	-	-	-	-	[A]
E. okaensis (Zachvatkin)	AK?	YK?	NT?	-	BC?	-	-	-	-	-	-	-	-	-	-	

* introduced; [] indoors; { } extirpated; ? uncertain record; @ unestablished adventive; $port-of-entry interception

Euscelis relativa (Gillette & Baker)	AK	-	-	-	BC	AB	SK	MB	ON	QC	-	-	NS	NF	-	
E. sibirica (Emeljanov)	AK?	YK?	-	-	-	-	-	-	-	-	-	-	-	-	-	
E. sordida (Zetterstedt)	AK	-	-	-	BC*	-	-	-	ON*	QC*	NB*	-	NS*	NF*	-	[A]

Genus EUTETTIX Van Duzee

E. borealis Hepner	-	-	-	-	-	-	-	-	ON	-	-	-	-	-	-	[r]
E. brunneus Osborn	-	-	-	-	-	-	-	-	ON	-	-	-	-	-	-	
E. glennanus Ball	-	-	-	-	BC	-	-	-	-	-	-	-	-	-	-	
E. marmoratus Van Duzee	-	-	-	-	-	-	-	-	ON	QC	-	-	-	-	-	
E. variabilis Hepner	-	-	-	-	-	-	-	-	ON	-	-	-	-	-	-	

Genus EXITIANUS Ball

E. exitiosus (Uhler)	-	-	-	-	BC	AB	SK	MB	ON	-	-	-	-	-	-	

Genus EXTRUSANUS Oman

E. extrusus (Van Duzee)	-	-	-	-	-	-	-	-	ON	QC	-	-	NS	-	-	
E. oryssus Hamilton	-	-	-	-	-	AB	SK	MB	-	-	-	-	-	-	-	
E. ovatus (Sanders & DeLong)	-	-	-	-	-	-	SK	-	-	-	-	-	-	-	-	

Genus FIEBERIELLA Signoret

F. florii (Stål) privet leafhopper	-	-	-	-	BC*	-	-	-	ON*	-	-	-	-	-	-	[O]

Genus FLEXAMIA DeLong

F. abbreviata (Osborn & Ball)	-	-	-	-	-	AB	SK	MB	-	-	-	-	-	-	-	
F. areolata (Ball)	-	-	-	-	-	-	-	-	ON	-	-	-	-	-	-	[r]
F. atlantica (DeLong)	-	-	-	-	-	-	-	MB	-	-	-	-	-	-	-	[r]
F. decora Beamer & Tuthill	-	-	-	-	-	AB	SK	MB	-	-	-	-	-	-	-	
F. delongi Ross & Cooley	-	-	-	-	-	-	-	MB	ON	-	-	-	-	-	-	
F. flexulosa (Ball)	-	-	-	-	-	AB	SK	MB	-	-	-	-	-	-	-	
F. graminea (DeLong)	-	-	-	-	-	-	-	MB	-	-	-	-	-	-	-	[r]
F. grammica (Ball)	-	-	-	-	-	AB	SK	-	-	-	-	-	-	-	-	
F. inflata (Osborn & Ball)	-	-	-	-	BC	-	SK	MB	ON	-	-	-	-	-	-	
F. prairiana DeLong	-	-	-	-	-	-	-	MB	ON	-	-	-	-	-	-	
F. reflexa (Osborn & Ball)	-	-	-	-	-	-	-	-	ON	-	-	-	-	-	-	[r]
F. serrata Beamer & Tuthill	-	-	-	-	BC	AB	SK	MB	-	-	-	-	-	-	-	
F. stylata (Ball)	-	-	-	-	-	AB	SK	MB	-	-	-	-	-	-	-	

Genus FRIGARTUS Oman

F. frigidus (Ball)	-	-	-	-	BC	AB	SK	-	-	-	-	-	-	-	-	

Genus GRAMINELLA DeLong

G. ampla Beamer	-	-	-	-	-	-	SK	MB	-	-	-	-	-	-	-	
G. fitchii (Van Duzee)	-	-	-	-	-	-	-	MB	ON	QC	NB	-	-	-	-	
G. mohri DeLong	-	-	-	-	-	-	-	-	ON	-	-	-	-	-	-	[r]
G. nigrifrons (Forbes) black-faced leafhopper	-	-	-	-	-	-	-	MB	ON	QC	NB	-	-	-	-	
G. oquaka DeLong	-	-	-	-	-	-	-	-	ON	-	-	-	-	-	-	
G. pallidula (Osborn)	-	-	-	-	-	-	-	-	ON	-	-	-	-	-	-	

Genus GRAPHOCRAERUS Thomson

G. ventralis (Fallén)	-	-	-	-	-	-	-	-	-	QC*	NB*	-	-	-	-	[A]

Genus HARDYA Edwards

H. dentata (Osborn & Ball)	-	-	-	-	BC	AB?	SK?	-	-	-	-	-	-	-	-	
H. youngi Beirne	AK	YK	-	-	-	-	-	-	-	-	-	-	-	-	-	

Genus HEBECEPHALUS DeLong

H. algidus DeLong & Davidson	AK	YK	NT	NU	BC	AB	-	MB	-	QC	-	-	-	-	LB	
H. beameri Hamilton & Ross	AK	-	-	-	-	-	-	-	-	-	-	-	-	-	-	
H. borealis DeLong & Davidson	-	-	-	-	BC	AB	SK	-	-	-	-	-	-	-	-	
H. crassus DeLong	-	-	-	-	BC	-	-	-	-	-	-	-	-	-	-	
H. occidentalis Beamer & Tuthill	AK	YK	-	-	BC	AB	SK	MB	-	-	-	-	-	-	-	
H. planaria Hamilton	-	-	-	-	BC	-	-	-	-	-	-	-	-	-	-	[r]
H. rostratus Beamer & Tuthill	AK	YK	-	-	-	AB	SK	MB	-	-	-	-	-	-	-	
H. sagittatus Beamer & Tuthill	-	YK	-	-	BC	-	-	-	-	-	-	-	-	-	-	
H. signatifrons (Van Duzee)	-	-	-	-	-	AB	-	-	-	-	-	-	-	-	-	
H. truncatus Beamer & Tuthill	-	YK	-	-	BC	AB	SK	MB	-	-	-	-	-	-	-	

Genus HECALUS Stål

H. atascaderus (Ball)	-	-	-	-	BC	-	-	-	-	-	-	-	-	-	-	
H. brunnea (Ball)	-	-	-	-	-	AB	SK	MB	-	-	-	-	-	-	-	
H. flavida (Signoret)	-	-	-	-	-	-	-	-	ON	-	-	-	-	-	-	[r]
H. grandis Shaw	-	-	-	-	-	AB	-	MB?	ON?	-	-	-	-	-	-	
H. major (Osborn)	-	-	-	-	BC	AB	SK	MB	ON	QC	NB	PE	-	-	-	
H. montanus (Ball)	AK	YK	NT	-	BC	AB	SK	MB	ON	QC	-	-	-	-	-	
H. viridis (Uhler)	-	-	-	-	BC	AB	SK	MB	ON	QC	-	-	-	-	-	

Genus IDIODONUS Ball

I. aurantiacus (Provancher)	AK	YK	NT	-	BC	AB	SK	-	ON	QC	-	-	-	-	-	
I. kennicottii (Uhler)	-	-	-	-	BC	AB	SK	MB	ON	QC	NB	-	NS	NF	-	
I. morsei (Osborn)	-	-	-	-	BC	AB	SK	MB	ON	QC	NB	-	NS	-	-	

Genus JAPANANUS Ball

J. hyalinus (Osborn)	-	-	-	-	BC*	-	-	-	ON*	-	-	-	-	-	-	[O]

Genus LAEVICEPHALUS DeLong

L. acus (Sanders & DeLong)	-	-	-	-	-	AB	SK	MB	ON	QC	NB	-	NS	NF	-	
L. exiguus Knull	-	-	-	-	-	-	SK	-	ON	-	-	-	-	-	-	[r]
L. melsheimerii (Fitch)	-	-	-	-	-	-	-	MB	ON	QC	NB	-	NS	-	-	
L. minimus (Osborn & Ball)	-	-	-	-	-	-	-	-	ON	-	-	-	-	-	-	
L. peronatus Ross & Hamilton	-	-	-	-	-	-	-	-	ON	-	-	-	-	-	-	[r]
L. poudris Tuthill	-	-	-	-	-	AB	-	MB	-	-	-	-	-	-	-	
L. salarius (Knull)	-	-	-	-	BC	-	-	-	-	-	-	-	-	-	-	
L. saskatchewanensis Hamilton & Ross	-	-	-	-	-	-	SK	MB	-	-	-	-	-	-	-	
L. sylvestris (Osborn & Ball)	-	-	-	-	-	-	-	-	ON	QC	NB	PE	NS	-	-	
L. unicoloratus (Gillette & Baker)	-	-	-	-	-	-	-	MB	ON	QC	-	-	-	-	-	

Genus LATALUS DeLong & Sleesman

L. curtus Beamer & Tuthill	-	-	-	-	BC	-	-	-	-	-	-	-	-	-	-	
L. distinguendus (Flor)	-	-	-	-	-	-	-	-	-	-	-	-	-	NF*	-	[A]
L. histrionicus Beirne	-	-	-	-	BC	-	-	-	-	-	-	-	-	-	-	
L. intermedius Ross & Hamilton	-	-	NT	-	BC	AB	-	-	-	-	-	-	-	-	-	
L. latidens (Sanders & DeLong)	-	-	-	-	-	-	-	MB	-	-	-	-	-	-	-	
L. missellus (Ball)	AK	YK	-	-	BC	AB	-	-	-	-	-	-	-	-	-	
L. mundus Beamer & Tuthill	-	-	-	-	BC	-	-	-	-	-	-	-	-	-	-	
L. occidentalis (DeLong)	-	-	-	-	BC	-	-	-	-	-	-	-	-	-	-	[r]
L. ocellaris (Fallén)	-	-	-	-	BC?	-	-	MB*	ON*	QC*	NB*	PE*	NS*	-	-	[A]
L. personatus Beirne	-	-	-	-	BC	AB	SK	MB	ON	QC	NB	PE	NS	-	-	
L. remotus Hamilton	-	-	-	-	-	-	-	-	ON	QC	-	-	NS	-	-	
L. sayii (Fitch)	-	-	-	-	-	AB	-	MB	ON	QC	NB	-	NS	-	-	
L. sobrinus DeLong & Sleesman	AK	YK	-	-	BC	AB	SK	MB	ON	QC	-	-	-	-	LB	
L. tatraensis (Heller)	AK	YK	NT	-	-	-	-	-	-	-	-	-	-	-	-	

* introduced; [] indoors; { } extirpated; ? uncertain record; @ unestablished adventive; $port-of-entry interception

Genus LIMOTETTIX Sahlberg

L. angustatus (Osborn)	AK	YK	-	-	BC	AB	SK	MB	ON	QC	-	-	NS	NF	LB	
L. anthracinus (Van Duzee)	-	-	-	-	-	-	-	-	-	QC	-	-	-	-	-	
L. arctostaphyli (Ball)	AK	YK	NT	-	BC	AB	SK	MB	ON	QC	-	-	-	NF	LB	
L. balli (Medler)	-	-	-	-	-	-	-	MB	ON	-	-	-	-	-	-	
L. beameri (Medler)	-	-	-	-	BC	-	-	-	-	-	-	-	-	-	-	[r]
L. bisoni Knull	-	-	-	-	-	AB	SK	MB	ON	-	-	-	-	-	-	
L. brooksi Hamilton	-	-	-	-	-	AB	-	-	-	-	-	-	-	-	-	
L. comptonianus (Ball)	-	-	-	-	-	-	-	-	ON	QC	NB	-	-	-	-	
L. conservatus Hamilton	-	-	-	-	-	AB	SK	-	ON	-	-	-	-	-	-	
L. cuneatus (Sanders & DeLong)	-	-	-	-	-	-	-	-	ON	-	-	-	-	-	-	[r]
L. dasidus (Medler)	AK	YK	-	-	BC	AB	SK	MB	-	QC	-	-	-	-	-	
L. divaricatus (Sanders & DeLong)	-	-	-	-	BC	-	SK	MB	ON	-	-	-	-	-	-	
L. elegans Hamilton	-	-	-	-	-	-	-	-	ON	-	-	-	-	-	-	[r]
L. ferganensis Dubovsky	AK	YK	-	-	BC	AB	SK	MB	ON	QC	NB	PE	NS	NF	-	
L. finitimus (Van Duzee)	-	-	-	-	BC	-	-	-	-	-	-	-	-	-	-	
L. glomerosus (Ball)	-	-	-	-	BC	-	-	-	-	-	-	-	-	-	-	
L. humidus (Osborn)	-	-	-	-	BC	-	-	-	ON	QC	NB	PE	NS	NF	-	
L. instabilis (Van Duzee)	AK	-	-	-	BC	AB	SK	MB	ON	QC	-	PE	NS	NF	LB	
L. kryptus (Medler)	-	-	-	-	-	AB	SK	MB	ON	QC	-	-	NS	-	-	
L. luteolus (Sleesman)	-	-	-	-	-	-	-	-	-	QC?	NB	-	NS	-	-	
L. medleri Hamilton	-	-	-	-	-	-	SK	-	-	-	-	-	-	-	-	
L. melastigmus (Medler)	-	-	-	-	BC	AB	-	MB	ON	QC	-	-	-	-	-	
L. myralis (Medler)	AK	-	-	-	-	AB	SK	-	-	-	-	-	-	-	-	
L. nigrax Medler	-	-	-	-	BC	-	-	-	ON	-	-	-	-	-	-	
L. nigristriatus Hamilton	-	-	-	-	-	-	-	-	ON	-	-	-	-	-	-	
L. obesura Hamilton	-	-	-	-	BC	-	-	-	-	-	-	-	-	-	-	[r]
L. omani (Medler)	-	-	-	-	-	AB	-	MB	-	-	-	-	-	-	-	
L. osborni (Ball)	-	YK	-	-	BC	AB	SK	MB	ON	QC	NB	-	NS	-	-	
L. paludosus (Boheman)	AK	YK	NT	-	BC	AB	SK	MB	ON	QC	-	PE	NS	NF	-	
L. parallelus (Van Duzee)	-	-	-	-	-	AB	SK	MB	ON	QC	NB	-	NS	-	-	
L. plutonius (Uhler)	AK	YK	-	-	BC	AB	SK	MB	-	-	-	-	-	-	-	
L. pseudosphagneticus Hamilton	-	-	-	-	-	-	-	-	ON	-	-	-	-	-	-	
L. schedia Hamilton	-	-	-	-	-	-	-	-	ON	-	NB	PE	NS	-	-	
L. scudderi Hamilton	-	YK	-	-	-	-	-	-	-	-	-	-	-	-	-	[r]
L. shastus (Ball)	-	-	-	-	BC	AB	SK	-	-	-	-	-	-	-	-	
L. sphagneticus Emeljanov	AK	YK	NT	-	BC	AB	SK	MB	ON	QC	NB	-	NS	NF	-	
L. symphoricarpae (Ball)	-	-	-	-	-	AB	-	MB	-	-	-	-	-	-	-	
L. taramus (Medler)	-	-	-	-	-	-	SK	-	-	-	-	-	-	-	-	
L. uhleri (Oman)	-	-	-	-	-	AB	SK	MB	ON	QC	-	-	-	-	-	
L. uneolus (Ball)	-	-	-	-	BC	AB	SK	MB	-	-	-	-	-	-	-	
L. urnura Hamilton	-	-	-	-	-	-	SK	MB	ON	-	-	-	-	-	-	
L. utahnus (Lawson)	-	-	-	-	BC	AB	SK	MB	-	-	-	-	-	-	-	
L. vaccinii (Van Duzee) bluntnosed cranberry leafhopper	AK?	-	-	-	BC?	AB	SK	MB	ON	QC	NB	PE	NS	NF	-	[A]
L. varus (Ball)	-	-	-	-	-	-	-	MB	ON	-	-	-	-	-	-	
L. xanthus Hamilton	-	-	-	-	BC	-	-	-	-	-	-	-	-	-	-	

Genus LONATURA Osborn & Ball

L. megalopa Osborn & Ball	-	-	-	-	-	-	SK	MB	-	-	-	-	-	-	-
L. melina DeLong	-	-	-	-	-	AB	SK	-	-	-	-	-	-	-	-
L. salsura Ball	-	-	-	-	BC	AB	SK	MB	-	-	-	-	-	-	-
L. teretis Beamer	-	-	-	-	-	AB	SK	MB	-	-	-	-	-	-	-

Genus MACROSTELES Fieber

M. alpinus (Zetterstedt)	AK	YK	NT	NU	BC	-	-	MB	ON	QC	-	-	-	-	LB
M. americanus (Van Duzee)	-	-	-	-	-	-	-	-	ON	QC	-	-	-	-	-

Species	AK	YK	NT	NU	BC	AB	SK	MB	ON	QC	NB	PE	NS	NF	LB	
Macrosteles bifurcatus Beirne	AK	-	-	-	-	-	-	MB	ON	QC	NB	-	NS	-	-	
M. binotatus (Sahlberg)	AK	YK	-	-	BC	AB	SK	MB	ON	QC	NB	-	NS	NF	-	
M. borealis (Dorst)	AK	YK	NT	-	BC	AB	SK	MB	ON	QC	-	-	-	NF	-	
M. curvata Beirne	-	-	-	-	-	-	-	MB	-	QC	-	-	-	NF	LB	
M. divisus (Uhler)	-	-	-	-	BC	AB	SK	-	-	-	-	-	-	-	-	
M. elongatus Beirne	-	-	-	-	BC	AB	SK	MB	-	QC	NB	-	-	NF	-	
M. fascifrons (Stål)	AK	YK	NT	NU	BC	-	-	MB	ON	QC	NB	-	NS	NF	LB	
M. fieberi (Edwards)	AK	YK	NT	-	BC	-	-	MB	-	QC	-	-	-	NF	-	
M. flavalis Hamilton	-	-	-	-	-	-	-	MB	-	-	-	-	-	-	-	
M. frontalis (Scott)	AK	-	-	-	-	-	-	-	-	-	-	-	-	-	-	
M. galeae Hamilton	-	-	-	-	-	-	-	-	ON	QC	NB	-	NS	NF	-	
M. inundatus Hamilton	AK	YK	-	-	BC	-	-	-	-	QC	-	-	NS	NF	-	
M. laevis (Ribaut) european grain leafhopper	AK	YK	-	-	BC	-	-	-	-	-	-	-	-	-	-	
M. lepidus (Van Duzee)	-	-	-	-	-	-	-	-	ON	QC	-	-	-	-	-	
M. lineatifrons (Stål)	AK	-	-	-	BC	-	-	-	-	-	-	-	-	-	-	
M. maculipes (Zetterstedt)	AK	YK	NT	-	BC	AB	-	-	ON	QC	-	-	NS	NF	LB	
M. major (Dorst)	-	-	-	-	-	AB	SK	MB	ON	-	-	-	-	-	-	
M. osborni (Dorst)	AK	-	-	-	-	-	-	-	-	-	-	-	-	-	-	
M. pallidus (Osborn)	-	-	-	-	-	-	-	MB	ON	QC	NB	PE	NS	NF	-	
M. potorius (Ball)	AK	-	-	-	-	AB	SK	MB	ON	QC	-	-	-	-	-	
M. quadrilineatus (Forbes) aster leafhopper; cicadelle de l'aster	-	YK	-	-	BC	AB	SK	MB	ON	QC	NB	PE	NS	NF	-	[A]
M. slossonae (Van Duzee)	-	YK	-	-	BC	AB	-	-	ON	QC	NB	-	NS	-	-	
M. snowi (Dorst)	-	-	-	-	-	AB	SK	MB	-	-	-	-	-	-	-	
M. sordidipennis (Stål)	AK	YK	NT	-	BC	AB	SK	-	-	-	-	-	-	-	-	
M. tessellatus Hamilton	AK	YK	NT	-	BC	AB	SK	MB	ON	QC	NB	-	NS	-	-	
M. urticae Moore & Ross	-	-	NT	-	BC	AB	-	MB	ON	QC	-	-	NS	-	-	
M. variatus (Fallén)	AK?	YK?	-	-	BC?	AB?	SK?	MB?	ON?	QC?	NB?	PE?	NS?	NF?	-	
M. vilbastei Hamilton	AK	-	-	-	BC	AB	SK	MB	ON	QC	NB	-	NS	-	-	

Genus MEMNONIA Ball

Species	AK	YK	NT	NU	BC	AB	SK	MB	ON	QC	NB	PE	NS	NF	LB	
M. consobrina Ball	-	-	-	-	-	AB	-	-	-	-	-	-	-	-	-	

Genus MENOSOMA Ball

Species	AK	YK	NT	NU	BC	AB	SK	MB	ON	QC	NB	PE	NS	NF	LB	
M. cincta (Osborn & Ball)	-	-	-	-	-	-	-	-	ON	QC	-	-	-	-	-	

Genus MESAMIA Ball

Species	AK	YK	NT	NU	BC	AB	SK	MB	ON	QC	NB	PE	NS	NF	LB	
M. frigida (DeLong & Hershberger)	-	-	-	-	BC	-	-	-	-	-	-	-	-	-	-	
M. ludovicia Ball	-	-	-	-	-	AB	SK	MB	-	-	-	-	-	-	-	
M. nigridorsum (Ball)	-	-	-	-	-	AB	SK	MB	ON	-	-	-	-	-	-	
M. straminea (Osborn)	-	-	-	-	-	AB	-	-	-	-	-	-	-	-	-	

Genus MOCUELLUS Ribaut

Species	AK	YK	NT	NU	BC	AB	SK	MB	ON	QC	NB	PE	NS	NF	LB	
M. americanus Emeljanov	-	-	-	-	-	AB	SK	MB	ON	-	-	-	-	-	-	
M. caprillus Ross & Hamilton	-	-	-	-	-	AB	SK	-	-	-	-	-	-	-	-	
M. larrimeri (DeLong)	-	-	-	-	BC	-	-	-	-	-	-	-	-	-	-	
M. strictus Ross & Hamilton	-	YK	-	-	BC	AB	-	-	-	-	-	-	-	-	-	

Genus NEOALITURUS Distant

Species	AK	YK	NT	NU	BC	AB	SK	MB	ON	QC	NB	PE	NS	NF	LB	
N. tenellus (Baker) beet leafhopper; cicadelle de la betterave	-	-	-	-	BC	AB?	-	-	-	-	-	-	-	-	-	[A]

Genus NEOHECALUS Linnavuori

Species	AK	YK	NT	NU	BC	AB	SK	MB	ON	QC	NB	PE	NS	NF	LB	
N. lineatus (Uhler)	-	-	-	-	-	-	-	MB?	-	QC	NB	PE	NS	-	-	

Genus NORVELLINA Ball

Species	AK	YK	NT	NU	BC	AB	SK	MB	ON	QC	NB	PE	NS	NF	LB	
N. chenopodii (Osborn)	-	-	-	-	BC	AB	SK	MB	ON	QC	-	-	-	-	-	
N. clarivida (Van Duzee)	-	-	-	-	-	AB	-	-	-	-	-	-	-	-	-	

Norvellina columbiana (Ball)	-	-	-	-	BC	AB	-	-	-	-	-	-	-	-	-	
N. novica Medler	-	-	-	-	-	-	-	MB	ON	QC	-	-	-	-	-	
N. rubida (Ball)	-	-	-	-	BC	-	-	-	-	-	-	-	-	-	-	
N. scitula (Ball)	-	-	-	-	BC	-	-	-	-	-	-	-	-	-	-	
N. seminuda (Say)	-	-	-	-	BC	AB	SK	MB	ON	-	-	-	-	-	-	

Genus OLLARIANUS Ball

O. insanus (Ball)	-	-	-	-	-	AB	-	-	-	-	-	-	-	-	-	

Genus OPSIUS Fieber

O. stactogalus Fieber	-	-	-	-	BC*	-	-	-	ON*	-	-	-	-	-	-	[O]

Genus ORIENTUS DeLong

O. ishidae (Matsumura)	-	-	-	-	-	-	-	-	ON*	-	-	-	-	-	-	[O]

Genus OROCASTUS Oman

O. labeculus (DeLong)	-	-	-	-	BC	AB	SK	-	-	-	-	-	-	-	-	
O. perpusillus (Ball & DeLong)	-	-	-	-	BC	AB	SK	MB	-	-	-	-	-	-	-	
O. pinnipenis Ross & Hamilton	-	-	-	-	BC	-	-	-	-	-	-	-	-	-	-	
O. tener (Beamer & Tuthill)	-	YK	-	-	-	AB	SK	-	-	-	-	-	-	-	-	

Genus OSBORNELLUS Ball

O. alatus Beamer	-	-	-	-	-	-	-	-	-	QC	-	-	-	-	-	
O. auronitens (Provancher)	-	-	-	-	-	-	-	-	ON	QC	-	PE	-	-	-	
O. borealis DeLong & Mohr	-	-	-	-	BC	AB	SK	-	-	-	-	-	-	-	-	
O. consors (Uhler)	-	-	-	-	-	-	-	MB	ON	QC	NB	-	-	-	-	
O. limosus DeLong	-	-	-	-	-	-	-	MB	ON	QC	-	-	-	-	-	
O. scalaris (Van Duzee)	-	-	-	-	BC	-	-	-	-	QC?	-	-	-	-	-	

Genus PALUDA DeLong

P. gladiola (Ball)	-	YK	-	-	BC	AB	SK	MB	ON	QC	NB	PE	NS	NF	-	

Genus PARAMESUS Fieber

P. major (Haupt)	-	-	-	-	BC?*	-	-	-	ON*	QC*	NB*	-	-	-	-	[A]

Genus PARAPHLEPSIUS Baker

P. apertinus (Osborn & Lathrop)	-	-	-	-	BC	-	-	-	-	-	-	-	-	-	-	
P. apertus (Van Duzee)	AK	YK	NT	-	BC	AB	SK	MB	ON	QC	NB	PE	NS	NF	LB	
P. bifidus (Sanders & DeLong)	-	-	-	-	BC	-	SK	MB	ON	QC	NB	-	-	-	-	
P. collitus (Ball)	-	-	-	-	-	-	-	-	ON	QC	-	-	-	-	-	
P. continuus (DeLong)	-	-	-	-	-	AB	SK	-	ON	-	-	-	-	-	-	
P. eburneolus (Osborn & Lathrop)	-	-	-	-	-	-	SK	MB	-	-	-	-	-	-	-	
P. electus (DeLong)	-	-	-	-	-	-	-	MB	-	-	-	-	-	-	-	[r]
P. excavatus Crowder	-	-	-	-	BC	-	-	-	-	-	-	-	-	-	-	[r]
P. fulvidorsum (Fitch)	-	-	-	-	BC?	-	-	-	ON	QC	NB	PE	NS	-	-	
P. fuscipennis (Van Duzee)	-	-	-	-	-	-	-	-	ON	-	NB	-	NS	NF	-	
P. hemicolor (Sanders & DeLong)	-	-	-	-	-	-	-	MB	-	-	-	-	-	-	-	
P. humidus (Van Duzee)	-	-	-	-	-	-	-	-	ON	QC	-	-	-	-	-	
P. incisus Van Duzee	-	-	-	-	-	-	-	-	ON	QC	-	-	-	-	-	
P. irroratus (Say) irrorate leafhopper	-	-	-	-	BC	AB	SK	MB	ON	QC	NB	PE	NS	NF	-	[A]
P. lascivius (Ball)	-	YK	-	-	BC	AB	SK	MB	-	-	-	-	-	-	-	
P. lobatus (Osborn)	-	-	-	-	-	-	-	MB	ON	QC?	-	-	-	-	-	
P. luxuria Hamilton	-	-	-	-	-	-	-	-	ON	QC	-	-	-	-	-	
P. occidentalis (Baker)	-	-	-	-	BC	-	-	-	-	-	-	-	-	-	-	
P. solidaginis (Walker)	-	-	-	-	-	AB	SK	MB	ON	QC	NB	-	NS	-	-	
P. tigrinus (Ball)	-	-	-	-	-	-	-	-	ON	-	-	-	-	-	-	

A agricultural, F forestry, G greenhouse, O ornamental pest; r rare; b biocontrol agent

Paraphlepsius truncatus (Van Duzee)	-	-	-	-	-	-	-	-	ON	-	-	-	-	-	-	
P. turpiculus (Ball)	-	-	-	-	-	AB	SK	MB	ON	-	-	-	-	-	-	
P. umbrosus (Sanders & DeLong)	-	-	-	-	-	-	SK	MB	ON	-	-	-	-	-	-	
P. varispinus Hamilton	-	-	-	-	-	AB	-	MB	ON	QC	NB	-	NS	NF	-	

Genus PASAREMUS Oman

P. concentricus (Van Duzee)	-	YK	-	-	BC	AB	SK	MB	-	QC	NB	PE	NS	-	-	

Genus PENDARUS Ball

P. franconianus (Ball)	-	-	-	-	-	-	-	-	ON	QC	NB	-	-	-	-	
P. magnus (Osborn & Ball)	-	-	-	-	-	AB	SK	MB	-	-	-	-	-	-	-	
P. punctiscriptus (Van Duzee)	-	-	-	-	-	-	-	MB	ON	-	-	-	-	-	-	

Genus PENTHIMIA Germar

P. americana Fitch	-	-	-	-	-	-	-	MB	ON	QC	-	-	-	-	-	
P. picta Provancher	-	-	-	-	-	-	-	-	-	QC	NB	-	NS	-	-	

Genus PINUMIUS Ribaut

P. sexmaculatus (Gillette & Baker)	-	-	-	-	BC	AB	SK	-	-	-	-	-	-	-	-	

Genus PLANAPHRODES Hamilton

P. bifasciatus (Linnaeus)	-	-	-	-	-	-	-	-	-	-	-	-	-	NF*	-	[A]

Genus PLATYMETOPIUS Linnavuori

P. twiningi (Uhler)	-	-	-	-	BC	AB	SK	MB	ON	QC	NB	-	NS	-	-	
P. vitellina (Fitch)	-	-	-	-	BC?	AB?	SK?	MB?	ON	QC	NB	-	NS	-	-	

Genus POLYAMIA DeLong

P. apicata (Osborn)	-	-	-	-	-	-	-	-	ON	QC	NB	-	NS	-	-	
P. caperata (Ball)	-	-	-	-	-	-	-	MB	ON	-	-	-	-	-	-	
P. compacta (Osborn & Ball)	-	-	-	-	-	-	-	-	ON	QC	-	-	NS	-	-	
P. obtecta (Osborn & Ball)	-	-	-	-	-	-	-	-	ON	QC	-	-	-	-	-	
P. similaris DeLong & Davidson	-	-	-	-	-	-	-	-	ON	-	-	-	-	-	-	[r]
P. weedi (Van Duzee)	-	-	-	-	-	-	-	-	-	QC	-	-	-	-	-	

Genus PRESCOTTIA Ball

P. lobata (Van Duzee)	-	-	-	-	-	-	-	MB	ON	QC	NB	PE	-	-	-	

Genus PSAMMOTETTIX Haupt

P. attenuens (DeLong & Davidson)	-	-	-	-	BC	-	-	-	-	-	-	-	-	-	-	
P. beirnei Greene	-	-	-	-	BC	-	-	-	-	-	-	-	-	-	-	[r]
P. confinis (Dahlbom)	AK	YK	NT	-	BC	-	-	-	-	-	-	-	-	-	-	
P. dentatus Knull	-	-	-	-	BC	-	-	-	-	-	-	-	-	-	-	[r]
P. knullae Greene	AK	YK	-	-	BC	AB	SK	MB	ON	-	-	-	-	-	-	
P. lapponicus (Ossiannilsson)	AK	YK	NT	NU	-	-	-	MB		-	-	-	-	-	-	
P. latipex (Sanders & DeLong)	AK	YK	NT	-	BC	AB	SK	MB	-	QC	-	-	-	-	-	
P. lividellus (Zetterstedt)	AK	YK	NT	NU	BC	AB	SK	MB	ON	QC	NB	PE	NS	NF	LB	

Genus ROSENUS Oman

R. acutus (Beamer)	-	-	-	-	BC	AB	-	MB	-	-	-	-	-	-	-	
R. cruciatus (Osborn & Ball)	-	-	-	-	BC	AB	SK	MB	-	-	-	-	-	-	-	
R. decurvus Hamilton & Ross	-	-	-	-	BC	-	-	-	-	-	-	-	-	-	-	[r]
R. obliquus (DeLong & Davidson)	-	-	-	-	BC	-	-	-	-	-	-	-	-	-	-	
R. pendulus Hamilton & Ross	AK	YK	-	-	BC	-	-	-	-	-	-	-	-	-	-	
R. transarcticus Hamilton & Ross	AK	YK	NT	-	BC	-	-	-	-	-	-	-	-	-	-	

* introduced; [] indoors; { } extirpated; ? uncertain record; @ unestablished adventive; $port-of-entry interception

Genus SANCTANUS Ball

S. sanctus (Say)	-	-	-	-	-	-	-	-	ON	-	-	-	-	-	-	[r]

Genus SCAPHOIDEUS Uhler

S. atlantus Ball	-	-	-	-	-	-	-	-	ON	-	-	-	-	-	-	
S. blennus Barnett	-	-	-	-	-	-	-	-	-	QC	-	-	-	-	-	
S. carinatus Osborn	-	-	-	-	-	-	-	-	ON	QC	NB	-	NS	-	-	
S. cinerosus Osborn	-	-	-	-	BC	-	-	MB	ON	QC	-	-	-	-	-	
S. crassus DeLong & Mohr	-	-	-	-	BC	-	-	MB	ON	QC	-	-	-	-	-	
S. curvatus DeLong & Mohr	-	-	-	-	-	-	-	-	ON	QC	-	-	-	-	-	
S. cylindratus DeLong & Beery	-	-	-	-	-	-	-	-	ON	QC	-	PE	NS	-	-	
S. cyprius Ball	-	-	-	-	-	-	-	-	ON	QC	NB	PE	-	-	-	
S. flavidus Barnett	-	-	-	-	-	-	-	-	-	QC	-	-	-	-	-	
S. immistus (Say)	-	-	-	-	BC	-	SK	-	ON	QC	-	-	-	-	-	
S. incisus Osborn	-	-	-	-	-	-	-	-	ON	QC	-	-	-	-	-	
S. incognitus Barnett	-	-	-	-	-	-	-	-	-	QC	-	-	-	-	-	
S. intricatus Uhler	-	-	-	-	-	-	-	-	ON	-	-	-	-	-	-	
S. lacyi Barnett	-	-	-	-	-	-	SK	-	ON	QC	-	-	-	-	-	
S. major Osborn	-	-	-	-	-	-	-	-	ON	QC	-	-	-	-	-	
S. mandevillei Barnett	-	-	-	-	-	-	-	-	-	QC	-	-	-	-	-	
S. melanotus Osborn	-	-	-	-	-	-	-	-	ON	QC	-	-	-	-	-	
S. merus DeLong & Beery	-	-	-	-	-	-	-	-	-	QC	-	-	-	-	-	
S. minor Osborn	-	-	-	-	-	-	-	MB	ON	QC	NB	PE	NS	-	-	
S. nigrellus DeLong & Mohr	-	-	-	-	-	-	-	-	ON	-	-	-	-	-	-	
S. ochraceus Osborn	-	-	-	-	-	-	-	-	ON	QC	-	-	-	-	-	
S. opalinus Osborn	-	-	-	-	-	-	-	-	ON	QC	-	-	-	-	-	
S. paludosus Ball	-	-	-	-	-	-	-	-	ON	-	-	-	-	-	-	
S. productus Osborn	-	-	-	-	-	-	-	-	-	QC	-	-	-	-	-	
S. pullus DeLong & Mohr	-	-	-	-	-	-	-	MB	ON	QC	-	-	-	-	-	
S. soleus DeLong	-	-	-	-	-	-	-	-	-	QC	-	-	-	-	-	
S. tergatus DeLong	-	-	-	-	-	-	-	-	ON	-	-	-	-	-	-	
S. titanus Ball	-	-	-	-	-	-	SK	MB	ON	QC	-	-	NS*	-	-	[A]
S. torqus DeLong & Beery	-	-	-	-	-	-	-	-	-	QC	-	-	-	-	-	
S. transeus DeLong & Beery	-	-	-	-	-	-	-	-	-	QC	-	-	-	-	-	
S. veterator DeLong & Beery	-	-	-	-	-	-	-	MB	ON	QC	NB	-	NS	-	-	

Genus SCAPHYTOPIUS Ball

S. acutus (Say) sharpnosed leafhopper	AK	YK	NT	-	BC	AB	SK	MB	ON	QC	NB	PE	NS	NF	-	
S. angustatus (Osborn)	-	-	-	-	-	-	-	-	ON	QC	-	-	-	-	-	
S. argutus (DeLong)	-	-	-	-	-	-	-	-	-	QC	-	-	-	-	-	
S. cinereus (Osborn & Ball)	-	-	-	-	-	AB?	-	-	ON	-	-	-	-	-	-	[r]
S. cinnamoneus (Osborn)	-	-	-	-	-	AB	SK	MB	ON	QC	-	PE	NS	-	-	
S. diabolus (Van Duzee)	-	-	-	-	BC	-	-	-	-	-	-	-	-	-	-	
S. frontalis (Van Duzee) yellowfaced leafhopper	-	-	-	-	BC*	-	-	-	ON	-	-	-	-	-	-	
S. fulvus (Osborn)	-	-	-	-	-	-	-	-	-	QC	-	-	-	-	-	
S. hastus (DeLong)	-	-	-	-	-	-	-	-	ON	QC	-	-	-	-	-	
S. latus (Baker)	-	-	-	-	BC	AB	SK	MB	ON	QC	-	-	-	-	-	
S. oregonensis (Baker)	-	-	-	-	BC	-	-	-	-	-	-	-	-	-	-	
S. vaccinium (DeLong) blueberry leafhopper	-	-	-	-	-	AB	-	MB	ON	QC	NB	PE	NS	NF	-	[A]

Genus SORHOANUS Ribaut

S. bimaculatus (Gillette & Baker)	AK	YK	NT	-	BC	AB	SK	MB	ON	QC	-	-	-	NF	-	
S. caricis (Gillette & Baker)	AK	YK	-	-	BC	AB	SK	MB	-	QC	-	-	-	-	-	
S. debilis (Uhler)	-	-	-	-	BC	AB	-	MB	ON	-	-	-	-	-	-	
S. flavovirens (Gillette & Baker)	AK	YK	NT	-	BC	AB	SK	MB	ON	QC	-	-	-	NF	LB	

Sorhoanus orientalis (DeLong & Davidson)	-	YK	-	-	BC	AB	SK	MB	ON	QC	NB	-	NS	-	LB	
S. pascuellus (Fallén)	-	-	-	-	BC*	-	-	-	ON*	QC*	NB*	PE*	NS*	NF*	-	[A]
S. uhleri (Oman)	-	YK	-	-	BC	AB	SK	MB	ON	QC	-	-	-	-	-	
S. xanthoneurus (Fieber)	AK	YK	NT	-	BC	AB	SK	MB	ON	QC	-	-	-	NF	-	

Genus STENOMETOPIELLUS Haupt

S. cookei (Gillette)	-	-	-	-	-	AB	SK	-	-	-	-	-	-	-	-	

Genus STIRELLUS Osborn & Ball

S. bicolor (Van Duzee)	-	-	-	-	-	-	-	MB	-	-	-	-	-	-	-	

Genus STROGGYLOCEPHALUS Flor

S. mixtus (Say)	AK	YK	-	-	BC	AB	SK	MB	ON	QC	-	-	-	-	LB	
S. placidus (Provancher)	-	-	-	-	-	-	-	-	ON	QC	-	-	-	-	-	

Genus TEXANANUS Ball

T. arctostaphylae (Ball)	-	-	-	-	BC	AB	SK	MB	ON	-	-	-	-	-	-	
T. cumulatus (Ball)	-	-	-	-	BC	AB	SK	MB	-	-	-	-	-	-	-	
T. decorus (Osborn & Ball)	-	-	-	-	BC?	-	-	MB	ON	QC	-	PE	-	-	-	
T. dolus DeLong	-	-	-	-	BC	-	SK	MB	-	-	-	-	-	-	-	
T. extremus (Ball)	-	-	-	-	BC	AB	-	-	-	-	-	-	-	-	-	
T. latipex (DeLong)	-	-	-	-	BC	-	-	-	-	-	-	-	-	-	-	[r]
T. longipennis Crowder	-	-	-	-	-	-	-	-	ON	-	-	-	-	-	-	
T. majestus (Osborn & Ball)	-	-	-	-	-	-	-	-	ON	-	-	-	-	-	-	
T. marmor (Sanders & DeLong)	-	-	-	-	-	AB	SK	MB	ON	-	-	-	-	-	-	
T. oregonus (Ball)	-	-	-	-	BC	-	-	-	-	-	-	-	-	-	-	
T. proximus Crowder	-	-	-	-	BC	-	-	-	-	-	-	-	-	-	-	[r]
T. superbus (Van Duzee)	-	-	-	-	BC	-	-	-	-	-	-	-	-	-	-	

Genus THAMNOTETTIX Zetterstedt

T. confinis (Zetterstedt)	AK	YK	NT	-	BC	AB	SK	MB	ON	QC	NB	-	NS	NF	-	

Genus TWININGIA Ball

T. fasciatus (Beamer)	-	-	-	-	BC	-	-	-	-	-	-	-	-	-	-	
T. pellucidus (Ball)	-	-	-	-	BC	-	-	-	-	-	-	-	-	-	-	
T. scrupulosus (Ball)	-	-	-	-	BC	-	-	-	-	-	-	-	-	-	-	[r]

Genus UNOKA Lawson

U. gillettei Metcalf	-	-	-	-	BC?	AB	-	-	-	-	-	-	-	-	-	

Genus XESTOCEPHALUS Van Duzee

X. brunneus Van Duzee	-	-	-	-	-	-	-	-	ON	QC	-	-	-	-	-	
X. coronatus Osborn & Ball	-	-	-	-	-	-	-	-	ON	QC	-	-	-	-	-	
X. fulvifrons Peters	-	-	-	-	BC	-	-	-	-	QC	-	-	-	-	-	[r]
X. fulvocapitatus Van Duzee	-	-	-	-	-	AB	-	MB	ON	QC	NB	-	NS	-	-	
X. nigrifrons Osborn	-	-	-	-	-	-	-	-	ON	QC	NB	-	-	-	-	
X. similis Peters	-	-	-	-	-	-	-	-	ON	-	-	-	-	-	-	
X. superbus (Provancher)	-	-	-	-	BC	-	SK	MB	ON	QC	NB	-	NS	-	-	

Subfamily CICADELLINAE

Genus CARSONUS Oman

C. aridus (Ball)	-	-	-	-	BC	-	-	-	-	-	-	-	-	-	-	

Genus CICADELLA Latreille

C. viridis (Linnaeus)	-	-	-	-	{BC}*	-	-	-	-	-	-	-	-	-	-	[A]

* introduced; [] indoors; { } extirpated; ? uncertain record; @ unestablished adventive; $port-of-entry interception

Genus CUERNA Melichar

C. alpina Oman & Beamer	-	-	-	-	-	AB	SK	-	-	-	-	-	-	-	-	
C. cuesta Hamilton	-	-	-	-	BC	-	-	-	-	-	-	-	-	-	-	
C. fenestrella Hamilton	-	-	-	-	-	-	-	MB	-	-	-	-	-	-	-	
C. nielsoni Hamilton	-	-	-	-	-	-	SK	MB	-	-	-	-	-	-	-	
C. sayi Nielson	-	-	-	-	-	AB	-	-	-	-	-	-	-	-	-	
C. septentrionalis (Walker)	AK	YK	NT	-	BC	AB	SK	MB	ON	QC	NB	PE	-	-	-	
C. striata (Walker)	-	-	-	-	-	AB	-	MB	ON	QC	NB	PE	NS	-	-	

Genus DRAECULACEPHALA Ball

D. angulifera (Walker)	AK	YK	NT	-	BC	AB	-	MB	ON	QC	NB	-	NS	NF	LB	
D. antica (Walker)	-	-	-	-	BC	AB	SK	MB	ON	QC	NB	PE	NS	-	-	
D. bivoltina Hamilton	-	-	-	-	BC	-	-	-	-	-	-	-	-	-	-	
D. borealis Hamilton	AK	YK	NT	-	BC	AB	-	MB	ON	QC	-	-	-	-	-	
D. constricta DeLong & Davidson	-	-	-	-	-	-	-	-	ON	QC	-	-	NS	-	-	
D. crassicornis Van Duzee	-	-	-	-	BC	-	-	-	-	-	-	-	-	-	-	
D. manitobiana Ball	-	-	-	-	BC	AB	SK	MB	ON	QC	NB	-	NS	-	-	
D. mollipes (Say)	-	-	-	-	-	-	-	-	ON	QC	NB	PE	NS	-	-	
D. noveboracensis (Fitch)	-	-	-	-	BC	AB	SK	MB	ON	QC	-	-	NS	-	-	
D. paludosa Ball & China	-	-	-	-	-	-	-	MB	ON	QC	-	-	-	-	-	
D. prasina (Walker)	-	-	-	-	BC	AB	SK	-	ON	QC	NB	PE	NS	-	-	
D. robinsoni Hamilton	-	-	-	-	BC	AB	SK	MB	ON	QC	NB	-	NS	-	-	
D. sphagneticola Hamilton	-	-	-	-	-	AB	-	MB	ON	-	NB	-	-	-	-	
D. zeae Hamilton sharpheaded sharpshooter	-	-	-	-	-	-	-	-	ON	QC	NB	PE	NS	-	-	[A]

Genus ERRHOMUS Oman

E. calvus Oman	-	-	-	-	BC	-	-	-	-	-	-	-	-	-	-	

Genus EVACANTHUS LePeletier & Serville

E. emilus (Bliven)	-	-	-	-	BC	-	-	-	-	-	-	-	-	-	-	
E. grandipes Hamilton	AK	-	-	-	BC	-	-	-	-	-	-	-	-	-	-	
E. interruptus (Linnaeus)	-	-	-	-	-	-	-	-	-	-	NB*	-	-	NF*	-	[A]
E. nigramericanus Hamilton	-	-	-	-	-	-	-	-	ON	-	-	-	-	-	-	[r]
E. orbitalis Fitch	-	-	-	-	BC	AB	SK	MB	ON	QC	NB	-	NS	-	-	

Genus GRAPHOCEPHALA Van Duzee

G. coccinea (Forster)	-	-	-	-	-	-	SK	MB	ON	QC	-	-	-	-	-	
G. fennahi Young rhododendron leafhopper	-	-	-	-	BC	-	-	-	-	-	-	-	-	-	-	[O]
G. picta (Walker)	-	-	-	-	-	-	-	-	ON	QC	NB	PE	NS	-	-	
G. teliformis (Walker)	-	-	-	-	-	-	-	-	ON	QC	-	-	-	-	-	

Genus HELOCHARA Fitch

H. communis Fitch bog leafhopper	-	-	NT	-	BC	AB	SK	MB	ON	QC	NB	PE	NS	-	-	
H. delta Oman	-	-	-	-	BC	-	-	-	-	-	-	-	-	-	-	

Genus HYLAIUS Oman & Musgrave

H. oregonensis (Baker)	-	-	-	-	BC	-	-	-	-	-	-	-	-	-	-	[r]

Genus NEOKOLLA Melichar

N. atra (Barber)	-	-	-	-	-	-	-	-	-	-	NB	-	-	-	-	[r]
N. atropunctata (Signoret)	-	-	-	-	{BC}	-	-	-	-	-	-	-	-	-	-	[r]
N. confluens (Uhler) purple sharpshooter	-	-	-	-	BC	-	-	-	-	-	-	-	-	-	-	
N. hieroglyphica (Say)	-	-	-	-	BC	AB	SK	MB	ON	QC	NB	PE	NS	-	-	

Neokolla surcula (DeLong & Currie)	-	-	-	-	-	AB	SK	-	-	-	-	-	-	-	-	
N. uhleri (Ball)	-	-	-	-	BC	-	-	-	-	-	-	-	-	-	-	

Genus ONCOMETOPIA Stål

O. orbona (Fabricius) broad-headed sharpshooter	-	-	-	-	-	-	-	-	ON	QC	-	-	-	-	-	

Genus PAGARONIA Ball

P. tredecimpunctata Ball	-	-	-	-	{BC}	-	-	-	-	-	-	-	-	-	-	[r]

Genus PARAULACIZES Young

P. irrorata (Fabricius)	-	-	-	-	-	-	-	-	ON	-	-	-	-	-	-	[r]

Genus PROVANCHERANA Hamilton

P. tripunctata (Fitch)	-	-	-	-	-	-	-	-	ON	QC	-	-	-	-	-	

Genus THATUNA Oman

T. gilletti Oman	-	-	-	-	BC	-	-	-	-	-	-	-	-	-	-	[r]

Genus TYLOZYGUS Fieber

T. bifidus (Say)	-	-	-	-	-	-	-	-	ON	QC	-	-	-	-	-	

Genus XYPHON Hamilton

X. triguttatum (Nottingham)	-	-	-	-	{BC}	-	-	-	-	-	-	-	-	-	-	[r]

Subfamily COELIDIINAE

Genus COELIDIA Germar

C. olitoria (Say)	-	-	-	-	-	-	-	-	ON	QC	-	-	-	-	-	

Subfamily EURYMELINAE

Genus AGALLIA Curtis

A. constricta (Van Duzee) constricted leafhopper	-	-	-	-	-	-	-	-	ON	-	-	-	-	-	-	[r]
A. quadripunctata (Provancher) four-spotted clover leafhopper	-	-	-	-	BC	AB	SK	MB	ON	QC	NB	PE	NS	-	-	[A]

Genus AGALLIOPSIS Kirkaldy

A. abietaria Oman	-	-	-	-	BC	-	-	-	-	-	-	-	-	-	-	
A. ancistra Oman	-	-	-	-	BC	AB	SK	MB	ON	QC	NB	-	NS	-	-	
A. peneoculata (Oman)	-	-	-	-	-	-	-	-	ON	QC	NB	-	-	-	-	

Genus BALCANOCERUS Maldonado-Capriles

B. crataegi (Van Duzee)	-	-	-	-	-	-	SK	-	ON	QC	-	-	-	-	-	
B. fitchi (Van Duzee)	-	-	-	-	-	AB	-	-	ON	QC	-	-	NS	-	-	
B. provancheri (Van Duzee)	-	-	-	-	BC	AB	SK	MB	ON	QC	NB	-	NS	-	-	

Genus CERATAGALLIA Kirkaldy

C. agricola Hamilton	-	-	-	-	-	-	-	-	ON	-	-	-	-	-	-	
C. arida (Oman)	-	-	-	-	BC	-	-	-	-	-	-	-	-	-	-	[r]
C. califa Oman	-	-	-	-	-	AB	-	-	-	-	-	-	-	-	-	
C. californica (Baker)	-	-	-	-	BC	-	-	-	-	-	-	-	-	-	-	
C. cerea Hamilton	-	-	-	-	-	AB	-	-	-	-	-	-	-	-	-	[r]
C. cinerea (Osborn & Ball)	-	-	-	-	-	AB	SK	-	-	-	-	-	-	-	-	
C. humilis (Oman) Canadian clover leafhopper	-	-	-	-	-	-	-	MB	ON	QC	NB	PE	NS	-	-	[A]

* introduced; [] indoors; { } extirpated; ? uncertain record; @ unestablished adventive; $port-of-entry interception

Ceratagallia nanella (Oman)	-	-	-	-	BC	-	-	MB	-	-	-	-	-	-	-	
C. okanagana Hamilton	-	-	-	-	BC	-	-	-	-	-	-	-	-	-	-	[r]
C. omani Hamilton	-	-	-	-	BC	-	-	-	-	-	-	-	-	-	-	[r]
C. robusta poudris (Oman)	-	-	-	-	BC	-	-	-	-	-	-	-	-	-	-	
C. sanguinolenta (Provancher)	AK	YK	NT	-	BC	AB	SK	MB	ON	QC	-	-	NS	-	-	
C. semiarida Hamilton	-	-	-	-	-	AB	SK	-	-	-	-	-	-	-	-	
C. siccifolia (Uhler)	AK	-	-	-	BC	AB	SK	MB	-	-	-	-	-	-	-	
C. uhleri (Van Duzee)	-	-	-	-	-	AB	SK	MB	-	-	-	-	-	-	-	
C. viator Hamilton	AK	YK	NT	-	BC	AB	SK	MB	ON	-	-	-	-	-	-	
C. vulgaris (Oman)	-	-	-	-	-	AB	-	-	ON	-	-	-	-	-	-	

Genus IDIOCERUS Lewis

I. albolinea Hamilton	-	-	-	-	-	-	SK	MB	ON	QC	-	-	-	-	-	
I. alternatus Fitch	-	-	-	-	BC	AB	SK	MB	ON	QC	-	-	NS	NF	-	
I. amabilis Ball	-	-	-	-	BC	-	SK	MB	ON	QC	-	-	-	NF	-	
I. amoenus Van Duzee	-	-	-	-	BC	-	-	-	-	-	-	-	-	-	-	
I. aureus Hamilton	-	-	-	-	BC	-	-	MB	ON	QC	-	-	-	-	-	
I. bovinus Ball & Parker	-	-	-	-	BC	-	-	-	-	-	-	-	-	-	-	
I. cabbottii Hamilton	-	-	-	-	-	-	-	-	-	-	-	-	NS	-	-	
I. canae Hamilton	-	-	-	-	-	AB	-	-	-	-	-	-	-	-	-	[r]
I. cedrus Hamilton	-	-	-	-	-	-	-	-	ON	-	-	-	-	-	-	
I. cinctus DeLong & Caldwell	-	-	-	-	BC	-	-	-	-	-	-	-	-	-	-	
I. cingulatus Ball	-	-	-	-	BC	-	-	-	-	-	-	-	-	-	-	
I. cognatus Fieber	-	-	-	-	BC*	-	-	-	ON*	QC*	-	PE*	NS*	-	-	[O]
I. concinnus Ball	-	-	-	-	BC	-	-	-	-	-	-	-	-	-	-	
I. couleanus Ball & Parker	AK	-	-	-	BC	-	-	-	-	-	-	-	-	-	-	
I. delongi Freytag	-	-	-	-	BC	-	-	-	-	-	-	-	-	-	-	
I. distinctus Gillette & Baker	-	-	-	-	BC	-	-	-	-	-	-	-	-	-	-	
I. duzeei Provancher	-	-	-	-	-	-	-	-	ON	QC	-	-	NS	-	-	
I. femoratus Ball	-	-	-	-	BC	AB	SK	-	-	-	-	-	-	-	-	
I. formosus Ball	-	-	-	-	BC	AB	SK	MB	ON	QC	NB	PE	NS	-	-	
I. freytagi Hamilton	-	-	-	-	BC	AB	SK	MB	ON	-	-	-	-	-	-	
I. fulgidus (Fabricius)	-	-	-	-	-	-	-	-	-	-	-	PE*	NS*	NF*	-	
I. gillettei Hamilton	-	-	-	-	-	AB	-	MB	ON	-	NB	-	-	-	-	
I. glacialis Hamilton	-	-	-	-	BC	-	-	-	-	-	-	-	-	-	-	
I. indistinctus Hamilton	-	-	-	-	BC	-	-	-	-	-	-	-	-	-	-	
I. interruptus Gillette & Baker	-	-	-	-	-	AB	-	-	-	-	-	-	-	-	-	
I. iodes Hamilton	-	-	-	-	BC	AB	SK	MB	-	QC	-	-	-	-	-	
I. lachrymalis Fitch	AK	YK	NT	-	BC	AB	SK	MB	ON	QC	-	PE	NS	-	LB	
I. levis DeLong & Caldwell	-	-	-	-	BC	AB	SK	MB	ON	QC	-	-	-	-	-	
I. lucidae Hamilton	-	-	-	-	-	AB	-	MB	ON	QC	NB	-	-	-	-	
I. lunaris Ball	-	-	-	-	BC	AB	SK	MB	ON	QC	NB	PE	NS	-	-	
I. maximus Freytag	AK	YK	-	-	BC	AB	SK	-	-	QC	-	-	-	-	-	
I. midas Hamilton	AK	YK	-	-	BC	AB	SK	MB	ON	QC	-	-	NS	-	-	
I. moniliferae Osborn & Ball	-	-	-	-	-	-	SK	MB	-	-	-	-	-	-	-	
I. morosus Ball	-	-	-	-	BC	-	-	-	-	-	-	-	-	-	-	
I. musteus Ball	-	-	-	-	BC	AB	SK	MB	ON	QC	NB	-	NS	NF	-	
I. nervatus Van Duzee	-	-	-	-	BC	-	-	-	ON	QC	-	-	-	-	-	
I. omani Freytag	-	-	-	-	BC	AB	-	-	-	-	-	-	-	-	-	
I. pallidus Fitch	-	-	NT	-	BC	AB	SK	MB	ON	QC	-	PE	NS	NF	LB	
I. pericallis Hamilton	-	-	-	-	BC	-	-	-	-	-	-	-	-	-	-	
I. productus Gillette & Baker	AK	YK	NT	-	BC	AB	-	-	ON	-	-	-	-	-	LB	
I. ramentosus Uhler	-	-	-	-	BC	AB	SK	MB	ON	-	-	-	-	-	-	
I. raphus Freytag	-	-	-	-	BC	AB	SK	MB	ON	QC	-	-	-	-	-	
I. rotundens DeLong & Caldwell	-	-	-	-	-	-	-	-	ON	-	-	-	-	-	-	
I. rufus Gillette & Baker	-	-	-	-	BC	-	-	-	-	-	-	-	-	-	-	
I. setaceus Hamilton	-	-	NT	-	BC	AB	SK	MB	ON	-	-	-	-	-	-	

Idiocerus snowi Gillette & Baker	-	-	-	-	BC	-	-	MB?	-	-	-	-	-	-	-	
I. stellaris DeLong & Caldwell	-	-	-	-	-	-	-	-	ON	QC	NB	-	NS	-	-	
I. stigmaticalis Lewis	-	-	-	-	-	-	-	-	ON*	-	-	PE*	NS*	NF*	-	[O]
I. striolus Fieber	AK	YK	NT	-	BC	AB	SK	MB	ON	-	-	-	-	NF	-	
I. subnitens Sanders & DeLong	-	-	-	-	-	-	-	MB	ON	QC	-	-	-	NF	-	
I. suturalis Fitch	-	-	-	-	BC	AB	SK	MB	ON	QC	NB	PE	NS	NF	-	
I. tahotus Ball & Parker	-	-	-	-	BC	-	-	-	-	-	-	-	-	-	-	
I. taiga Hamilton	-	-	-	-	-	AB	-	-	ON	-	-	-	-	-	-	
I. unicolor Hamilton	-	-	-	-	-	-	-	MB	ON	-	NB	-	-	-	-	
I. utahnus Ball & Parker	-	-	-	-	BC	-	-	-	-	-	-	-	-	-	-	
I. vanduzeei Hamilton	-	-	-	-	BC	-	-	-	-	-	-	-	-	-	-	[r]
I. varians Hamilton	-	-	-	-	-	-	-	-	ON	-	-	-	-	-	-	
I. venosus Hamilton	-	YK	NT	-	BC	AB	SK	MB	ON	QC	NB	PE	NS	NF	-	
I. xanthiops Hamilton	AK	YK	NT	-	BC	AB	-	MB	ON	QC	-	-	-	NF	-	

Genus MACROPSIS Lewis

M. aurifera Breakey	-	-	-	-	BC	AB	SK	MB	ON	QC	NB	-	-	-	-	
M. basalis (Van Duzee)	AK	YK	NT	-	BC	AB	SK	MB	ON	QC	NB	-	NS	-	-	
M. bifasciata (Van Duzee)	-	-	-	-	BC	-	-	MB	ON	QC	NB	-	NS	-	-	
M. borealis Hamilton	AK	YK	NT	-	BC	AB	SK	MB	ON	-	-	-	-	-	-	
M. canadensis (Van Duzee)	-	-	-	-	BC	AB	SK	MB	ON	QC	NB	-	NS	-	-	
M. cinerea Breakey	-	-	-	-	-	AB	SK	MB	ON	-	-	-	-	-	-	
M. confusa Breakey	-	-	-	-	-	-	-	-	ON	QC	-	-	NS	-	-	
M. decisa Hamilton	-	-	-	-	-	-	-	-	-	-	NB	-	-	-	-	[r]
M. deviridis Hamilton	-	-	-	-	BC	AB	SK	MB	ON	QC	-	-	-	-	-	
M. feminis Hamilton	-	-	-	-	-	AB	SK	MB	-	-	-	-	-	-	-	
M. ferrugineoides (Van Duzee)	-	-	-	-	-	AB	SK	MB	ON	-	-	-	-	-	-	
M. fumipennis (Gillette & Baker)	-	-	-	-	-	-	-	-	ON	-	-	-	-	-	-	
M. fuscula (Zetterstedt) raspberry leafhopper	-	-	-	-	BC*	-	-	-	ON*	-	-	PE*	-	-	-	[O]
M. graminea (Fabricius)	-	-	-	-	BC*	-	-	-	ON*	QC*	-	PE*	NS*	NF*	-	[A]
M. hesperia Breakey	-	-	-	-	BC	AB	SK	MB	ON	QC	-	-	-	-	-	
M. insignis (Van Duzee)	-	-	-	-	-	-	-	-	ON	-	-	-	-	-	-	
M. jocosa Hamilton	-	-	-	-	-	-	SK	MB	ON	QC	-	-	-	-	-	
M. mendax (Fieber)	-	-	-	-	-	-	-	-	ON*	-	-	-	-	-	-	[O]
M. notata Prohaska	-	-	-	-	-	-	-	-	-	QC*	NB*	-	-	-	-	
M. ocellata (Provancher)	-	-	-	-	-	-	-	-	ON*	QC*	-	-	NS*	-	-	[O]
M. osborni Breakey	-	-	-	-	-	AB	-	MB	ON	-	-	-	-	-	-	
M. robusta Breakey	AK	YK	-	-	BC	AB	SK	MB	ON	QC	NB	-	NS	-	-	
M. rufescens Hamilton	-	-	-	-	-	AB	SK	MB	-	-	-	-	-	-	-	
M. rufocephala Osborn	-	-	-	-	-	AB	SK	MB	ON	-	-	-	-	-	-	
M. sauroni Hamilton	-	-	-	-	-	-	-	MB	ON	-	-	-	-	-	-	
M. sordida (Van Duzee)	-	-	-	-	BC	AB	SK	MB	ON	QC	NB	-	-	-	-	
M. trimaculata (Fitch) plum leafhopper; cicadelle du prunier	-	-	-	-	-	-	-	-	ON	-	-	-	NS*	-	-	[A]
M. tristis (Van Duzee)	-	-	-	-	-	-	-	MB	-	-	-	-	-	-	-	[rA]
M. trivialis (Ball)	-	-	-	-	-	AB	SK	MB	ON	QC	-	-	-	-	-	
M. tunicata Hamilton	-	-	-	-	-	-	SK	MB	ON	QC	-	-	-	-	-	
M. vicina (Horváth)	-	-	-	-	-	-	-	-	ON*	-	-	PE*	-	-	-	[O]
M. vinea Hamilton	AK	YK	NT	-	BC	AB	SK	MB	ON	QC	-	-	-	-	-	
M. viridis (Fitch)	-	-	-	-	-	AB	SK	MB	ON	QC	NB	-	-	-	-	

Genus ONCOPSIS Burmeister

O. abietis (Fitch)	-	-	-	-	BC	AB	SK	MB	ON	QC	NB	-	-	NF	-	
O. albicollis Hamilton	AK	YK	NT	-	BC	AB	-	-	-	-	-	-	-	-	-	
O. californica Van Duzee	AK	-	-	-	BC	-	-	-	-	-	-	-	-	-	-	
O. cinctifrons (Provancher)	AK	YK	NT	-	BC	AB	SK	MB	ON	QC	-	-	-	-	-	

* introduced; [] indoors; { } extirpated; ? uncertain record; @ unestablished adventive; $port-of-entry interception

Oncopsis citra Hamilton	-	-	-	-	BC	-	-	-	ON	QC	NB	-	NS	NF	-	
O. citrella Hamilton	-	-	-	-	-	-	-	-	ON	QC	NB	-	NS	-	-	
O. coloradensis (Baker)	-	-	-	-	-	AB	-	-	-	-	-	-	-	-	-	
O. concurrens Hamilton	-	-	-	-	-	-	-	-	ON	QC	NB	-	NS	-	-	
O. crispae Hamilton	AK	YK	NT	-	BC	AB	SK	MB	ON	QC	NB	PE	NS	NF	LB	
O. dorsalis (Provancher)	-	-	-	-	-	-	-	MB	ON	QC	NB	PE	NS	-	-	
O. ferrosa Hamilton	-	-	-	-	BC	-	-	-	-	-	-	-	-	-	-	[r]
O. flavidorsum (Amyot & Serville)	-	-	-	-	-	-	-	MB	ON	QC	NB	PE	NS	NF	LB	
O. incidens Hamilton	-	-	-	-	BC	-	SK	MB	-	-	-	-	-	-	-	
O. interior Hamilton	AK	YK	NT	-	BC	AB	-	-	-	-	-	-	-	-	-	
O. juno Hamilton	-	-	-	-	BC	-	SK	-	-	-	-	-	-	-	-	
O. marilynae Hamilton	-	-	-	-	BC	AB	SK	MB	-	-	-	-	-	-	-	
O. minor (Fitch)	-	-	-	-	BC	AB	SK	MB	ON	QC	NB	PE	NS	NF	-	
O. monticola Hamilton	AK	-	-	-	BC	-	-	-	-	-	-	-	-	-	-	
O. nigrinasi (Fitch)	-	-	-	-	-	-	-	-	ON	QC	-	-	-	-	-	
O. prolixa Hamilton	-	-	-	-	BC	-	-	-	ON	QC	NB	-	NS	NF	-	
O. quebecensis Hamilton	-	-	-	-	-	-	-	-	-	QC	-	-	NS	NF	-	
O. sobria (Walker)	-	-	-	-	BC	-	SK	MB	ON	QC	NB	-	NS	NF	-	
O. speciosa Hamilton	-	-	-	-	-	-	-	-	-	-	-	-	NS	NF	-	
O. tenuifoliae Hamilton	-	-	-	-	BC	-	-	-	-	-	-	-	-	-	-	
O. tristis (Zetterstedt)	-	-	-	-	BC*	-	-	-	-	-	-	-	-	-	-	[A]
O. truncata (Baker)	-	-	-	-	-	-	-	-	ON	-	-	-	-	-	-	[r]
O. variabilis (Fitch)	-	-	-	-	BC	AB	SK	MB	ON	QC	NB	PE	NS	NF	-	
O. vartyi Hamilton	-	-	-	-	-	-	-	-	ON	QC	NB	PE	NS	NF	-	

Genus PEDIOPSIS Burmeister

P. tiliae (Germar)	-	-	-	-	-	-	-	-	ON*	QC*	-	-	-	-	-	[O]

Genus PEDIOPSOIDES Matsumura

P. distinctus (Van Duzee)	-	-	-	-	-	-	-	-	ON	-	-	-	-	-	-	

Genus RHYTIDODUS Fieber

R. decimaquartus (Schrank)	-	-	-	-	BC*	-	-	-	ON*	-	-	-	NS*	NF*	-	[O]

Genus TIAJA Oman

T. arenaria Oman	-	-	-	-	BC	-	-	-	-	-	-	-	-	-	-	[r]

Subfamily KOEBELIINAE

Genus KOEBELIA Baker

K. californica Baker	-	-	-	-	BC	-	-	-	-	-	-	-	-	-	-

Subfamily LEDRINAE

Genus XEROPHLOEA Germar

X. major Baker	-	-	-	-	-	-	-	-	ON	QC	-	-	-	-	-
X. peltata (Uhler)	-	-	-	-	-	AB	SK	MB	-	-	-	-	-	-	-
X. zionis Lawson	-	-	-	-	BC	-	-	MB	-	-	-	-	-	-	-

Subfamily NEOCOELIDIINAE

Genus NEOCOELIDIA Gillette & Baker

N. lineata Baker	-	-	-	-	BC	-	-	-	-	-	-	-	-	-	-	
N. tuberculata (Baker)	-	-	-	-	-	-	-	-	-	QC	-	-	-	-	-	[r]
N. tumidifrons Gillette & Baker	-	-	-	-	-	AB	SK	MB	ON	QC	-	PE	-	-	-	

Subfamily SCARINAE

Genus **GYPONA** Germar

G. contana DeLong	-	-	-	-	-	AB	SK	MB	-	-	-	-	-	-	-	
G. melanota Spångberg	-	-	-	-	-	-	-	-	ON	-	-	-	-	-	-	

Genus **GYPONANA** Ball

G. acia DeLong	-	-	-	-	-	-	-	-	ON	QC	-	-	-	-	-	
G. avara DeLong	-	-	-	-	-	-	-	-	-	QC	-	-	-	-	-	
G. brevita DeLong	-	-	-	-	-	-	-	-	ON	-	-	-	-	-	-	
G. cacumina DeLong	-	-	-	-	-	-	-	-	ON	QC	NB	PE	NS	-	-	
G. expanda DeLong	-	-	-	-	-	-	-	-	ON	-	-	-	-	-	-	
G. extenda DeLong	-	-	-	-	-	-	-	-	ON	QC	-	-	-	-	-	
G. geminata (Osborn)	-	-	-	-	-	-	SK	MB	ON	QC	NB	-	NS	-	-	
G. gladia DeLong	-	-	-	-	-	-	-	-	ON	-	-	-	-	-	-	[r]
G. hasta DeLong	-	-	-	-	BC	-	-	-	-	-	-	-	-	-	-	
G. mali DeLong	-	-	-	-	-	-	-	-	ON	-	-	-	-	-	-	
G. octolineata (Say)	-	-	-	-	-	-	-	-	ON	-	-	-	-	-	-	
G. ortha DeLong	-	-	-	-	-	-	-	-	ON	-	NB	-	-	-	-	
G. praelonga DeLong	-	-	-	-	-	-	-	-	ON	-	-	-	-	-	-	[r]
G. pruinosa (Spångberg)	-	-	-	-	-	-	-	-	-	QC	-	-	-	-	-	
G. quebecensis (Provancher)	-	-	-	-	BC	-	SK	MB	ON	QC	NB	-	-	-	-	
G. querci DeLong	-	-	-	-	-	-	-	-	ON	-	-	-	-	-	-	
G. salsa DeLong	-	-	-	-	BC	AB	SK	MB	ON	-	-	-	-	-	-	
G. serpenta DeLong	-	-	-	-	BC	AB	SK	MB	ON	QC	NB	PE	NS	-	-	
G. striata (Burmeister)	-	-	-	-	BC?	-	-	-	ON	QC	-	-	NS	-	-	
G. tubera DeLong	-	-	-	-	-	-	-	-	ON	QC	-	-	-	-	-	
G. vasta DeLong	-	-	-	-	-	-	-	-	ON	-	-	-	-	-	-	
G. vincula DeLong	-	-	-	-	-	AB	SK	MB	-	-	-	-	-	-	-	

Genus **PONANA** Ball

P. albosignata DeLong	-	-	-	-	-	-	-	-	-	QC	-	-	-	-	-	
P. pectoralis (Spångberg)	-	-	-	-	-	-	-	MB	ON	QC	-	-	-	-	-	
P. puncticollis (Spångberg)	-	-	-	-	-	-	-	-	ON	QC	-	-	-	-	-	
P. quadralaba DeLong	-	-	-	-	-	-	-	-	ON	-	-	-	-	-	-	
P. rubida DeLong	-	-	-	-	-	-	-	MB	ON	QC	-	-	-	-	-	
P. scarlatina (Fitch)	-	-	-	-	-	-	-	-	ON	QC	-	-	NS	-	-	

Genus **PRAIRIANA** Ball

P. angustens DeLong	-	-	-	-	-	-	-	-	ON	-	-	-	-	-	-	[r]
P. cinerea (Uhler)	-	-	-	-	BC?	AB	SK	MB	ON	-	-	-	-	-	-	
P. kansana (Ball)	-	-	-	-	-	-	-	MB?	ON	-	-	-	-	-	-	[r]

Genus **STRAGANIA** Stål

S. alabamensis (Baker)	-	-	-	-	-	-	-	-	ON	-	-	PE*	-	-	-	
S. apicalis (Osborn & Ball)	-	-	-	-	-	-	-	-	ON	QC	-	-	-	-	-	
S. atra (Baker)	-	-	-	-	BC	AB	SK	MB	-	-	-	-	-	-	-	
S. robusta (Uhler)	-	-	-	-	-	AB	-	-	ON?	-	-	-	-	-	-	
S. rufoscutellata (Baker)	-	-	-	-	-	AB	SK	-	-	-	-	-	-	-	-	

Subfamily TYPHLOCYBINAE

Genus **AGURIAHANA** Distant

A. stellulata (Burmeister)	-	-	-	-	-	-	-	-	ON*	-	-	-	NS*	-	-	[O]

* introduced; [] indoors; { } extirpated; ? uncertain record; @ unestablished adventive; $port-of-entry interception

Genus ALEBRA Fieber

A. aurea (Walsh)	-	-	-	-	-	-	-	-	ON	QC	-	-	-	-	-	
A. bicincta DeLong	-	-	-	-	-	-	-	-	ON	-	-	-	-	-	-	
A. fumida Gillette	-	-	-	-	-	-	-	-	ON	-	-	-	-	-	-	
A. thoracica Hamilton	-	-	-	-	-	-	-	-	ON	QC	-	-	-	-	-	
A. wahlbergi (Boheman)	-	-	-	-	-	-	-	-	-	-	-	-	NS*	-	-	

Genus ALNETOIDIA Dlabola

A. alneti (Dahlbom)	-	-	-	-	-	-	-	-	-	-	-	-	NS*	-	-	[O]

Genus ARBORIDIA Zachvatkin

A. betulaspera (Richards & Varty)	-	-	-	-	BC	AB	SK	-	ON	QC	NB	PE	NS	NF	-	
A. bitincta (McAtee)	-	-	-	-	-	-	-	-	ON	-	NB	-	-	-	-	
A. cornipes (Beamer)	-	-	-	-	-	-	-	-	ON	-	-	-	-	-	-	
A. crataegi (Johnson)	-	-	-	-	-	-	-	-	ON	-	-	-	-	-	-	
A. dorsalis (Gillette)	-	-	-	-	-	-	-	-	ON	-	-	-	-	NF*	-	
A. enata (Knull)	-	-	-	-	-	-	-	-	ON	-	-	-	-	-	-	
A. fumida (Gillette)	-	-	-	-	-	-	-	MB	ON	QC	NB	-	-	-	-	
A. insigna (Beamer & Griffith)	-	-	-	-	BC	-	SK	MB	ON	-	NB	PE	NS	-	-	
A. nitida (Beamer)	-	-	-	-	-	-	-	-	ON	-	NB	-	NS	-	-	
A. noeva (Gillette)	-	-	-	-	-	-	-	-	ON	-	-	-	-	-	-	
A. obliqua (Say)	-	-	-	-	-	-	-	-	ON	QC	-	-	NS	-	-	
A. obvia (Beamer)	-	-	-	-	-	-	-	-	ON	QC	-	-	-	-	-	
A. plena (Beamer)	-	-	-	-	BC	AB	-	-	ON	-	NB	-	NS	-	-	
A. sinua (Johnson)	-	-	-	-	BC	-	SK	-	ON	-	NB	-	NS	-	-	
A. stolata (McAtee)	-	-	-	-	-	-	-	-	ON	QC	NB	PE	-	-	-	
A. varia (McAtee)	-	-	-	-	-	-	-	-	-	-	-	-	NS	-	-	

Genus CHLORITA Fieber

C. nearctica Hamilton	-	YK	-	-	BC	-	-	-	-	-	-	-	-	-	-	

Genus DIKRANEURA Hardy

D. abnormis (Walsh)	-	-	-	-	BC?	-	-	-	-	QC	-	-	-	-	-	
D. absenta DeLong & Caldwell	-	-	-	-	BC	-	-	-	-	-	-	-	-	-	-	
D. angustata Ball & DeLong	-	-	-	-	-	-	-	-	ON	-	-	-	-	-	-	
D. carneola (Stål)	AK	YK?	-	-	BC	AB	SK?	MB?	-	-	-	-	-	-	-	
D. etiolata Knull	-	-	-	-	-	-	-	-	ON	-	-	-	-	-	-	
D. hungerfordi Lawson	AK	YK	NT	-	-	AB	SK	MB	ON	-	-	-	NS	NF	-	
D. mali (Provancher)	AK	-	-	-	BC	AB	SK	MB	ON	QC	NB	-	NS	-	-	
D. omani Knight	-	-	-	-	BC	-	-	-	-	-	-	-	-	-	-	[r]
D. rubrala DeLong & Caldwell	-	-	-	-	-	AB	-	MB	ON	-	-	-	NS	-	-	
D. rufula Gillette	-	-	-	-	BC	-	-	-	-	QC?	-	-	-	-	-	
D. urbana Ball & DeLong	-	-	-	-	-	-	-	-	ON	-	NB	-	NS	-	-	
D. variata Hardy	-	YK	-	-	BC	AB	SK	MB	-	-	-	-	-	-	-	

Genus DIKRELLA Oman

D. californica (Lawson)	-	-	-	-	BC	-	-	-	-	-	-	-	-	-	-	
D. cruentata (Gillette)	-	-	-	-	-	-	-	-	ON	QC	-	-	-	-	-	

Genus EMPOASCA Walsh

E. abbreviata DeLong	-	-	-	-	BC	-	-	-	-	-	-	-	-	-	-	
E. abrupta DeLong	-	-	-	-	BC	-	-	-	-	-	-	-	-	-	-	
E. albolinea Gillette	-	-	-	-	-	-	-	-	ON	-	-	-	-	-	-	
E. alboneura Gillette	-	-	-	-	BC	-	SK	MB	ON	-	-	-	-	-	-	
E. alexanderae Ross	-	-	-	-	BC	-	-	-	ON	QC	-	-	-	-	-	
E. andresia Ross	AK	YK	-	-	BC	AB	-	MB	ON	QC	NB	PE	NS	-	-	
E. angustella DeLong	-	-	-	-	-	-	-	-	ON	-	-	-	-	-	-	

Empoasca angustifoliae Ross	-	-	-	-	-	AB	-	-	-	-	-	-	-	-	-	[r]
E. aspersa Gillette	-	-	-	-	BC	-	-	-	-	-	-	-	-	-	-	
E. aureoviridis (Uhler)	-	-	-	-	BC	-	-	-	ON?	QC?	-	-	-	-	-	
E. banksianae Hamilton	-	-	-	-	BC	AB	-	-	ON	-	-	-	-	-	-	
E. biarca Davidson & DeLong	-	-	-	-	BC	-	-	-	-	-	-	-	-	-	-	
E. bifurcata DeLong	-	-	-	-	-	-	-	MB	ON	QC	NB	-	NS	-	-	
E. brunneus Osborn	-	-	-	-	-	-	-	-	ON	-	-	-	-	-	-	
E. caesarsi Hamilton	-	-	-	-	BC	-	-	-	-	-	-	-	-	-	-	
E. carsona DeLong & Davidson	-	-	-	-	BC	-	-	-	-	-	-	-	-	-	-	
E. caverna Davidson & DeLong	-	-	-	-	BC	-	-	-	ON	-	-	-	-	-	-	
E. cerea DeLong & Davidson	-	-	-	-	-	-	SK	-	-	-	NB	-	-	-	-	
E. clinata Ross	AK	YK	-	-	BC	AB	-	-	ON	-	-	-	-	NF	-	
E. clypeata Gillette & Baker	-	-	-	-	BC	AB	-	-	-	-	-	-	-	-	-	
E. coccinea Fitch	-	-	-	-	-	-	-	-	ON	-	-	-	-	NF	-	
E. columbiana Hamilton	-	-	-	-	BC	-	-	-	-	-	-	-	-	-	-	
E. convergens Davidson & DeLong	-	-	-	-	BC	-	-	-	ON	-	NB	-	NS	NF	-	
E. copula DeLong	AK	YK	NT	-	BC	-	SK	MB	ON	-	-	-	-	NF	-	
E. coronata Hamilton	-	YK	-	-	BC	AB	-	-	ON	QC	-	-	-	-	-	
E. crystola Ross	AK	YK	NT	-	BC	AB	-	-	-	-	-	-	-	-	-	
E. curvata Poos	-	-	-	-	BC	AB	-	MB	ON	-	NB	-	NS	-	-	
E. curvatura Davidson & DeLong	-	-	-	-	BC	-	-	-	-	QC	-	-	-	-	-	
E. deluda DeLong	AK	-	NT	-	BC	-	SK	MB	ON	QC	NB	-	-	-	-	
E. digita DeLong	-	-	-	-	BC	-	SK	MB	-	-	-	-	-	-	-	
E. dissimilaris Hamilton	-	-	-	-	BC	AB	-	-	-	-	-	-	-	-	-	
E. distracta Caldwell & DeLong	-	-	-	-	-	AB	-	-	-	-	-	-	-	-	-	
E. ditata Caldwell & DeLong	-	-	-	-	BC	-	-	-	-	-	-	-	NS	-	-	
E. elongella DeLong	-	-	-	-	BC	-	-	-	-	-	-	-	-	-	-	
E. empusa Hamilton	-	-	-	-	BC	-	SK	MB	ON	QC	NB	-	-	-	-	
E. ensiformis Oman & Wheeler	-	-	-	-	BC	-	-	-	-	-	-	-	-	-	-	
E. erigeron DeLong	-	-	-	-	BC	-	-	-	ON	QC	NB	-	NS	NF	-	
E. esuma Goding	-	-	-	-	BC	-	-	-	ON	-	-	-	-	-	-	
E. exiguae Ross	-	-	-	-	BC	AB	SK	MB	ON	-	-	-	-	-	-	
E. fabae Harris potato leafhopper; cicadelle de la pomme de terre	-	-	-	-	BC?	-	SK	MB	ON	QC	NB	PE	NS	NF	-	[A]
E. falca Davidson & DeLong	-	-	-	-	-	-	SK	MB	-	-	-	-	-	-	-	
E. filamenta DeLong	-	YK	-	-	BC	AB	SK	MB	ON	-	-	-	-	-	-	[A]
E. flavescens (Fabricius)	-	-	-	-	BC*	-	-	-	-	-	-	-	-	-	-	
E. fontana Ross	-	-	-	-	BC	-	-	-	-	-	-	-	-	-	-	
E. gelbata DeLong & Davidson	-	-	-	-	BC	AB	SK	-	ON	QC	-	-	-	-	-	
E. hartzelli Baker	-	-	-	-	BC	-	-	-	-	-	-	-	-	-	-	
E. incida DeLong	-	-	-	-	BC	-	-	-	-	-	-	-	-	-	-	
E. junipera DeLong	-	-	-	-	BC	-	-	-	ON	QC	-	-	-	-	-	
E. livingstonii Gillette	-	YK	-	-	BC	AB	-	-	-	-	-	-	-	-	-	
E. lucidae Ross	-	YK	-	-	BC	AB	-	-	-	-	-	-	-	-	-	
E. luda Davidson & DeLong	-	-	-	-	BC*	-	-	-	ON*	-	-	-	-	NF*	-	[O]
E. medora DeLong	-	-	-	-	-	AB	SK	MB	-	-	-	-	-	-	-	
E. mexicana Gillette	-	-	-	-	-	-	SK	-	-	-	-	-	-	-	-	
E. neaspersa Oman & Wheeler	-	-	-	-	BC	-	-	-	-	-	-	-	-	-	-	
E. neclya Davidson & DeLong	-	-	-	-	-	-	SK	-	-	-	-	-	-	-	-	
E. nema Davidson & DeLong	-	-	-	-	-	-	-	-	ON	-	-	-	-	-	-	
E. nigra Gillette & Baker	-	-	-	-	BC	-	-	-	-	-	-	-	-	-	-	[r]
E. nigroscuta Gillette & Baker	AK	YK	-	-	BC	AB	SK	-	-	-	-	-	-	-	-	
E. obtusa Walsh	-	-	-	-	-	-	SK	MB	-	QC	-	-	NS	-	-	
E. osborni Hartzell	-	-	-	-	-	-	-	-	-	QC	-	-	-	-	-	
E. pallida Gillette	-	-	-	-	-	-	-	-	ON	-	-	-	-	-	-	[r]
E. panda DeLong	-	-	-	-	BC	-	-	-	-	-	-	-	-	-	-	
E. patula DeLong	-	-	-	-	BC	-	SK	MB	ON	-	NB	-	-	NF	-	

* introduced; [] indoors; { } extirpated; ? uncertain record; @ unestablished adventive; $port-of-entry interception

Empoasca pelecana Oman & Wheeler	-	-	-	-	BC	-	-	-	-	-	-	-	-	-	-	
E. pergrada Davidson & DeLong	-	-	-	-	-	-	SK	-	ON	-	-	-	-	NF	-	
E. perlonga Davidson & DeLong	-	-	-	-	-	-	-	-	ON	-	-	-	-	-	-	[r]
E. petiolaridis Ross	-	-	-	-	BC	AB	SK	MB	ON	QC	NB	PE	NS	-	-	
E. pinella Davidson & DeLong	-	-	-	-	BC	-	-	-	-	-	-	-	-	-	-	
E. populi Edwards	-	-	-	-	-	-	-	-	-	-	-	-	NS*	NF*	-	[O]
E. portola DeLong	AK	YK	-	-	BC	AB	-	-	-	-	-	-	-	-	-	
E. pulchella Gillette & Baker	-	-	-	-	-	-	-	-	-	-	-	-	NS?	-	-	
E. pura Stål	AK	YK	NT	-	BC	AB	SK	-	ON	-	-	-	-	NF	-	
E. racina DeLong & Liles	-	-	-	-	-	-	-	MB	-	-	-	-	-	-	-	
E. ratio Davidson & DeLong	AK	YK	-	-	BC	AB	-	-	-	QC	-	-	NS	-	-	
E. recta Caldwell & DeLong	-	-	-	-	BC	-	-	-	ON	-	-	-	-	-	-	
E. recurvata DeLong	-	-	-	-	-	-	-	-	ON	-	-	-	-	-	-	
E. rossi Hamilton	-	-	-	-	BC	-	-	-	-	-	-	-	-	-	-	
E. salicis Wheeler	-	-	-	-	-	-	-	MB	ON	-	-	-	-	-	-	
E. smaragdula (Fallén)	-	-	-	-	BC*	-	-	-	-	-	-	-	-	-	-	[O]
E. sordidula Ossiannilsson	AK	YK	-	-	BC	AB	-	-	-	QC	-	-	-	-	-	
E. spira Caldwell & DeLong	-	YK	-	-	-	-	-	-	-	-	-	-	NS	-	-	
E. tigris Hamilton	-	-	-	-	BC	-	-	-	-	-	-	-	-	-	-	
E. tolana DeLong & Guevara	-	-	-	-	BC	-	-	-	-	-	-	-	-	-	-	
E. torqua Davidson & DeLong	-	-	-	-	-	-	-	MB	ON	QC	-	-	NS	-	-	
E. trifasciata Gillette	-	-	-	-	BC	-	-	-	ON	-	-	-	-	-	-	
E. typhlocyboides Gillette & Baker	-	-	-	-	BC	AB	SK	-	-	-	-	-	-	-	-	
E. unica Provancher	-	-	-	-	-	-	-	-	-	QC	-	-	-	-	-	
E. venusta Davidson & DeLong	-	-	-	-	-	-	-	-	-	QC	NB	-	NS	-	-	
E. vergena Caldwell & DeLong	-	-	-	-	-	-	SK	MB	ON	-	-	-	NS	-	-	
E. vincula DeLong	-	-	-	-	BC	AB	SK	MB	ON	-	-	-	-	-	-	
E. volsella Hamilton	-	-	-	-	-	-	-	-	-	QC	-	-	NS	-	-	
E. yukonensis Ross	AK	YK	-	-	BC	AB	-	-	-	-	-	-	-	-	-	
E. zaisanensis (Mitjaev)	AK	-	NT	-	-	AB	SK	-	ON	QC	NB	-	-	-	-	
E. zanclus Hamilton	-	-	-	-	-	-	SK	-	ON	QC	-	-	NS	-	-	

Genus ERYTHRONEURA Fitch

E. abjecta Beamer	-	-	-	-	-	-	-	-	ON	-	-	-	-	-	-	
E. acicularis Beamer	-	-	-	-	BC	-	-	-	-	-	-	-	-	-	-	
E. ador McAtee	-	-	-	-	-	-	-	-	-	-	-	-	NS	-	-	
E. affinis Fitch	-	-	-	-	-	-	-	MB	-	-	-	-	-	-	-	
E. ardens McAtee	-	-	-	-	-	-	-	-	ON	-	-	-	-	-	-	
E. aspera Beamer & Griffith	-	-	-	-	BC	-	-	-	-	-	-	-	-	-	-	
E. basilaris (Say)	-	-	-	-	-	-	-	-	ON	QC	-	-	-	-	-	
E. beameri Robinson	-	-	-	-	-	-	-	-	ON	-	-	-	-	-	-	
E. bigemina McAtee	-	-	-	-	BC	-	-	-	ON	QC	-	-	-	-	-	
E. bispinosa Beamer	-	-	-	-	-	-	-	-	-	-	NB	-	-	-	-	
E. brevipes Beamer	-	-	-	-	-	-	-	-	-	-	NB	-	-	-	-	
E. calva Beamer	-	-	-	-	-	AB	SK	MB	-	-	-	-	-	-	-	
E. calycula McAtee	-	-	-	-	-	-	-	-	ON?	-	-	-	-	-	-	
E. campora Robinson	-	-	-	-	-	-	-	MB	ON	QC	NB	-	NS	-	-	
E. carbonata McAtee	-	-	-	-	-	AB	SK	MB	-	-	-	-	-	-	-	
E. carmini Beamer	-	-	-	-	-	-	-	MB	ON	-	NB	-	-	-	-	
E. certa Beamer	-	-	-	-	-	-	-	-	ON	-	-	-	-	-	-	
E. coloradensis (Gillette)	-	-	-	-	-	-	-	-	ON	-	-	-	-	-	-	
E. comes (Say) grape leafhopper; cicadelle du raisin	-	-	-	-	BC*	-	-	-	ON	QC	-	-	-	-	-	[A]
E. complicata Johnson	-	-	-	-	BC?	-	-	-	-	-	-	-	-	-	-	
E. compta McAtee	-	-	-	-	-	-	-	-	ON	-	-	-	-	-	-	
E. concisa Beamer	-	-	-	-	-	-	-	-	ON	QC	-	-	-	-	-	
E. continua Knull & Auten	-	-	-	-	-	-	-	-	-	QC	-	-	-	-	-	

Erythroneura corni Robinson	-	-	-	-	BC*	-	-	-	-	-	-	-	NS	-	-
E. crevecoeuri Gillette	-	-	-	-	-	-	-	-	-	QC	-	-	-	-	-
E. cuneata Beamer	-	-	-	-	-	-	-	-	-	QC	-	-	-	-	-
E. cymbium McAtee	-	-	-	-	-	-	-	-	ON	-	-	-	-	-	-
E. delicata McAtee	-	-	-	-	-	-	-	-	ON	-	-	-	-	-	-
E. dira Beamer	-	-	-	-	-	-	-	-	ON	QC	-	-	-	-	-
E. divisa McAtee	-	-	-	-	-	-	-	-	ON	-	-	-	-	-	-
E. dolosa Beamer & Griffith	-	-	-	-	BC	-	-	-	-	-	-	-	-	-	-
E. electa McAtee	-	-	-	-	BC	-	-	-	-	-	-	-	-	-	-
E. elegans McAtee	-	-	-	-	-	-	-	-	ON	-	-	-	-	-	-
E. elegantula Osborne	-	-	-	-	BC	-	-	-	-	-	-	-	-	-	-
E. eluta McAtee	-	-	-	-	-	-	SK	-	ON	QC	NB	-	-	-	-
E. eversi Ross & DeLong	-	-	-	-	-	-	-	-	ON	-	-	-	-	-	-
E. fulvocephala Robinson	-	-	-	-	-	-	SK	-	ON	-	-	-	-	-	-
E. funesta Beamer	-	-	-	-	-	-	-	-	ON	-	-	-	-	-	-
E. gemina McAtee	-	-	-	-	-	-	-	-	ON	-	-	-	-	-	-
E. hamata Beamer	-	-	-	-	-	-	-	-	ON	-	NB	-	NS	-	-
E. harpola Ross	-	-	-	-	-	-	-	-	ON	QC	-	-	-	-	-
E. haysensis Hepner	-	-	-	-	-	-	-	-	ON	-	-	-	-	-	-
E. inconspicua Johnson	-	-	-	-	-	-	-	-	ON?	-	-	-	-	-	-
E. infuscata (Gillette)	-	-	-	-	-	-	-	-	ON	-	-	-	-	-	-
E. jocosa Beamer	-	-	-	-	-	-	-	-	ON?	-	-	-	-	-	-
E. kanwakae Robinson	-	-	-	-	-	AB	-	-	-	QC	-	-	-	-	-
E. latapex Beamer	-	-	-	-	-	-	-	-	ON	-	-	-	-	-	-
E. lawsoniana Baker	-	-	-	-	-	-	-	-	ON	-	-	-	-	-	-
E. ligata McAtee	-	-	-	-	-	-	-	-	-	QC	-	-	-	-	-
E. linea Beamer	-	-	-	-	-	-	-	MB	ON	-	NB	-	-	-	-
E. macra Beamer	-	-	-	-	-	-	-	-	ON	QC	NB	-	NS	-	-
E. maculata Gillette	-	-	-	-	-	-	-	-	ON	QC	-	-	-	-	-
E. magnacalx Beamer	-	-	-	-	-	-	-	-	ON	-	-	-	-	-	-
E. manus Beamer	-	-	-	-	-	-	-	-	ON	-	-	-	-	-	-
E. maritima Hamilton	-	-	-	-	-	-	-	-	-	-	NB	-	NS	-	-
E. marra Beamer	-	-	-	-	-	-	-	-	ON	-	-	-	-	-	-
E. masonae Knull	-	-	-	-	-	-	-	-	ON	-	-	-	-	-	-
E. mimica Ross	-	-	-	-	-	-	-	-	ON	-	-	-	-	-	-
E. minima Johnson	-	-	-	-	-	-	-	-	ON?	-	-	-	-	-	-
E. minor Beamer	-	-	-	-	-	-	-	-	ON	-	-	-	-	-	-
E. nigra (Gillette)	-	-	-	-	-	-	-	-	ON	QC	-	-	-	-	-
E. nondescripta Johnson	-	-	-	-	-	-	-	MB	ON	QC	NB	-	-	-	-
E. nudata McAtee	-	-	-	-	-	-	-	-	ON	-	-	-	-	-	-
E. octonotata Walsh	-	-	-	-	-	-	-	-	ON	QC	-	-	-	-	-
E. ontari Robinson	-	-	-	-	BC	AB	-	-	ON	QC	-	PE	NS	-	-
E. parallela McAtee	-	-	-	-	-	-	-	-	ON	-	-	-	-	-	-
E. patris Ross & DeLong	-	-	-	-	-	-	-	-	ON	-	-	-	-	-	-
E. penesica Beamer	-	-	-	-	-	-	-	-	ON	-	-	-	-	-	-
E. perita Beamer	-	-	-	-	-	-	-	-	ON	-	-	-	-	-	-
E. prima Beamer	-	-	-	-	-	-	-	-	ON	QC	NB	-	-	-	-
E. reflecta McAtee	-	-	-	-	-	-	-	-	ON	QC	-	-	-	-	-
E. restricta Beamer	-	-	-	-	-	-	-	-	ON	-	-	-	-	-	-
E. rosa Robinson	-	-	-	-	BC	-	SK	-	ON	QC	NB	-	-	-	-
E. rubens Beamer	-	-	-	-	-	-	-	-	-	QC	-	-	-	-	-
E. rubra (Gillette)	-	-	-	-	-	-	-	-	ON?	-	-	-	-	-	-
E. rubrella McAtee	-	-	-	-	BC	-	SK	-	ON	-	-	-	-	-	-
E. rubroscuta Gillette	-	-	-	-	-	-	SK?	-	ON?	-	-	-	-	-	-
E. sancta Beamer	-	-	-	-	-	-	-	-	ON	-	-	-	-	-	-
E. separata Beamer	-	-	-	-	-	-	-	-	ON	-	-	-	-	-	-
E. septima Beamer	-	-	-	-	BC*	-	-	-	ON	QC	-	-	NS	-	-

* introduced; [] indoors; { } extirpated; ? uncertain record; @ unestablished adventive; $port-of-entry interception

Erythroneura smithi Ross	-	-	-	-	-	-	SK	-	-	-	-	-	-	-	-	
E. solita Beamer	-	-	-	-	-	-	-	-	ON	-	-	-	-	-	-	
E. tacita Beamer	-	-	-	-	BC	-	-	-	-	-	-	-	-	-	-	
E. tenuispica Beamer	-	-	-	-	-	-	-	-	ON	-	NB	-	-	-	-	
E. trautmanae Knull	-	-	-	-	-	-	-	-	ON	QC	-	-	NS	-	-	
E. tricincta Fitch threebanded grape leafhopper; cicadelle à trois bandes	-	-	-	-	-	-	-	-	ON	QC	-	-	-	-	-	[A]
E. tridens Beamer	-	-	-	-	-	-	-	-	ON	-	NB	-	-	-	-	
E. tumida Knull	-	-	-	-	-	-	-	-	ON	-	NB	-	-	-	-	
E. uncinata Beamer	-	-	-	-	-	AB	-	MB	ON	QC	NB	-	NS	-	-	
E. victorialis Knull	-	-	-	-	-	-	-	-	ON	-	-	-	NS	-	-	
E. vitifex Fitch	-	-	-	-	BC*	-	-	-	ON	QC	-	-	-	-	-	
E. vitis (Harris) grapevine leafhopper; cicadelle de la vigne	-	-	-	-	-	-	-	-	ON	QC	-	-	-	-	-	[A]
E. vulgaris Hepner	-	-	-	-	-	-	-	-	ON	-	-	-	-	-	-	
E. vulnerata Fitch	-	-	-	-	-	-	-	MB	ON	QC	-	-	-	-	-	[A]
E. ziczac Walsh virginiacreeper leafhopper; cicadelle de la vigne-vierge	-	-	-	-	BC	AB	SK	MB	ON	QC	NB	-	-	-	-	

Genus EUPTERYX Curtis

E. atropunctata (Goeze)	-	-	-	-	-	-	-	-	ON*	QC*	NB*	-	-	-	-	[A]
E. aurata (Linnaeus)	-	-	-	-	-	-	-	-	-	QC*	-	-	-	-	-	[A]
E. clavalis McAtee	-	-	-	-	-	-	-	-	-	QC	-	-	-	-	-	
E. flavoscuta Gillette	-	-	-	-	-	-	-	-	ON	QC	-	PE	NS	-	-	
E. furcata (Beamer)	-	-	-	-	BC	-	-	-	-	-	-	-	-	-	-	[r]
E. melissae Curtis	-	-	-	-	BC*	-	-	-	ON*	QC*	-	-	-	-	-	[A]
E. nigra Osborn	-	-	-	-	-	-	-	-	ON	QC	-	-	NS	-	-	
E. vanduzeei Gillette	-	-	-	-	-	-	-	-	-	QC	-	-	NS	NF	-	
E. vittata (Linnaeus)	-	-	-	-	-	-	-	-	-	QC*	-	-	-	-	-	[A]

Genus FAGOCYBA Dlabola

F. douglasi (Edwards)	-	-	-	-	BC*	-	-	-	-	-	-	-	-	NF*	-	[O]

Genus FORCIPATA DeLong & Caldwell

F. acclina DeLong & Caldwell	AK	-	-	-	BC	-	SK	MB	ON	-	-	-	NS	NF	-	
F. calipera DeLong & Caldwell	-	-	-	-	-	-	-	-	ON	QC	-	PE	-	-	-	
F. citrinella (Zetterstedt)	AK	-	-	-	-	-	-	-	-	-	-	-	-	-	-	
F. frigida (Beirne)	-	-	-	-	-	AB	-	MB	-	-	-	-	-	-	-	
F. ips Hamilton	AK	YK	NT	-	-	-	-	-	-	-	-	-	-	-	-	
F. lobata Bearmer	-	-	-	-	-	-	-	-	ON	-	-	-	-	-	-	
F. loca DeLong & Caldwell	AK	YK	-	-	BC	-	SK	MB	ON	QC	NB	PE	NS	NF	-	
F. montana Hamilton	-	-	-	-	BC	-	-	-	-	-	-	-	-	-	-	
F. sicula DeLong & Caldwell	-	-	-	-	BC	-	-	-	-	-	-	-	-	-	-	
F. triquetra DeLong & Caldwell	AK	YK	NT	-	BC	-	SK	MB	-	-	-	-	NS	NF	-	

Genus HYMETTA McAtee

H. balteata McAtee	-	-	-	-	-	-	-	-	ON	-	-	-	-	-	-	
H. trifasciata (Say)	-	-	-	-	-	-	-	-	-	QC	-	-	-	-	-	

Genus KYBOASCA Zachvatkin

K. acodens (DeLong)	-	-	-	-	-	-	-	-	ON	QC	-	-	-	-	-	
K. atrolabes (Gillette)	AK	YK	NT	-	BC	AB	SK	MB	ON	QC	NB	-	NS	NF	LB	
K. bipunctata (Oshanin)	-	-	-	-	BC*	AB*	-	MB*	ON*	-	-	-	-	-	-	[O]
K. confusa (DeLong & Davidson)	-	-	-	-	BC	-	-	-	-	-	-	-	-	-	-	
K. denticula (Gillette)	-	-	-	-	-	-	-	MB	-	-	-	-	-	-	-	
K. maligna (Walsh) apple leafhopper; cicadelle du pommier	-	-	-	-	BC	AB	SK	-	ON	QC	NB	-	NS	-	-	[A]
K. papyriferae Hamilton	-	-	-	-	-	-	-	-	ON	-	-	-	NS	-	-	

Kyboasca pergandei (Gillette)	-	-	-	-	BC	-	-	-	ON	QC	NB	-	NS	-	-	
K. simplex (DeLong & Davidson)	AK	YK	NT	-	BC	AB	SK	-	-	-	-	-	-	-	-	
K. splendida (Gillette)	-	-	-	-	BC?	-	-	-	ON	QC	NB	-	NS	-	-	
K. trilobata (DeLong)	-	-	-	-	-	-	-	-	ON	QC	NB	PE	NS	NF	-	
Genus NOTUS Fieber																
N. alta (DeLong & Caldwell)	-	-	-	-	-	-	-	-	ON	-	-	-	-	-	-	[r]
N. isolatus Hamilton	-	-	-	-	-	-	-	-	-	QC	-	-	-	-	-	[r]
N. sitka (DeLong & Caldwell)	AK	YK	NT	-	-	-	-	MB	-	-	-	-	-	-	-	
Genus OSSIANNILSSONOLA Christian																
O. antigone (McAtee)	-	-	-	-	-	-	-	-	ON	-	-	-	-	-	-	
O. appendiculata (Malloch)	-	-	-	-	-	-	-	-	ON	-	-	-	-	-	-	
O. australis (Walsh)	-	-	-	-	-	-	-	-	ON	-	-	-	-	-	-	
O. bangsoni Christian	-	-	-	-	-	-	-	-	ON	-	-	-	-	-	-	
O. berenice (McAtee)	-	-	-	-	-	-	-	-	ON	-	NB	-	-	-	-	
O. clymene (McAtee)	-	-	-	-	-	-	-	-	ON	QC	NB	PE	NS	-	-	
O. danae (McAtee)	-	-	-	-	-	-	-	-	ON	-	-	-	-	-	-	
O. duplicata (McAtee)	-	-	-	-	-	-	-	-	ON	QC	-	-	-	-	-	
O. macateei Christian	-	-	-	-	-	-	-	-	ON	-	-	-	-	-	-	
O. phryne (McAtee)	-	-	-	-	-	-	-	-	ON	-	-	-	-	-	-	
O. quadrata (DeLong & Johnson)	-	-	-	-	-	-	-	-	ON	QC	-	PE	NS	-	-	
O. serrula (Ross & DeLong)	-	-	-	-	-	-	-	-	ON	-	-	-	NS	-	-	
O. tunicarubra (Gillette)	-	-	-	-	-	-	-	-	ON	-	-	-	-	-	-	
O. ulmi Hamilton	-	-	-	-	-	-	-	MB	ON	QC	-	-	-	-	-	
O. volans (McAtee)	-	-	-	-	BC	AB	-	-	ON	-	NB	PE	NS	-	-	
Genus RIBAUTIANA Dlabola																
R. foliosa (Knull)	-	-	-	-	BC	-	-	MB	ON	-	NB	PE	NS	-	-	
R. multispinosa Christian	-	-	-	-	-	-	-	-	ON	-	-	-	-	-	-	
R. parapiscator Christian	-	-	-	-	-	-	-	-	ON	QC	-	-	-	-	-	
R. piscator (McAtee)	-	-	-	-	-	-	-	-	ON	-	-	-	-	-	-	
R. tenerrima (Herrich-Schaeffer) bramble leafhopper; cicadelle des petits fruits	-	-	-	-	BC*	-	-	-	-	-	-	-	-	-	-	[O]
R. ulmi (Linnaeus)	-	-	-	-	BC*	-	-	-	ON*	-	-	PE*	NS*	NF*	-	[A]
R. unca (McAtee)	-	-	NT	-	BC	-	SK	-	ON	QC	NB	-	NS	NF	-	
Genus TYPHLOCYBA Germar																
T. albicans (Walsh)	-	-	-	-	-	-	-	-	ON	QC	-	PE	NS	NF	-	
T. alcorni (Christian)	AK	YK	NT	-	BC	AB	-	-	-	-	-	-	-	-	-	
T. andromache McAtee	-	-	-	-	-	-	-	-	ON	-	NB	-	-	-	-	
T. apicata McAtee	-	-	-	-	-	-	-	-	ON	QC	-	-	-	-	-	
T. ariadne McAtee	AK	YK	-	-	BC	-	-	-	ON	QC	NB	-	NS	NF	LB	
T. ariste McAtee	-	-	-	-	BC	-	-	-	-	-	-	-	-	-	-	
T. arsinoe McAtee	-	-	-	-	-	-	-	-	ON	-	-	-	-	-	-	
T. athene McAtee	-	-	-	-	-	-	-	-	ON	-	-	-	-	-	-	
T. avellanae Edwards	-	-	-	-	BC*	-	-	-	-	-	-	-	-	-	-	[A]
T. barbata Ribaut	-	-	-	-	BC*	-	-	-	-	-	-	-	-	-	-	[O]
T. bella Hamilton	-	-	-	-	-	-	-	-	-	-	NB	PE	NS	-	-	
T. candidula Kirschbaum	-	-	-	-	BC*	-	-	-	ON*	-	-	-	-	-	-	[O]
T. cassiopeia Knull	-	-	-	-	BC	AB	-	-	ON	-	-	-	-	-	-	
T. casta McAtee	-	-	-	-	-	-	-	MB	ON	QC	-	PE	-	-	-	
T. commissuralis Stål	AK	YK	NT	-	BC	-	-	-	-	-	-	-	-	-	-	
T. crassa DeLong & Johnson	-	-	-	-	BC	-	-	-	ON	-	-	-	-	-	-	
T. dejecta (Christian)	-	-	-	-	BC	-	-	-	ON	-	-	-	-	-	-	
T. dorsti Ossiannilsson	-	-	-	-	BC	-	-	-	-	-	-	-	-	-	-	
T. euphrante McAtee	-	-	NT	-	BC	AB	-	-	-	-	-	-	-	-	-	

* introduced; [] indoors; { } extirpated; ? uncertain record; @ unestablished adventive; $port-of-entry interception

Typhlocyba expanda DeLong & Johnson	AK	YK	-	-	-	-	-	-	-	-	-	-	-	-	-	
T. froggatti Baker	-	-	-	-	BC*	-	-	-	ON*	-	-	-	-	-	-	[A]
yellow apple leafhopper; cicadelle jaune du pommier																
T. frustrator Edwards	-	-	-	-	BC*	-	-	-	-	-	-	-	-	-	-	[A]
T. gillettei (Van Duzee)	AK	YK	NT	-	BC	AB	-	-	ON	QC	NB	PE	NS	NF	LB	
T. hippocastani Edwards	-	-	-	-	-	-	-	-	-	QC*	-	-	NS*	-	-	[O]
T. hockingensis Knull	AK	-	-	-	BC	-	-	-	ON	QC	NB	PE	NS	NF	-	
T. hollandi Hamilton	-	-	-	-	-	-	-	-	-	-	-	-	NS	-	-	[r]
T. inflata Christian	-	-	-	-	BC	-	-	-	ON	-	-	-	-	-	-	
T. latifasciata (Christian)	-	YK	-	-	BC	AB	-	-	ON	QC	NB	PE	NS	NF	-	
T. lethierryi Edwards	-	-	-	-	BC*	-	-	-	ON*	QC*	-	-	NS*	-	-	[O]
T. medleri Christian	-	-	-	-	-	-	-	-	ON	-	-	-	-	-	-	
T. melite McAtee	-	-	-	-	-	-	-	MB	ON	-	-	-	-	-	-	
T. modesta Gibson	-	-	-	-	-	-	-	-	ON	QC	-	-	-	-	-	
T. nigriloba Edwards	-	-	-	-	-	-	-	-	-	-	-	-	NS*	-	-	[O]
T. niobe McAtee	-	-	-	-	-	-	-	-	ON	-	-	-	-	-	-	
T. oneka Knull	-	-	-	-	-	-	-	-	ON	-	-	-	-	-	-	
T. oregonensis Beamer	-	-	-	-	BC	-	-	-	-	-	-	-	-	-	-	
T. persephone McAtee	-	-	-	-	-	-	-	-	ON	QC	-	-	NS	-	-	
T. plebeja Edwards	-	-	-	-	BC*	-	-	-	-	-	-	-	-	NF*	-	[O]
T. pomaria McAtee	AK	YK	-	-	BC	-	SK	-	ON	QC	NB	-	NS	NF	-	[A]
white apple leafhopper; cicadelle blanche du pommier																
T. prunicola Edwards	-	-	-	-	BC*	-	-	-	-	-	-	-	-	-	-	[A]
T. pseudocommissuralis (Christian)	-	-	-	-	-	-	-	-	ON	QC	NB	-	NS	NF	-	
T. putmani Knull	-	-	-	-	BC	-	-	-	ON	QC	-	-	-	-	-	
T. querci (Fitch)	-	-	-	-	-	-	-	-	ON	QC	-	-	NS	-	-	
T. quercus (Fabricius)	-	-	-	-	BC*	-	-	-	-	-	-	PE*	NS*	-	-	[O]
T. rosae (Linnaeus)	-	-	-	-	BC*	-	-	-	ON*	QC*	NB*	-	NS*	-	-	[O]
rose leafhopper																
T. rubricola (Christian)	-	-	-	-	-	-	-	-	ON	-	-	-	-	-	-	
T. saffrana McAtee	-	-	-	-	-	-	-	-	ON	QC	-	PE	-	-	-	
T. scripta McAtee	-	-	-	-	BC?	-	-	-	-	QC	NB	PE	NS	NF	-	
T. spinosa Beamer	AK	YK	-	-	BC	AB	-	MB	ON	-	NB	-	-	NF	-	
T. tersa Edwards	AK	YK	-	-	-	AB	-	-	-	-	-	-	-	NF	-	
T. tortosa Ross & DeLong	-	-	-	-	-	-	-	-	ON	-	-	-	-	-	-	
T. transviridis Christian	-	-	-	-	-	-	-	-	ON	-	-	-	-	-	-	
T. unicorn Hamilton	-	-	-	-	-	-	-	-	-	-	-	-	-	NF	-	[r]
T. venusta McAtee	-	-	-	-	-	-	-	MB	ON	QC	-	-	-	-	-	
T. vestita McAtee	-	-	-	-	-	-	-	-	ON	QC	-	-	-	-	-	

Genus ZYGINA Fieber

Z. flammigera (Fourcroy)	-	-	-	-	BC*	-	-	-	-	-	-	-	-	-	-	[O]

Family MEMBRACIDAE

Keys to genera and species: Beirne 1961 [Canada], Kopp & Yonke (1973a, b, c, 1974) [eastern USA].

Treehoppers; membracides. Treehoppers are mainly a tropical group best developed in the new world. As their name implies, many genera are restricted to trees, although shrubs and forbs are also attacked. Both adults and nymphs are sluggish and gregarious when not disturbed; females may show maternal instinct. Nymphs have a terminal tube for presenting droplets of honeydew to ants, which may "herd" and protect them from enemies. The pronotum of the adults is much enlarged, often spined, horned or carinate, and extending backwards over the abdomen, sometimes concealing the wings.

A number of tropical groups of treehoppers have been assigned separate family status, but all Canadian treehoppers are now considered to belong to a single family. Within the family, subfamilies and tribes have been stabilized, but generic boundaries (traditionally based almost exclusively on pronotal shape; recently becoming based almost exclusively on male genitalia) remain unsatisfactory. The highly ornamented bodies of nymphs offer additional useful generic characters that are not yet explored. For the present, a conservative generic classification is recommended (e.g., *Ceresa sensu lato*).

There are over 100 estimated species in Canada as compared to the previous estimate of 79 species (Scudder 1979), but this may be higher than the actual number as many "species" of the genus *Cyrtolobus* may in fact be merely colour forms of a very few species.

Subfamily CENTROTINAE

Genus GARGARA Amyot & Serville

G. genistae Fabricius	-	-	-	-	BC*	-	-	-	ON?*	-	-	-	-	-	-	

Subfamily MEMBRACINAE

Genus CAMPYLENCHIA Stål

C. latipes (Say)	-	-	-	-	-	-	-	MB	ON	QC	NB	PE	NS	-	-	
C. rugosa (Fowler)	-	-	-	-	BC	AB	SK	MB	-	-	-	-	-	-	-	

Genus ENCHENOPA Amyot & Serville

E. binotata (Say)	-	-	-	-	-	-	-	MB	ON	QC	-	-	NS	-	-	
E. brevis Walker	-	-	-	-	-	-	-	MB	ON	-	-	-	NS?	-	-	
E. porrecta Buckton	-	-	-	-	-	-	-	MB	ON	-	-	-	-	-	-	

Genus LEIOSCYTA Fowler

L. ferruginipennis (Goding)	-	-	-	-	-	AB	SK	-	-	-	-	-	-	-	-	

Genus PLATYCOTIS Stål

P. humilis (Walker)	-	-	-	-	BC	-	-	-	-	-	-	-	-	-	-	
P. lineosa (Walker)	-	-	-	-	BC	-	-	-	-	-	-	-	-	-	-	[r]

Subfamily SMILIINAE

Genus ACUTALIS Fairmaire

A. brunnea (Provancher)	-	-	-	-	-	-	-	-	ON	QC	-	-	-	-	-	
A. tartarea (Say)	-	-	-	-	-	-	-	-	ON	-	-	-	-	-	-	

Genus ARCHASIA Stål

A. auriculata (Fitch)	-	-	-	-	-	-	-	-	ON	QC	-	-	-	-	-	
A. belfragei Stål	-	-	-	-	-	-	-	-	ON	QC	-	-	-	-	-	

Genus ATYMNA Stål

A. castaneae (Fitch)	-	-	-	-	-	-	-	-	{ON}	-	-	-	-	-	-	
A. helena (Woodruff)	-	-	-	-	-	-	-	MB	-	QC	-	-	-	-	-	
A. inornata (Say)	-	-	-	-	-	-	-	-	ON	QC	-	-	-	-	-	
A. querci (Fitch)	-	-	-	-	-	-	-	-	ON	QC	-	-	-	-	-	

Genus CARYNOTA Fitch

C. marmorata (Say)	-	-	-	-	-	-	-	MB	ON	QC	NB	PE	NS	-	-	
C. mera (Say)	-	-	-	-	-	-	-	-	ON	QC	NB	-	-	-	-	
C. stupida (Walker)	-	-	-	-	-	-	SK	MB	ON	QC	NB	-	NS	-	-	

Genus CERESA Amyot & Serville

C. albescens Van Duzee	-	-	-	-	-	-	-	MB	ON	QC	-	-	-	-	-	
C. albidosparsa Stål	-	-	-	-	{BC}	-	-	-	-	-	-	-	-	-	-	
C. alta Walker buffalo treehopper; cérèse buffle	-	-	-	-	BC	AB	SK	MB	ON	QC	NB	PE	NS	-	-	

* introduced; [] indoors; { } extirpated; ? uncertain record; @ unestablished adventive; $port-of-entry interception

Ceresa basalis Walker	-	-	NT	-	BC	AB	SK	MB	ON	QC	NB	PE	NS	-	-	
C. borealis Fairmaire	-	-	-	-	-	-	-	-	ON	QC	-	-	-	-	-	
C. brevicornis Fitch	-	-	-	-	-	AB	SK	MB	ON	QC	-	-	-	-	-	
C. brevis Walker	-	-	-	-	-	AB	-	-	-	QC	-	-	-	-	-	
C. brevitylus Van Duzee	-	-	-	-	-	-	-	MB	-	QC	-	-	-	-	-	
C. constans (Walker)	-	-	-	-	-	-	-	-	ON	QC	-	-	-	-	-	
C. diceros (Say)	-	-	-	-	BC	AB	SK	MB	ON	QC	NB	-	NS	-	-	
C. festina (Say)	-	-	-	-	BC	-	-	-	{ON}	-	-	-	-	-	-	[r]
C. inermis (Fabricius)	-	-	-	-	-	AB	SK	MB	ON	QC	-	-	-	-	-	
C. palmeri Van Duzee	-	-	-	-	-	-	-	-	ON	QC	-	-	-	-	-	
C. taurina (Harris)	-	-	-	-	-	-	-	MB	ON	QC	-	-	NS	-	-	
C. wickhami Van Duzee	-	-	-	-	BC	AB	-	-	-	-	-	-	-	-	-	

Genus **CYRTOLOBUS** Goding

C. cinctus Van Duzee	-	-	-	-	-	-	-	-	-	QC	-	-	-	-	-	
C. cinereus (Emmons)	-	-	-	-	-	-	-	-	ON	QC	-	-	-	-	-	
C. discoidalis (Emmons)	-	-	-	-	-	-	-	-	ON	QC	-	-	-	-	-	
C. dixianus Woodruff	-	-	-	-	-	-	-	-	ON	QC	-	-	-	-	-	
C. fenestratus (Fitch)	-	-	-	-	-	-	-	-	ON	QC	-	-	-	-	-	
C. fulginosus (Emmons)	-	-	-	-	-	-	-	-	ON	QC	-	-	-	-	-	
C. funkhouseri Woodruff	-	-	-	-	-	-	-	-	ON?	-	-	-	-	-	-	
C. fuscipennis Van Duzee	-	-	-	-	-	-	-	-	-	QC	-	-	-	-	-	
C. gramatanus Woodruff	-	-	-	-	-	-	-	-	-	QC	-	-	-	-	-	
C. gratiosus Woodruff	-	-	-	-	-	-	-	-	-	QC	-	-	-	-	-	
C. griseus Van Duzee	-	-	-	-	-	-	-	MB	ON	QC	-	-	-	-	-	
C. intermedius (Emmons)	-	-	-	-	-	-	-	-	ON	QC	-	-	-	-	-	
C. maculifrontis (Emmons)	-	-	-	-	-	-	-	MB	ON	QC	-	-	-	-	-	
C. muticus (Fabricius)	-	-	-	-	-	-	-	MB	ON	QC	-	-	-	-	-	
C. ovatus Van Duzee	-	-	-	-	-	-	-	-	ON	-	-	-	-	-	-	
C. pallidifrontis (Emmons)	-	-	-	-	-	-	-	MB	ON	QC	-	-	-	-	-	
C. puritanus Woodruff	-	-	-	-	-	-	-	-	ON	QC	-	-	-	-	-	
C. rufulus Woodruff	-	-	-	-	-	-	-	-	-	QC	-	-	-	-	-	
C. tuberosus (Fairmaire)	-	-	-	-	-	-	-	MB	ON	-	-	-	-	-	-	
C. tumidus (Walker)	-	-	-	-	-	-	-	-	-	QC	-	-	-	-	-	
C. vau (Say)	-	-	-	-	-	-	-	MB	ON	QC	-	-	-	-	-	

Genus **ENTYLIA** Germar

E. carinata (Forster)	-	-	-	-	-	-	-	-	ON	QC	-	PE	NS	-	-	

Genus **GLOSSONOTUS** Butler

G. acuminatus (Fabricius)	-	-	-	-	-	-	-	-	ON	QC	-	-	-	-	-	
G. crataegi (Fitch)	-	-	-	-	-	-	-	MB	ON	QC	-	-	NS	-	-	
G. nimbatulus (Ball)	-	-	-	-	-	-	SK	MB	ON	QC	NB	-	-	-	-	
G. turriculatus (Emmons)	-	-	-	-	-	-	-	MB	-	QC	-	-	-	-	-	
G. univittatus (Harris)	-	-	-	-	BC	-	SK	MB	ON	QC	-	-	-	-	-	

Genus **MICRUTALIS** Fowler

M. calva (Say)	-	-	-	-	-	-	-	-	ON	-	-	-	-	-	-	[r]
M. dorsalis (Fitch)	-	-	-	-	-	-	-	-	ON	QC	-	-	-	-	-	

Genus **OPHIDERMA** Fairmaire

O. definita Woodruff	-	-	-	-	-	-	-	-	ON	QC	-	-	-	-	-	
O. evelyna Woodruff	-	-	-	-	-	-	-	-	ON	QC	-	-	-	-	-	
O. flava Goding	-	-	-	-	-	-	-	-	ON	QC	-	-	-	-	-	
O. flavicephala Goding	-	-	-	-	-	-	-	-	ON	QC	-	-	-	-	-	
O. pubescens (Emmons)	-	-	-	-	-	-	-	-	ON	QC	-	-	-	-	-	
O. salamandra Fairmaire	-	-	-	-	-	-	-	-	ON	QC	NB	-	NS	-	-	

Genus PUBLILIA Stål

P. brunnea (Ball)	-	-	-	-	-	AB	SK	MB	-	-	-	-	-	-	-	
P. concava (Say)	-	-	-	-	-	-	SK	MB	ON	QC	NB	-	NS	-	-	

Genus SMILIA Germar

S. camelus (Fabricius)	-	-	-	-	-	-	-	-	ON	QC	-	-	-	-	-	

Genus STICTOCEPHALA Stål

S. lutea (Walker)	-	-	-	-	-	-	-	-	ON	QC	-	-	-	-	-	
S. pacifica Van Duzee	-	-	-	-	BC	AB	-	-	-	-	-	-	-	-	-	
S. tauriniformis Caldwell	-	-	-	-	-	-	-	-	ON	QC	-	-	-	-	-	

Genus TELAMONA Fitch

T. ampelopsidis (Harris)	-	-	-	-	-	-	-	MB	ON	QC	-	-	NS	-	-	
T. barbata Van Duzee	-	-	-	-	{BC}	-	-	-	-	-	-	-	-	-	-	
T. compacta Ball	-	-	-	-	-	-	-	-	-	-	-	-	NS	-	-	[r]
T. concava Fitch	-	-	-	-	-	-	-	-	ON	QC	-	-	-	-	-	
T. cristata (Fairmaire)	-	-	-	-	-	-	-	-	ON	-	-	-	-	-	-	
T. decorata Ball	-	-	-	-	-	AB	-	-	ON	QC	-	-	-	-	-	
T. dubiosa Van Duzee	-	-	-	-	-	-	SK	-	ON	QC	-	-	-	-	-	
T. gemma (Ball)	-	-	-	-	-	-	-	-	ON	-	-	-	-	-	-	
T. maculata Van Duzee	-	-	-	-	-	-	-	-	ON	QC	-	-	-	-	-	[r]
T. molaris Butler	-	-	-	-	-	-	SK	MB	ON	-	-	-	-	-	-	
T. monticola (Fabricius)	-	-	-	-	-	-	SK	MB	ON	QC	-	-	-	-	-	
T. praealta Fowler	-	-	-	-	-	AB	SK	-	-	-	-	-	-	-	-	
T. pyramidata Uhler	-	-	NT	-	BC	AB	SK	MB	ON	QC	-	-	-	-	-	
T. reclivata Fitch	-	-	-	-	-	AB	-	-	ON	QC	-	-	-	-	-	
T. scalaris (Fairmaire)	-	-	-	-	-	AB	SK	MB	ON	QC	NB	-	NS	-	-	
T. spreta Goding	-	-	-	-	-	AB	SK	MB	ON	QC	-	-	-	-	-	
T. strombergi (Goding)	-	-	-	-	-	-	-	-	ON	-	NB	-	-	-	-	
T. tiliae Ball	-	-	-	-	-	-	-	-	ON	QC	-	-	-	-	-	
T. tremulata (Ball)	-	-	-	-	BC	AB	SK	MB	ON	QC	NB	-	NS	-	-	
T. tristis (Fitch)	-	-	-	-	-	-	-	MB	ON	QC	NB	-	-	-	-	
T. unicolor Fitch	-	-	-	-	-	-	-	-	ON	QC	-	-	-	-	-	
T. viridia Ball	-	-	-	-	-	AB	-	MB	-	QC	-	-	-	-	-	
T. westcotti Goding	-	-	-	-	-	-	-	MB	ON	QC	-	-	-	-	-	

Genus THELIA Amyot & Serville

T. bimaculata (Fabricius)	-	-	-	-	-	-	-	-	ON	QC	-	-	-	-	-	
T. uhleri Stål	-	-	-	-	-	-	-	MB	ON	QC	-	-	-	-	-	

Genus TORTISTYLUS Caldwell

T. minutus Caldwell	-	-	-	-	-	-	-	MB	-	-	-	-	-	-	-	

Genus VANDUZEA Goding

V. arquata (Say)	-	-	-	-	-	-	-	-	ON	-	-	-	NS	-	-	
V. triguttata (Burmeister)	-	-	-	-	-	-	SK	MB	-	-	-	-	-	-	-	

Subfamily STEGASPIDINAE

Genus MICROCENTRUS Stål

M. caryae (Fitch)	-	-	-	-	-	-	-	-	ON	QC	-	-	-	-	-	

Suborder Archaeorrhyncha

Planthoppers. Mouthparts arising from posterior part of head capsule; antenna short, bristle-like, arising below eyes on sharply defined sides of head; middle coxae separated, elongate; tarsi with 3 segments.

Superfamily Fulgoroidea

Keys to genera: Metcalf 1923, Dozier 1926 [both eastern North America].

Planthoppers, lanternflies; fulgorides. Planthoppers are mainly a tropical group with whole subfamilies exclusively tropical and subtropical. Family Delphacidae is the only family of economic importance in the temperate zone, while in the tropics it encompasses some of the most devastating pests of rice, sugar, taro and other (mainly graminaceous) crops. Their adults and nymphs feed actively above ground as do leafhoppers, but (unlike most leafhoppers) many delphacids, particularly in the tropics, are migratory and disperse widely (some subtropical species have been recorded well into Canada). Other Fulgoroidea have sessile nymphs that live in rotten wood or in pockets of waxy secretions on their hosts.

The superfamily has been variously separated into two families (Fulgoridae, Delphacidae) or many; their exact relationships, subfamilies and boundaries are still debatable. The families of Fulgoroidea presented here are those of Scudder (1979). Kinnaridae are considered a subfamily of Cixiidae, as they share tegmina with a stigma on the first apical cell. Caliscelidae are removed from Issidae on the basis of very different wing venation, and allied to the tropical Lophopidae by their swollen hind basal tarsal segments; they also have very different acoustic signals (Tishechkin 1998). Acanaloniidae are a subfamily of Issidae (Fennah 1954). Dictyopharidae and the tropical Tropiduchidae are closely related (Emeljanov 1979) and are united with Fulgoridae on the basis of their bicarinate pronotal margins. The familiar genera of "Dictyopharidae" in Canada, *Scolops* and *Phylloscelis*, both have nymphs with the characteristic chevron-shaped abdominal segments of the subfamily Orgeriinae. They are less closely allied to the tropical and subtropical Fulgorinae than are true Dictyopharinae (*e.g. Rhynchomitra*), which, like Fulgorinae, have the terminal segments of the nymphal abdomen divided by a recessed telson. Thus, "Dictyopharidae" are paraphyletic with respect to Fulgorinae.

There are 213 species in Canada, compared to the previous estimate of 192 species (Scudder 1979), but many undescribed species of Delphacidae will greatly increase this number. A rough estimate is 260 species.

Family ACHILIDAE

Genus CATONIA Uhler																
C. lunata Metcalf	-	-	-	-	-	-	-	-	ON	-	-	-	-	-	-	[r]
C. nava (Say)	-	-	-	-	-	-	-	-	ON	-	-	-	-	-	-	
Genus EPIPTERA Metcalf																
E. brittoni Metcalf	-	-	-	-	BC	-	SK	MB	ON	QC	NB	-	-	-	-	
E. confusa Beirne	-	YK	-	-	BC	AB	SK	MB	-	QC	NB	-	-	-	LB	
E. floridae (Walker)	-	-	-	-	-	-	-	-	ON	QC	-	-	-	-	-	
E. fusca (Walker)	-	-	-	-	-	-	-	-	ON	QC	-	-	-	-	-	
E. fusiformis (Van Duzee)	-	-	-	-	BC	-	-	-	-	-	-	-	-	-	-	
E. henshawi (Van Duzee)	-	-	-	-	BC	AB	-	-	-	-	-	-	-	-	-	
E. manitobiana Beirne	-	YK	-	-	BC	AB	SK	MB	-	QC	NB	-	-	-	-	
E. opaca (Say)	-	-	-	-	-	-	-	-	ON	QC	-	-	-	-	-	
E. pallida (Say)	-	YK	NT	-	BC	AB	SK	MB	-	QC	-	-	-	-	LB	
E. septentrionalis (Provancher)	-	-	-	-	BC	AB	SK	MB	ON	QC	NB	-	-	-	LB	
E. slossonae (Van Duzee)	-	-	NT?	-	BC?	-	-	-	-	QC	-	-	-	-	-	
E. variegata (Van Duzee)	-	-	-	-	-	-	-	-	ON	QC	NB	-	-	-	-	

Genus **SYNECDOCHE** Doering

S. constellata (Ball)	-	-	-	-	BC	-	-	-	-	-	-	-	-	-	-	
S. dimidiata (Van Duzee)	-	-	-	-	-	-	-	-	ON	-	-	-	NS	-	-	
S. grisea (Van Duzee)	-	-	-	-	-	-	-	-	ON	QC	-	-	-	-	-	
S. impunctata (Fitch)	-	-	-	-	-	-	-	-	ON	QC	-	-	-	-	-	
S. nemoralis (Van Duzee)	-	-	-	-	BC	-	-	-	-	-	-	-	-	-	-	

Family CALISCELIDAE

Genus **APHELONEMA** Uhler

A. simplex Uhler	-	-	-	-	-	-	SK	MB	ON	-	-	-	-	-	-	

Genus **BRUCHOMORPHA** Newman

B. beameri Doering	-	-	-	-	BC	AB	SK	MB	-	-	-	-	-	-	-	
B. dorsata Fitch	-	-	-	-	-	-	SK	MB	ON	-	-	-	-	-	-	
B. jocosa Stål	-	-	-	-	-	-	SK	MB	-	-	-	-	-	-	-	
B. keidensia Doering	-	-	-	-	-	AB	SK	MB	-	-	-	-	-	-	-	
B. oculata Newman	-	-	-	-	-	-	-	-	ON	QC	-	PE	-	-	-	
B. tristis Stål	-	-	-	-	-	AB	SK	MB	ON	-	-	-	-	-	-	

Genus **FITCHIELLA** Van Duzee

F. robertsoni (Fitch)	-	-	-	-	-	-	-	-	ON	-	-	-	-	-	-	[r]

Genus **PELTONOTELLUS** Puton

P. bivittatus Ball	-	-	-	-	-	AB	SK	-	-	-	-	-	-	-	-	
P. histrionicus Stål	-	YK	-	-	BC	AB	SK	MB	ON	QC	NB	PE	NS	NF	-	
P. rugosus (Ball)	-	-	-	-	-	AB	SK	-	-	-	-	-	-	-	-	

Family CIXIIDAE

Genus **CIXIUS** Latreille

C. apicalis Metcalf	-	-	-	-	-	-	-	MB	ON	-	NB	-	-	-	-	
C. basalis Van Duzee	-	-	-	-	-	-	SK	MB	ON	QC	NB	-	NS	-	-	
C. coelopium Fitch	-	-	-	-	BC?	-	-	-	ON	QC	NB	-	-	-	-	
C. cultus Ball	-	-	-	-	-	AB	SK	-	-	-	-	-	-	-	-	
C. guttulatus Walley	-	-	-	-	-	-	-	-	ON	QC	-	-	-	-	-	
C. meridionalis Beirne	-	YK	NT	-	BC	AB	SK	MB	-	QC	-	-	-	-	LB	
C. misellus Van Duzee	-	-	-	-	BC	AB	-	MB	ON	QC	NB	-	NS	NF	-	
C. narke Kramer	-	-	-	-	BC?	-	-	-	-	-	-	-	-	-	-	[r]
C. nervosus (Linnaeus)	AK	-	-	-	BC	AB	-	-	-	-	-	-	-	-	-	
C. nike Kramer	-	-	-	-	-	-	-	MB	ON	QC	NB	-	NS	-	-	
C. pini Fitch	-	-	-	-	-	-	-	MB	ON	QC	-	-	-	-	-	
C. praecox Van Duzee	-	-	-	-	BC	-	-	-	-	-	-	-	-	-	-	
C. quebecensis Beirne	-	-	-	-	-	-	-	-	ON	QC	NB	PE	NS	-	-	
C. stigmatus (Say)	-	-	-	-	-	AB	SK	-	ON	-	-	-	-	-	-	
C. umbrosus Walley	-	-	-	-	BC	-	-	-	ON	-	-	-	-	-	-	

Genus **MYNDUS** Stål

M. collinus Ball	-	-	-	-	-	-	SK	-	-	-	-	-	-	-	-	
M. fulvus Osborn	-	-	-	-	-	-	-	-	ON	-	-	-	-	-	-	
M. glyphus Kramer	-	-	-	-	-	-	-	-	-	QC	NB	-	-	-	-	
M. pictifrons Stål	-	-	-	-	-	-	-	-	ON	-	-	-	-	-	-	
M. xyron Kramer	-	-	-	-	-	-	-	-	ON	QC	-	-	-	-	-	

* introduced; [] indoors; { } extirpated; ? uncertain record; @ unestablished adventive; $ port-of-entry interception

Genus OECLEUS Stål

O. borealis Van Duzee	-	-	-	-	-	-	-	-	ON	-	-	-	-	-	-	[r]
O. obtusus Ball	-	-	-	-	BC	-	-	-	-	-	-	-	-	-	-	
O. venosus Van Duzee	-	-	-	-	BC	-	-	-	-	-	-	-	-	-	-	[r]

Genus OECLIDIUS Van Duzee

O. brickellus Ball	-	-	-	-	BC	-	-	-	-	-	-	-	-	-	-	[r]

Genus OLIARUS Stål

O. aridus Ball	-	-	-	-	-	-	-	-	ON	-	-	-	-	-	-	
O. beirnei Meade & Kramer	-	-	-	-	BC	-	-	-	-	-	-	-	-	-	-	
O. cinnamoneus Provancher	-	-	-	-	-	-	-	-	ON	QC	NB	-	-	-	-	
O. coconinus Ball	-	-	-	-	BC	-	-	-	-	-	-	-	-	-	-	[r]
O. dondonius Ball	-	-	-	-	BC	AB	SK	-	-	-	-	-	-	-	-	
O. exoptatus Van Duzee	-	-	-	-	BC	-	-	-	-	-	-	-	-	-	-	
O. humilis (Say)	-	-	-	-	-	-	SK	-	ON	QC	-	-	-	-	-	
O. placitus Van Duzee	-	-	-	-	-	-	-	-	ON	-	-	-	-	-	-	[r]
O. quinquelineatus (Say)	-	-	-	-	-	-	-	-	ON	QC	-	-	NS	-	-	
O. sablensis Caldwell	-	-	-	-	-	-	-	-	ON	-	NB	-	-	-	-	
O. zyxus Caldwell	-	-	-	-	BC	-	-	-	-	-	-	-	-	-	-	[r]

Family DELPHACIDAE

Genus ACHROTILE Fieber

A. acuta Scudder	-	YK	-	-	-	AB	SK	-	-	QC	-	-	-	-	-	
A. distincta Scudder	-	-	-	-	BC	AB	SK	-	-	-	-	-	-	-	-	
A. stylata Scudder	-	YK	-	-	BC	AB	-	-	-	-	-	-	-	-	-	
A. subarctica Scudder	AK	YK	NT	-	BC	AB	-	-	-	-	-	-	-	-	LB	
A. transbaicalica Kuznezov	-	-	NT?	-	-	-	-	-	-	-	-	-	-	-	-	

Genus ASCHEDELPHAX Wilson

A. hochae Wilson	-	YK	-	-	BC	AB	SK	-	-	-	-	-	-	-	-	

Genus BAKERELLA Crawford

B. muscotana (Beamer)	-	-	-	-	BC	-	-	-	-	-	-	-	-	-	-	
B. penefusca (Beamer)	-	-	-	-	-	-	-	-	ON	-	-	-	-	-	-	

Genus CRIOMORPHUS Curtis

C. conspicuus Metcalf	-	-	-	-	-	-	-	-	ON	QC	-	-	NS	-	-	
C. inconspicuus (Uhler)	-	YK	-	-	-	AB	SK	-	-	-	-	-	-	-	-	
C. wilhelmi Anufriev & Averkin	AK	YK	NT	-	-	-	-	-	-	-	-	-	-	-	-	

Genus DELPHACODES Fieber

D. albocarinata (Stål)	-	YK	NT	-	-	-	-	MB	-	-	-	-	-	-	LB	
D. albostriata (Fieber)	-	YK	-	-	BC	AB	SK	MB	-	-	-	-	-	-	-	
D. analis (Crawford)	-	YK	-	-	BC	AB	SK	MB	ON	QC	-	-	-	-	LB	
D. anufrievi Wilson	-	YK	-	-	BC	-	SK	-	-	-	-	-	-	-	-	
D. arcanistyla (Beamer)	-	-	-	-	BC	-	-	-	ON	QC	-	-	-	NF	LB	
D. balli Muir & Gifford	-	-	-	-	-	-	-	-	ON	-	-	-	-	-	-	[r]
D. basivitta (Van Duzee)	-	-	-	-	-	-	-	-	ON	-	-	-	-	-	-	
D. campestris (Van Duzee)	-	YK	-	-	BC	AB	SK	MB	ON	QC	NB	PE	NS	NF	-	[A]
D. consimilis (Van Duzee)	[r]	-	-	-	BC	AB	-	-	-	-	-	-	-	-	-	
D. detecta (Van Duzee)	-	-	-	-	-	-	-	-	-	-	-	PE	NS	-	-	
D. emeljanovi Wilson	-	YK	NT	-	BC	AB	-	-	-	-	-	-	-	-	-	
D. foveata (Van Duzee)	-	-	-	-	BC	AB	SK	MB	ON	-	-	-	-	-	-	

Delphacodes gillettei (Van Duzee)	-	-	-	-	BC	AB	SK	-	-	-	-	-	-	-	-	
D. incurva Beamer	-	-	-	-	BC	-	-	-	-	-	-	-	-	-	-	
D. indistincta (Crawford)	-	-	-	-	-	AB	SK	-	-	-	-	-	-	-	-	
D. kilmani (Van Duzee)	-	YK	NT	-	BC	AB	SK	MB	ON	QC	-	-	-	NF	-	
D. kolophon (Kirkaldy)	-	-	-	-	-	-	-	-	-	-	-	-	NS*	-	-	[r]
D. lineatipes (Van Duzee)	-	-	-	-	BC	AB	SK	MB	ON	-	-	-	NS	-	-	
D. litoralis (Reuter)	-	YK	NT	-	BC	-	-	-	-	-	-	-	-	-	-	
D. lutulenta (Van Duzee)	-	-	-	-	BC	AB	SK	-	ON	QC	-	-	NS	-	-	
D. lutulentoides Beamer	-	-	-	-	BC	-	SK	-	-	-	-	-	-	-	-	
D. magna (Crawford)	-	YK	-	-	BC	AB	-	-	-	QC	-	-	-	NF	-	
D. munda Beamer	-	-	-	-	-	AB	-	-	-	-	-	-	-	-	-	[r]
D. neocclusa Muir & Gifford	-	-	-	-	BC	-	-	-	-	-	-	-	-	-	-	
D. nigridorsum Crawford	-	YK	-	-	-	AB	SK	-	ON	QC	-	-	-	-	LB	[r]
D. nigriscutellata Beamer	-	-	-	-	-	AB	-	MB	ON	-	-	-	-	-	-	[r]
D. occlusa (Van Duzee)	-	-	-	-	BC	-	-	-	-	-	-	-	-	-	-	
D. paransera Beamer	-	-	-	-	-	AB	SK	MB	-	-	-	-	-	-	-	
D. parvula Ball	-	-	-	-	-	-	-	MB	ON	-	-	-	-	-	-	
D. pedaforma Beamer	-	-	-	-	BC	AB	-	-	-	-	-	-	-	-	-	
D. propinqua (Fieber)	-	-	-	-	BC*	-	-	-	-	-	-	-	-	-	-	
D. puella (Van Duzee)	-	-	-	-	-	-	-	-	ON	QC	-	-	-	-	-	
D. sagae Beamer	-	-	-	-	-	-	-	-	ON	-	-	-	-	-	-	[r]
D. serrata Beamer	-	YK	NT	-	-	AB	SK	MB	ON	-	-	-	-	-	-	
D. slossonae Ball	-	-	-	-	-	-	-	-	ON	-	-	-	-	-	-	[r]
D. stramineosa Beamer	-	-	-	-	-	-	-	-	-	QC	-	-	-	-	-	[r]
D. stricklandi (Metcalf)	-	-	-	-	-	AB	SK	MB	-	-	-	-	-	-	-	

Genus DICRANOTROPIS Fieber

D. incerta (Van Duzee)	-	-	-	-	-	-	-	-	-	QC	-	-	-	-	-	[r]

Genus EUIDES Fieber

E. gerhardi (Metcalf)	-	-	-	-	BC	-	SK	MB	ON	QC	-	-	-	-	-	

Genus EURYSA Fieber

E. magnifrons (Crawford)	-	YK	-	-	BC	AB	SK	-	-	-	-	-	-	-	-	
E. montana Beamer	-	-	-	-	-	AB	-	-	-	-	-	-	-	-	-	
E. obesa Beamer	-	-	-	-	BC	-	-	-	-	-	-	-	-	-	-	

Genus JAVASELLA Fennah

J. atrata (Osborn)	-	YK	-	-	-	AB	-	MB	ON	QC	-	-	-	-	-	
J. beringiaca Emeljanov	-	YK	-	-	-	-	SK	MB	-	-	-	-	-	-	-	
J. discolor (Boheman)	AK	YK	NT	-	-	-	-	-	-	-	-	-	-	-	-	
J. dolera (Spooner)	-	-	-	-	-	AB	-	-	ON	-	-	-	-	-	-	
J. lutulentella (Muir & Gifford)	-	-	-	-	BC	-	-	-	-	-	-	-	-	-	-	
J. obscurella (Boheman)	-	YK	NT	-	-	-	-	-	-	-	-	-	-	-	-	
J. opaca (Beamer)	-	-	-	-	-	-	-	-	-	QC	-	-	-	-	-	
J. pellucida (Fabricius)	AK	YK	NT	NU	BC	AB	SK	MB	ON	QC	NB	-	NS	NF	LB	[A]
J. simillima (Linnavuori)	-	YK	NT	-	-	-	-	MB	-	-	-	-	-	-	-	

Genus KELISIA Fieber

K. axialis Van Duzee	-	-	-	-	-	-	-	-	ON	-	-	PE	NS	-	-	
K. flava Beamer	-	-	-	-	-	-	-	-	ON	-	-	-	-	-	-	
K. hyalina Beamer	-	-	-	-	-	-	-	-	ON	-	-	-	-	-	-	[r]
K. parvicurvata Beamer	-	-	-	-	-	-	-	-	-	-	-	-	-	NF	-	
K. pectinata Beamer	-	-	-	-	-	-	-	-	ON	QC	-	-	-	-	-	
K. retrorsa Beamer	-	-	-	-	-	-	-	-	ON	-	-	-	-	-	-	
K. spinosa Beamer	-	-	-	-	-	-	-	-	ON	-	-	PE	-	-	-	
K. vesiculata Beamer	-	-	-	-	-	-	SK	-	-	-	-	-	-	-	-	

* introduced; [] indoors; { } extirpated; ? uncertain record; @ unestablished adventive; $ port-of-entry interception

Genus KUSNEZOVIELLA Vilbaste

K. macleani Wilson	-	YK	NT	-	-	-	-	-	-	QC	-	-	-	-	-	

Genus LACCOCERA Van Duzee

L. canadensis Beirne	-	-	-	-	-	AB	SK	-	-	-	-	-	-	-	-	
L. flava Crawford	-	-	-	-	-	AB	-	-	-	-	-	-	-	-	-	
L. lineata Scudder	-	-	-	-	BC	AB	SK	-	-	-	-	-	-	-	-	[r]
L. obesa Van Duzee	-	-	-	-	-	AB	SK	-	-	-	-	-	-	-	-	
L. oregonensis Penner	-	-	-	-	BC	-	-	-	-	-	-	-	-	-	-	[r]
L. vanduzeei Penner	-	-	-	-	BC	-	-	-	-	-	-	-	-	-	-	
L. vittipennis Van Duzee	-	-	-	-	BC	AB	SK	MB	ON	QC	-	-	-	-	-	

Genus LIBURNIELLA Crawford

L. ornata (Stål)	-	-	-	-	BC*	-	-	-	ON	-	-	-	-	-	-	

Genus MEGAMELANUS Ball

M. bicolor Ball	-	-	-	-	-	-	-	MB?	-	-	-	-	-	-	-	[r]

Genus MEGAMELUS Fieber

M. aestus Metcalf	-	-	-	-	-	-	-	-	ON	QC	-	-	-	-	-	
M. bifidus Beamer	-	-	-	-	-	-	-	-	ON	-	-	-	-	-	-	[r]
M. distinctus Metcalf	-	-	-	-	-	-	SK	-	ON	QC	NB	PE	NS	NF	-	
M. flavus Crawford	-	YK	NT	-	BC	AB	SK	MB	-	-	-	-	-	-	-	
M. inflatus Metcalf	-	-	-	-	-	-	-	-	-	-	-	-	NS	-	-	[r]
M. metzaria Crawford	-	-	-	-	-	-	SK	MB	-	QC	NB	PE	NS	-	-	
M. recurvatus Beamer	-	-	-	-	-	-	-	MB	ON	-	-	-	-	NF	-	
M. uncus Metcalf	-	YK	-	-	BC	AB	SK	MB	ON	QC	NB	-	NS	NF	-	

Genus MUELLERIANELLA Wagner

M. laminalis (Van Duzee)	-	-	-	-	-	-	-	-	ON	QC	-	-	-	-	-	

Genus NOTHODELPHAX Fennah

N. eburneocarinata (Anufriev)	AK	YK	NT	-	-	-	-	-	-	-	-	-	-	-	-	
N. glacia Wilson	-	YK	-	-	-	-	-	MB	-	-	-	-	-	-	-	
N. guentheri (Dlabola)	-	YK?	-	-	BC	AB?	SK?	-	-	-	-	-	-	-	-	
N. umbrata Emeljanov	-	YK	-	-	-	-	-	-	-	-	-	-	-	-	-	

Genus PARKANA Beamer

P. alata Beamer	-	-	-	-	BC	-	-	-	-	-	-	-	-	-	-	

Genus PENTAGRAMMA Van Duzee

P. variegata Penner	-	-	-	-	BC	-	SK	MB	ON	-	-	-	NS	-	-	
P. vittatifrons (Uhler)	-	-	-	-	-	AB	SK	MB	-	-	-	-	-	-	-	

Genus PHYLLODINUS Van Duzee

P. nervatus Van Duzee	-	-	-	-	BC	AB	SK	MB	ON	QC	-	-	-	-	-	

Genus PISSONOTUS Van Duzee

P. aphidioides Van Duzee	-	-	-	-	-	-	-	MB	ON	QC	-	PE?	-	-	-	
P. basalis Van Duzee	-	-	-	-	-	AB?	-	-	ON	QC	NB	-	NS	-	-	
P. brunneus Van Duzee	-	-	-	-	-	AB?	-	-	ON	QC	-	-	NS	-	-	
P. delicatus Van Duzee	-	-	-	-	BC	AB	SK	-	-	-	-	-	-	-	-	
P. dorsalis Van Duzee	-	-	-	-	-	-	-	-	ON	QC	NB	PE	NS	NF	-	
P. flabellatus (Ball)	-	-	-	-	-	-	-	-	ON	-	-	-	-	-	-	
P. guttatus Spooner	-	-	-	-	-	-	SK	-	ON	QC	-	-	NS	-	-	
P. marginatus Van Duzee	-	-	-	-	-	-	-	-	ON	QC	-	-	-	-	-	
P. niger Morgan & Beamer	-	-	-	-	-	-	-	MB	-	-	-	-	-	-	-	

Pissonotus piceus Van Duzee	-	-	-	-	-	-	-	-	ON	-	-	-	-	-	-	
P. rubrilatus Morgan & Beamer	-	-	-	-	BC	-	-	-	-	-	-	-	-	-	-	
P. tumidus Morgan & Beamer	-	-	-	-	-	-	-	MB	-	-	-	-	-	-	-	

Genus PROKELISIA Osborn

P. carolae Wilson	-	-	-	-	BC	-	-	-	-	-	-	-	-	-	-	[r]
P. crocea (Van Duzee)	-	-	-	-	-	-	SK	MB	ON	QC	NB	PE	-	-	-	
P. dolus Wilson	-	-	-	-	-	-	-	-	ON	QC	NB	-	NS	-	-	
P. frontalis (Crawford)	-	-	-	-	-	-	SK	-	-	-	-	-	-	-	-	
P. salina (Ball)	-	-	-	-	BC	-	SK	MB	ON	-	-	-	-	-	-	

Genus RIBAUTIDELPHAX Wagner

R. bidentata Anufriev	-	-	-	-	-	-	-	-	ON	QC	-	-	-	NF	LB	
R. pusilla Emeljanov	-	YK	-	-	BC	AB	SK	-	ON	QC	-	-	-	-	-	

Genus STENOCRANUS Fieber

S. dorsalis (Fitch)	-	-	-	-	-	AB?	-	-	ON	QC	-	-	-	-	-	
S. felti Van Duzee	-	-	-	-	BC	AB	SK	MB	ON	QC	-	PE	-	-	-	
S. lautus Van Duzee	-	-	-	-	BC?	-	-	-	ON	QC	-	-	-	-	-	
S. pallidus Beamer	-	-	-	-	BC	AB	SK	MB	ON	QC	NB	PE	-	-	-	
S. sandersoni Beamer	-	-	-	-	BC	-	SK	-	ON	QC	-	-	-	-	-	
S. unipunctatus (Provancher	-	-	-	-	-	-	-	-	ON	QC	-	-	-	-	-	

Genus STOBAERA Stål

S. tricarinata (Say)	-	-	-	-	BC	-	-	-	ON	QC	-	-	-	-	-	

Genus UNKANODES Fennah

U. excisa (Melichar)	AK	-	-	-	-	-	-	-	-	-	-	-	-	-	-	

Genus YUKONODELPHAX Wilson

Y. kendallae Wilson	-	YK	-	-	-	AB	-	-	-	-	-	-	-	-	-	

Family DERBIDAE

Genus AMALOPOTA Van Duzee

A. uhleri Van Duzee	-	-	-	-	-	-	-	-	ON	QC	-	-	-	-	-	

Genus ANOTIA Kirby

A. burnetii Fitch	-	-	-	-	-	-	-	-	ON	-	-	-	-	-	-	[r]
A. kirkaldyi Ball	-	-	-	-	-	-	-	-	ON	-	-	-	-	-	-	

Genus CEDUSA Fowler

C. hedusa McAtee	-	-	-	-	-	-	-	-	ON	QC	NB	-	-	-	-	
C. incisa (Metcalf)	-	-	-	-	-	-	-	MB	ON	QC	NB	-	-	-	-	
C. maculata (Van Duzee)	-	-	-	-	-	-	-	-	ON	QC	-	-	-	-	-	
C. obscura (Ball)	-	-	-	-	-	-	-	MB	ON	QC	-	-	-	-	-	
C. vulgaris (Fitch)	-	-	-	-	-	-	-	-	ON	QC	-	-	-	-	-	

Genus OTIOCERUS Kirby

O. coquebertii Kirby	-	-	-	-	-	-	-	-	ON	QC	NB	PE	-	-	-	
O. degeerii Kirby	-	-	-	-	BC	AB	-	-	ON	QC	-	-	NS	-	-	
O. kirbyii Fitch	-	-	-	-	-	-	-	-	ON	-	-	-	NS	-	-	[r]
O. signoretii Fitch	-	-	-	-	-	-	-	-	ON	-	-	-	-	-	-	
O. wolfei Kirby	-	-	-	-	-	-	-	-	ON	-	-	-	-	-	-	

* introduced; [] indoors; { } extirpated; ? uncertain record; @ unestablished adventive; $ port-of-entry interception

Genus PATARA Westwood

P. vanduzeei Ball	-	-	-	-	-	-	-	-	ON	-	-	-	-	-	-	

Family FLATIDAE

Genus METCALFA Caldwell & Martorell

M. pruinosa (Say)	-	-	-	-	-	-	-	-	ON	QC	-	-	-	-	-	

Family FULGORIDAE

Genus PHYLLOSCELIS Germar

P. atra Germar	-	-	-	-	-	-	-	-	ON	-	-	-	-	-	-	
P. pallescens Germar	-	-	-	-	-	-	-	-	ON	-	-	-	-	-	-	

Genus SCOLOPS Schaum

S. abnormis Ball	-	-	-	-	BC	-	-	-	-	-	-	-	-	-	-	
S. angustatus Uhler	-	-	-	-	BC	AB	SK	MB	ON	-	-	-	-	-	-	
S. grossus Uhler	-	-	-	-	-	AB	-	-	-	-	-	-	-	-	-	
S. hesperius Uhler	-	-	-	-	-	AB	SK	MB	-	-	-	-	-	-	-	
S. pungens (Germar)	-	-	-	-	-	-	SK?	MB?	-	-	-	-	-	-	-	
S. sulcipes (Say)	-	-	-	-	BC	-	SK	MB	ON	QC	NB	-	-	-	-	

Family ISSIDAE

Genus ACANALONIA Spinola

A. bivittata (Say)	-	-	-	-	-	-	-	-	ON	-	-	-	-	-	-	
A. conica (Say)	-	-	-	-	-	-	-	-	ON	-	-	-	-	-	-	[r]

Genus THIONIA Stål

T. bullata (Say)	-	-	-	-	-	-	-	-	ON	-	-	-	-	-	-	

Suborder Prosorrhyncha

Key to infraorders, families, subfamilies and some tribes: Schuh & Slater 1995 [world]

Key to genera: Slater & Baranowski 1978 [North America (most genera)].

Catalogue: Henry & Froeschner 1988 [Canada and Continental United States].

Regional guide: Larochelle 1983 [Québec].

Regional lists: Scudder 1997 [Yukon], Downes 1927, 1935, Scudder 1961, 1986 [British Columbia], Strickland 1953 [Alberta], Moore 1950 [Québec], Parshley 1923 [Nova Scotia], Lindberg 1958 [Newfoundland].

True bugs. Punaises. Prosorrhyncha is equivalent to Heteroptera plus Pelordidiidae (= Coleorrhyncha, a group not occurring in Canada or Alaska). Characterized by the forward position of the rostrum and (usually) closure of the head capsule behind it; antennae of 3-5 segments (flagellum with 1 to 3 segments, scape and pedicel often elongate, similar to flagellar segments); forewing usually with differentiated basal thickened and apical membranous area, folded flat and overlapping over the abdomen.

Infraorder Enicocephalomorpha

Revision: Wygodzinsky & Schmidt 1991 [New World].

Superfamily Enicocephaloidea

Family ENICOCEPHALIDAE

Unique-headed bugs. Small, frail predators, with head characteristically elongate with a marked transverse impression, the eyes situated laterally on the anterior lobe and the ocelli dorsal on the posterior lobe. Fore tibia flattened with distinct distal spines; fore tarsus opposable to apex of fore tibia. Pronotum with three apparent lobes. Forewing, when present, uniform in texture and not differentiated with distinct corium and membrane, claval commissure and costal fracture absent.

Most species fly readily and some perform prenuptial flight-dance in patches of sunlight. Most abundant in humid tropics and subtropics. Usually found in leaf litter or loose soil, in moss, rotting wood and under bark.

Subfamily ENICOCEPHALINAE

Genus SYSTELLODERES Blanchard

Key to species: Kritsky 1978 [North America].

S. biceps (Say)	-	-	-	-	-	-	-	-	ON	QC	-	-	NS	-	-

Infraorder Dipsocomorpha

Superfamily Dipsocoidea

Family CERATOCOMBIDAE

Key to species: McAtee & Malloch 1925 [North America].

Ceratocombids. Very small, cryptic predatory bugs recognized by long slender antennae, two-segmented tarsi on all legs, and forewing with a short but distinct fracture at middle of costal margin. Usually found in leaf litter, at base of grass clumps, or in rotting wood.

Subfamily CERATOCOMBINAE

Tribe Ceratocombini

Genus CERATOCOMBUS Signoret
Key to species: McAtee & Malloch 1925 [New World].

C. vagans McAtee & Malloch - - - - BC AB? - - ON QC - - - - -

Infraorder Nepomorpha

Regional guide: Brooks & Kelton 1967 [prairie provinces].
Regional list: Scudder 1977 [B.C.], Chagnon & Fournier 1948 [Québec].

Superfamily Nepoidea

Family BELOSTOMATIDAE

Key to subfamilies and genera: Lauck & Menke 1961.

Giant water bugs (electric light bugs, toe biters). Flattened and elongate-oval in shape, uniform dull brown colour with short hidden antennae, greatly enlarged and raptorial forelegs, and with middle and hind legs adapted for swimming. Voracious aquatic predators, usually found in freshwater ponds, lakes and rivers. Although prey is smaller aquatic organisms such as immature insects and tadpoles, at times prey include fish, frogs and even small birds. Victims are subdued by powerful, hydrolytic enzymes injected via the short beak: the "bite" can produce a painful swelling in humans. In *Lethocerus*, egg masses are laid on emergent vegetation, but in *Belostoma*, they are laid on the backs of males. Adults regularly come to light at night.

Subfamily BELOSTOMATINAE

Genus BELOSTOMA Latreille
Key to species: Menke 1958 [North America].

B. flumineum Say - - - - BC - - MB ON QC NB PE NS - -

* introduced; [] indoors; { } extirpated; ? uncertain record; @ unestablished adventive; $ port-of-entry interception

Subfamily LETHOCERINAE

Genus LETHOCERUS Mayr
Key to species: Menke 1963.

(Subgenus BENACUS Stål)

L. griseus (Say) - - - - - - - MB ON QC - - - - -

(Subgenus LETHOCERUS Mayr)

L. americanus (Leidy) - - - - BC AB SK MB ON QC NB PE NS NF -
giant water bug; léthocère

L. uhleri (Montandon) - - - - - - - - ON - - - - - -

Family NEPIDAE

Key to species: Sites & Polhemus 1994 [North America].

Water scorpions. Aquatic predators with characteristic elongate breathing tube extending from end of abdomen like a slender tail, and raptorial forelegs with elongate femur. Tarsi on all legs with single segment; middle and hind legs slender and not modified for swimming. Found in ponds and streams, usually clinging to sticks and weeds.

Subfamily NEPINAE

Tribe Nepini

Genus NEPA Linnaeus

N. apiculata Uhler - - - - - - - MB ON QC - - - - -

Subfamily RANATRINAE

Genus RANATRA Fabricius

R. fusca Palisot - - - - BC - - MB ON QC NB PE NS - -

R. kirkaldyi Torre Bueno - - - - - - - - ON - - - - - - [r]

R. nigra Herrich-Schaeffer - - - - - - - - ON - - - - - - [r]

Superfamily Ochteroidea

Family GELASTOCORIDAE

Key: Todd 1955.

Toad bugs. Ovoid, flattened, dull coloured, predaceous bugs with prominent, bulging eyes, concealed antennae, and raptorial forelegs. Usually found along margins of ponds, rivers and streams. Predominantly a northern hemisphere group.

Subfamily GELASTOCORINAE

Genus GELASTOCORIS Kirkaldy

G. oculatus (Fabricius) - - - - BC - - MB ON - - - - - -

Family OCHTERIDAE

Key to taxa: Schell 1943 [Western Hemisphere].

Velvety shore bugs. Small, oval, blackish bugs with velvety texture. Ocelli present, antennae relatively long, filiform and partially hidden in dorsal view; eyes sessile and legs cursorial. Predaceous and found in damp littoral habitats.

Genus OCHTERUS Latreille

O. americanus (Uhler)	-	-	-	-	-	-	-	-	ON	-	-	-	-	-	-	[r]

Superfamily Corixoidea

Family CORIXIDAE

Key to taxa: Hungerford 1948 [Western Hemisphere].

Water boatmen. Aquatic predators, living mostly in ponds and at the margins of lakes. Streamlined body has a broad head with large eyes, small and hidden antennae, and a short triangular "beak". Forelegs are scoop-like (palae), middle legs are long and slender with a single tarsal segment, and hind legs are flattened and oarlike, fringed with long hairs.

Subfamily CORIXINAE

Tribe Corixini

Genus ARCTOCORISA Wallengren

A. chanceae Hungerford	AK	YK	NT	-	-	-	-	MB	-	QC	-	-	-	NF	LB	
A. convexa (Fieber)	-	-	NT	-	BC	AB	-	MB	-	QC	-	-	-	NF	LB	
A. planifrons (Kirby)	AK	YK	NT	-	-	-	-	-	-	-	-	-	-	NF	LB	
A. sutilis (Uhler)	AK	YK	NT	-	BC	AB	SK	MB	ON	QC	NB	-	-	-	-	

Genus CALLICORIXA White

C. alaskensis Hungerford	AK	YK	NT	-	BC	AB	SK	MB	ON	QC	NB	-	NS	NF	LB	
C. audeni Hungerford	AK	YK	NT	-	BC	AB	SK	MB	ON	QC	NB	PE	NS	NF	-	
C. producta noorvikensis Hungerford	AK	YK	NT	-	-	-	-	MB	-	-	-	-	-	-	-	
C. scudderi Jansson	-	-	-	-	BC	-	-	-	-	-	-	-	-	-	-	
C. vulnerata (Uhler)	AK	-	-	-	BC	-	-	-	-	-	-	-	-	-	-	

Genus CENOCORIXA Hungerford

Key to species: Jansson 1972.

C. andersoni Hungerford	-	-	-	-	BC	-	-	-	-	-	-	-	-	-	-	[r]
C. bifida bifida (Hungerford)	-	-	-	-	-	AB	SK	MB	-	-	-	-	-	-	-	
C. bifida hungerfordi Lansbury	-	-	-	-	BC	-	-	-	-	-	-	-	-	-	-	
C. blaisdelli (Hungerford)	-	-	-	-	BC	-	-	-	-	-	-	-	-	-	-	[r]
C. dakotensis (Hungerford)	-	-	NT	-	BC	AB	SK	MB	-	-	-	-	-	-	-	
C. expleta (Uhler)	-	-	-	-	BC	AB	SK	MB	-	-	-	-	-	-	-	
C. utahensis (Hungerford)	-	-	-	-	BC	AB	SK	MB	ON	-	-	-	-	-	-	

Genus CORISELLA Lundblad

C. decolor (Uhler)	-	-	-	-	BC	-	-	-	-	-	-	-	-	-	-	[r]
C. inscripta (Uhler)	-	-	-	-	BC	-	-	-	-	-	-	-	-	-	-	[r]
C. tarsalis (Fieber)	-	-	-	-	-	AB	SK	MB	ON	-	-	-	-	-	-	

Genus HESPEROCORIXA Kirkaldy

H. atopodonta (Hungerford)	-	-	NT	-	BC	AB	SK	MB	ON	QC	NB	PE	NS	NF	-	
H. harrisi (Uhler)	-	-	-	-	-	-	-	-	ON	QC	-	PE	NS	-	-	

* introduced; [] indoors; { } extirpated; ? uncertain record; @ unestablished adventive; $ port-of-entry interception

Hesperocorixa interrupta (Say)	-	-	-	-	-	-	-	-	ON	QC	NB	-	NS	-	-	
H. kennicotti (Uhler)	-	-	-	-	-	-	-	-	ON	QC	-	-	NS	-	-	
H. laevigata (Uhler)	-	-	-	-	BC	AB	SK	MB	ON	-	-	-	-	-	-	
H. lucida (Abbott)	-	-	-	-	-	-	-	-	ON	-	-	-	-	-	-	
H. michiganensis (Hungerford)	-	-	-	-	BC	AB	SK	MB	ON	QC	NB	-	NS	-	-	
H. minorella (Hungerford)	-	-	NT	-	-	AB	-	MB	ON	QC	NB	-	NS	NF	-	
H. obliqua (Hungerford)	AK	-	-	-	-	-	-	-	ON	-	-	-	-	-	-	
H. scabricula (Walley)	-	-	-	-	-	-	SK	MB	ON	QC	NB	PE	-	-	-	
H. semilucida (Walley)	-	-	-	-	-	-	-	-	ON	QC	-	-	-	-	-	
H. vulgaris (Hungerford)	-	-	-	-	BC	AB	SK	MB	ON	QC	NB	PE	-	-	-	
Genus PALMACORIXA Abbott																
P. buenoi Abbott	-	-	-	-	-	AB	SK	MB	ON	QC	-	-	-	NF	-	
P. gillettei confluens Walley	-	-	-	-	-	-	-	-	ON	-	-	-	-	-	-	
P. gillettei gillettei Abbott	-	-	-	-	-	-	SK	MB	-	QC	-	-	-	-	-	
P. janeae Brooks	-	-	-	-	-	-	SK	MB	-	-	-	-	-	-	-	
P. nana nana Walley	-	-	-	-	-	-	-	-	ON	QC	-	-	-	-	-	
Genus RAMPHOCORIXA Abbott																
R. acuminata (Uhler)	-	-	-	-	-	-	-	-	ON	-	-	-	-	-	-	
Genus SIGARA Fabricius																
(Subgenus ALLOSIGARA Hungerford)																
S. decorata (Abbott)	-	-	-	-	-	AB	SK	-	-	-	-	-	-	-	-	
(Subgenus ARCTOSIGARA Hungerford)																
S. bicoloripennis (Walley)	AK	-	-	-	BC	AB	SK	MB	ON	QC	NB	PE	NS	NF	-	
S. conocephala (Hungerford)	-	-	NT	-	BC	AB	SK	MB	ON	QC	-	PE	NS	NF	-	
S. decoratella (Hungerford)	AK	YK	NT	-	BC	AB	SK	MB	ON	QC	NB	PE	NS	NF	LB	
S. penniensis (Hungerford)	AK	YK	NT	-	BC	AB	SK	MB	ON	QC	NB	PE	-	NF	LB	
(Subgenus LASIOSIGARA Hungerford)																
S. lineata (Forster)	-	-	-	-	-	-	SK	MB	ON	QC	-	-	-	-	-	
S. trilineata (Provancher)	-	-	NT	-	-	AB	SK	MB	ON	QC	NB	-	NS	NF	LB	
(Subgenus PHAEOSIGARA Hungerford)																
S. compressoidea (Hungerford)	-	-	-	-	-	-	-	-	ON	QC	NB	-	NS	NF	-	
S. dolabra Hungerford & Sailer	-	-	-	-	BC	-	-	-	ON	QC	-	-	NS	NF	-	
S. mackinacensis (Hungerford)	-	-	-	-	-	-	-	-	ON	QC	NB	-	-	-	-	
S. macropala Hungerford	-	-	-	-	-	-	-	-	-	-	-	-	-	NF	-	
S. quebecensis (Walley)	-	-	-	-	-	-	-	-	-	QC	-	-	-	-	-	[r]
S. signata (Fieber)	-	-	-	-	-	-	-	MB	ON	QC	NB	PE	NS	NF	LB	
S. variabilis (Hungerford)	-	-	-	-	-	-	-	-	ON	QC	NB	PE	NS	-	-	
(Subgenus PILEOSIGARA Hungerford)																
S. douglasensis (Hungerford)	-	-	-	-	-	-	-	-	ON	-	NB	-	NS	-	-	
(Subgenus SUBSIGARA Stichel)																
S. fallenoidea (Hungerford)	AK	YK	NT	-	-	AB	SK	MB	-	-	-	-	-	-	-	
(Subgenus VERMICORIXA Walton)																
S. alternata (Say)	-	-	-	-	BC	AB	SK	MB	ON	QC	NB	PE	NS	-	-	
S. defecta Hungerford	-	-	-	-	-	-	-	-	ON	-	-	-	-	-	-	
S. grossolineata Hungerford	-	-	-	-	BC	AB	SK	MB	ON	QC	NB	-	-	-	-	
S. hubbelli (Hungerford)	-	-	-	-	-	-	-	-	ON	-	-	-	-	-	-	
S. johnstoni Hungerford	-	-	-	-	-	-	-	-	ON	QC	NB	-	-	-	-	
S. knighti Hungerford	-	-	-	-	-	-	-	-	ON	QC	NB	-	NS	-	-	
S. mathesoni Hungerford	-	-	-	-	-	AB	SK	MB	ON	-	NB	-	NS	NF	-	
S. modesta (Abbott)	-	-	-	-	BC	-	-	-	ON	-	-	-	-	-	-	

A agricultural, F forestry, G greenhouse, O ornamental pest; r rare; b biocontrol agent

Sigara mullettensis (Hungerford)	-	-	-	-	BC	-	SK	MB	ON	QC	NB	-	-	-	-	
S. omani (Hungerford)	-	-	-	-	BC	-	-	-	-	-	-	-	-	-	-	
S. solensis (Hungerford)	AK	-	-	-	BC	AB	SK	MB	ON	QC	NB	PE	-	-	-	
S. transfigurata (Walley)	-	-	-	-	-	-	-	-	ON	QC	-	-	-	-	-	
S. washingtonensis Hungerford	-	-	-	-	BC	AB	SK	MB	-	-	-	-	-	-	-	

(Subgenus XENOSIGARA Hungerford)

S. ornata (Abbott)	-	-	-	-	-	-	-	-	ON	QC	NB	-	-	-	-	

Genus TRICHOCORIXA Kirkaldy

Key to species: Sailer 1948.

T. borealis Sailer	-	-	NT	-	-	AB	SK	MB	ON	QC	-	-	-	-	-	
T. calva (Say)	-	-	-	-	-	-	-	-	ON	-	-	-	-	-	-	
T. kanza Sailer	-	-	-	-	-	-	-	-	ON	-	-	-	-	-	-	
T. louisianae Jaczewski	-	-	-	-	-	-	-	-	-	-	-	-	NS	-	-	
T. sexcincta (Champion)	-	-	-	-	-	AB	SK	MB	ON	QC	-	PE	NS	NF	-	
T. verticalis californica Sailer	-	-	-	-	BC	-	-	-	-	-	-	-	-	-	-	[r]
T. verticalis fenestrata Walley	-	-	-	-	-	-	-	-	-	QC	-	-	-	NF	-	
T. verticalis interiores Sailer	-	-	-	-	-	AB	SK	MB	-	-	-	-	-	-	-	
T. verticalis verticalis (Fieber)	-	-	-	-	-	-	-	-	-	-	-	PE	-	-	-	

Tribe Glaenocorisini

Genus DASYCORIXA Hungerford

D. hybrida (Hungerford)	-	YK	-	-	BC	AB	SK	-	-	QC	-	-	-	-	-	
D. johanseni (Walley)	-	-	NT	-	BC	AB	SK	MB	-	QC	-	-	-	-	-	
D. rawsoni Hungerford	-	-	NT	-	BC	AB	SK	MB	-	-	-	-	-	-	-	

Genus GLAENOCORISA Thomson

G. propinqua cavifrons (Thomson)	AK	YK	NT	-	-	-	-	-	-	QC	-	-	-	-	LB	
G. quadrata Walley	AK	-	NT	-	-	-	-	-	-	QC	-	-	-	NF	-	

Subfamily CYMATINAE

Genus CYMATIA Flor

C. americana Hussey	AK	YK	NT	-	BC	AB	SK	MB	ON	QC	-	-	-	-	-	

Superfamily Naucoroidea

Family NAUCORIDAE

Creeping water bugs. Oval, brownish aquatic predators with enlarged raptorial forelegs, middle and hind legs with two-segmented tarsi and long claws, and leathery forewings without veins in the membrane. These bugs lack ocelli and the single Canadian species occurs in ponds with abundant vegetation. Mainly a tropical family, often in rapidly flowing streams and rivers.

Subfamily NAUCORINAE

Genus PELOCORIS Stål

Key to species: La Rivers 1948 [USA].

P. femoratus (Palisot)	-	-	-	-	-	-	-	-	ON	-	-	-	-	-	-	[r]

* introduced; [] indoors; { } extirpated; ? uncertain record; @ unestablished adventive; $ port-of-entry interception

Superfamily Notonectoidea

Family NOTONECTIDAE

Key to genera: Hungerford 1933.

Backswimmers. Active predators in ponds, lakes and other aquatic habitats, characteristically swimming and resting upside-down under water. Forelegs raptorial and not scoop like, rostrum short but distinctly segmented, and hind legs flattened and oarlike, fringed with long hairs for swimming.

Subfamily ANISOPINAE

Genus BUENOA Kirkaldy

Key to species: Truxal 1953.

B. confusa Truxal	-	-	-	-	BC	AB	SK	MB	ON	QC	-	PE	NS	-	-
B. limnocastoris Hungerford	-	-	-	-	-	-	-	-	ON	QC	-	PE	NS	-	-
B. macrotibialis Hungerford	-	-	-	-	BC	-	SK	MB	ON	QC	-	-	NS	-	-
B. margaritacea Torre Bueno	-	-	-	-	-	-	-	MB	ON	QC	-	-	-	-	-

Subfamily NOTONECTINAE

Tribe Notonectini

Genus NOTONECTA Linnaeus
Key to species: Hungerford 1933.

(Subgenus NOTONECTA Linnaeus)

N. borealis Hussey	-	-	NT	-	BC	AB	SK	MB	ON	QC	-	-	-	NF	-	
N. irrorata Uhler	-	-	-	-	-	-	SK	MB	ON	QC	NB	PE	NS	-	-	

(Subgenus PARANECTA Hutchinson)

N. insulata Kirby	-	-	-	-	-	-	-	MB	ON	QC	NB	PE	NS	NF	-	
N. kirbyi Hungerford	-	YK	NT	-	BC	AB	SK	MB	-	-	-	-	-	-	-	
N. lunata Hungerford	-	-	-	-	-	-	-	-	ON	QC	NB	PE	NS	-	-	
N. spinosa Hungerford	-	-	-	-	BC	-	-	-	-	-	-	-	-	-	-	[r]
N. undulata Say	-	-	NT	-	BC	AB	SK	MB	ON	QC	NB	PE	NS	NF	-	
N. unifasciata andersoni Hungerford	-	-	-	-	BC	-	-	-	-	-	-	-	-	-	-	[r]

Family PLEIDAE

Key to species: Drake & Chapman 1953 [western hemisphere].

Pygmy Backswimmers. Aquatic predators, small, with strongly convex body form, coleopteriform hemelytra, and hind legs with long hairs for swimming. They swim upside down, inhabiting weedy ponds.

Genus NEOPLEA Esaki & China

N. striola (Fieber)	-	-	-	-	-	-	-	MB	ON	QC	-	-	-	-	-

Infraorder Gerromorpha

Keys to families, subfamilies, tribes and genera: Andersen 1982.

Regional guides: Brooks & Kelton 1967 [Prairie Provinces], Chagnon & Fournier 1948 [Québec].

Regional list: Scudder 1977 [British Columbia].

Superfamily Mesovelioidea

Family MESOVELIIDAE

Water treaders. Predators and scavengers on water surface or floating vegetation. Body greenish, without dense velvety pile; often apterous. Head porrect, longer than wide, with ocelli present in winged forms; eyes not kidney-shaped; venter of head without rostral groove; labium three-segmented. Mesonotum exposed. Membrane of hemelytra, when present, without cells or veins. Coxae of hind legs not fused to thorax; tibiae with stiff black spines; tarsi three-segmented with apical claws.

Subfamily MESOVELIINAE

Genus MESOVELIA Mulsant & Rey

M. amoena Uhler	-	-	-	-	-	-	-	-	-	QC	NB	-	-	-	-	[r]
M. mulsanti White	-	-	NT	-	BC	AB	SK	MB	ON	QC	NB	-	NS	NF	-	

Superfamily Hebroidea

Family HEBRIDAE

Velvet water bugs. Small and inconspicuous predators and scavengers in semi-aquatic habitats, occurring either in moist detritus near shore or among floating aquatic plants. Body with thick velvety pile; often apterous. Head with ocelli; antennae four- or five-segmented; venter of head with prominent buccula, a deep rostral groove continued onto thoracic venter; labium four-segmented. Pronotum large, but small scutellum visible. Hemelytra, when present, lacking cuneus and veins to membrane. Coxae of hind legs cylindrical and not fused to thorax; tibiae without stiff black spines; tarsi two-segmented with apical claws.

Subfamily HEBRINAE

Genus HEBRUS Curtis

H. buenoi Drake & Harris	-	-	-	-	-	-	-	-	ON	-	-	-	-	-	-
H. burmeisteri Lethierry & Severin	-	-	-	-	-	-	SK	-	ON	QC	-	-	NS	-	-
H. concinnus Uhler	-	-	-	-	-	-	-	-	ON	QC	-	-	-	-	-
H. sobrinus Uhler	-	-	-	-	-	-	-	-	ON?	-	-	-	-	-	-

Genus MERRAGATA White

M. brunnea Drake	-	-	-	-	-	-	-	-	ON	-	-	-	-	-	-
M. hebroides White	-	-	NT	-	BC	AB	SK	MB	ON	QC	-	-	-	-	-

* introduced; [] indoors; { } extirpated; ? uncertain record; @ unestablished adventive; $ port-of-entry interception

Superfamily Hydrometroidea

Family HYDROMETRIDAE

Key to species: Hunderford & Evans 1934 [world].

Marsh treaders. Predators and scavengers along margins or on emergent or floating vegetation of still or slowly moving water. Elongate and stick-like, with long antennae and long thread-like legs with terminal claws. Head very elongate with eyes located midway along length; head longer than thorax. Usually apterous, but when winged, membrane with three closed cells.

Subfamily HYDROMETRINAE

Genus HYDROMETRA Latreille

H. martini Kirkaldy	-	-	-	-	BC	-	-	MB	ON	QC	-	-	NS	-	-

Superfamily Gerroidea

Family VELIIDAE

Key to genera and species: Smith & Polhemus 1978 [North America].

Small water striders. Active predators on the water surface, in either still or running water. Short-bodied and often apterous. Head dorsally with distinct median impression, and with a pair of deep pits near posterior angle of eyes; ocelli absent. Pronotum large and obscuring meso- and metanotum; scent gland openings with laterally extended channels. Forewings when present with fewer than six closed cells. Legs with subapical claws; fore tibiae of male usually with grasping comb of short spines along inner margin; fore tarsi either one- or three-segmented; middle tarsi two- or three-segmented; hind femora shorter than middle femora.

Subfamily MICROVELIINAE

Tribe Microveliini

Genus MICROVELIA Westwood
Key to species: Smith 1980 [North America].
(Subgenus KIRKALDYA Torre-Bueno)

M. americana (Uhler)	-	-	-	-	-	-	-	-	ON	QC	NB	PE	NS	-	-	

(Subgenus MICROVELIA Westwood)

M. albonotata Champion	-	-	-	-	-	-	-	-	-	-	NB	-	-	-	-	[r]
M. buenoi Drake	AK	YK	NT	-	BC	AB	SK	MB	ON	QC	NB	PE	NS	NF	-	
M. hinei Drake	-	-	-	-	-	-	-	-	ON	-	-	-	NS	-	-	[r]
M. pulchella Westwood	AK	-	-	-	BC	AB	-	MB	ON	QC	NB	-	NS	NF	-	

Subfamily RHAGOVELIINAE

Genus RHAGOVELIA Mayr
Key to species: Bacon 1956 [Western Hemisphere], Smith & Polhemus 1978 [North America].
(Subgenus RHAGOVELIA Mayr)

R. obesa Uhler	-	-	-	-	-	-	-	MB	ON	QC	NB	-	NS	-	-

Subfamily VELIINAE

Genus STEINOVELIA Polhemus & Polhemus

S. stagnalis (Burmeister)	-	-	-	-	-	-	-	-	ON	-	-	-	-	-	-	

Family GERRIDAE

Key to species: Scudder 1971 [adults, British Columbia], Scudder & Jamieson 1972, Spence & Scudder 1978 [immatures, British Columbia], Cheng & Fernando 1970 [Ontario].

Water striders. Active predators on the surface of lakes, ponds, rivers, streams, temporary puddles and the open ocean. Body elongate with venter clothed with dense silvery pubescence. Head shorter than thorax, without distinct median impression; ocelli absent. Thorax without scutellum; venter of metathorax with median scent gland opening or omphalium. Often apterous, but when forewings present with never more than four cells. Legs with claws subterminal, and tarsi two-segmented; fore tibiae of male without grasping comb; hind femora usually more slender than middle femora, and extending beyond end of abdomen.

Subfamily GERRINAE

Tribe Gerrini

Genus AQUARIUS Schellenberg

Key to species: Andersen 1990.

A. conformis (Uhler)	-	-	-	-	-	-	-	-	-	QC	-	-	NS	-	-	
A. nyctalis (Drake & Hottes)	-	-	-	-	BC	AB	SK	-	-	QC	-	-	-	NF	-	
A. remigis (Say)	-	-	-	-	BC	AB	SK	MB	ON	QC	NB	PE	NS	NF	-	

Genus GERRIS Fabricius

Key to species: Drake & Harris 1934 [Western Hemisphere].

G. alacris Hussey	-	-	-	-	-	-	-	-	ON	-	-	-	-	-	-	[r]
G. argenticollis Parshley	-	-	-	-	-	-	-	-	ON	QC	-	-	-	-	-	
G. buenoi Kirkaldy	AK	YK	NT	-	BC	AB	SK	MB	ON	QC	NB	PE	NS	NF	LB	
G. comatus Drake & Hottes	-	-	-	-	BC	AB	SK	MB	ON	QC	NB	PE	NS	NF	-	
G. gillettei Lethierry & Severin	-	-	-	-	BC?	-	-	-	-	-	-	-	-	-	-	
G. incognitus Drake & Hottes	-	-	-	-	BC	AB	SK	MB	ON	QC	-	-	NS	-	LB	
G. incurvatus Drake & Hottes	-	-	-	-	BC	-	-	-	-	-	-	-	-	-	-	
G. insperatus Drake & Hottes	-	-	-	-	-	-	-	-	ON	QC	NB	PE	NS	-	-	
G. marginatus Say	-	-	-	-	-	-	-	MB	ON	QC	-	-	NS	-	-	
G. pingreensis Drake & Hottes	AK	YK	NT	-	BC	AB	SK	MB	ON	QC	-	-	-	-	LB	

Genus LIMNOPORUS Stål

Key to species: Andersen & Spence 1992 [Holarctic].

L. canaliculatus (Say)	-	-	-	-	-	-	-	-	ON	-	-	-	-	-	-	[r]
L. dissortis (Drake & Harris)	-	-	NT	-	BC	AB	SK	MB	ON	QC	-	PE	NS	NF	LB	
L. notabilis (Drake & Harris)	-	-	-	-	BC	AB	-	-	-	-	-	-	-	-	-	
L. rufoscutellatus (Latreille)	AK	YK	NT	-	-	-	-	-	-	-	-	-	-	-	-	

Subfamily RHAGODOTARSINAE

Genus RHEUMATOBATES Bergroth

Key to species: Hungerford 1954.

R. palosi Blatchley	-	-	-	-	-	-	SK	MB	-	-	-	-	-	-	-	
R. rileyi Bergroth	-	-	-	-	-	-	-	-	ON	QC	-	-	-	-	-	

* introduced; [] indoors; { } extirpated; ? uncertain record; @ unestablished adventive; $ port-of-entry interception

Subfamily TREPOBATINAE

Genus METROBATES Uhler
Key to species: Anderson 1932.

M. hesperius Uhler	-	-	-	-	-	-	-	MB	ON	QC	NB	-	-	-	-	

Genus TREPOBATES Uhler
Key to species: Kittle 1977.

T. inermis Esaki	-	-	-	-	-	-	-	-	ON	-	-	-	-	-	-	[r]
T. pictus (Herrich-Schaeffer)	-	-	-	-	-	-	-	-	ON	-	-	-	-	-	-	[r]
T. subnitidus Esaki	-	-	-	-	-	-	-	-	ON	-	-	-	-	-	-	[r]

Infraorder Leptopodomorpha

Catalogue of world species: Schuh *et al.* 1987.

Superfamily Saldoidea

Family SALDIDAE

Key to genera and species: Brooks & Kelton 1967 [prairie provinces].

Shore bugs. Active predators, usually living in damp areas or near water. Oval, blackish in colour, with large eyes, 3-segmented rostrum, and with 4 or 5 closed cells in membrane of forewing.

Subfamily CHILOXANTHINAE

Genus CHILOXANTHUS Reuter

C. arcticus (Sahlberg)	AK	-	NT	-	-	-	-	-	-	-	-	-	-	-	-
C. stellatus (Curtis)	AK	YK	NT	NU	-	-	-	MB	-	-	-	-	-	-	-

Genus PENTACORA Reuter

P. hirta (Say)	-	-	-	-	-	-	-	-	-	QC	NB	PE	-	-	-
P. ligata (Say)	-	-	-	-	-	-	-	MB	ON	QC	-	-	NS	NF	-
P. signoreti (Guérin-Méneville)	-	-	-	-	-	-	SK	MB	-	-	-	-	-	-	-
P. sphacelata (Uhler)	-	-	-	-	-	-	-	-	-	QC	NB	PE	NS	NF	-

Subfamily SALDINAE

Key to genera: Polhemus 1985 [New World].

Tribe Saldini

Genus LAMPRACANTHIA Reuter

L. crassicornis (Uhler)	AK	-	-	-	BC	AB	SK	MB	ON	QC	-	-	-	NF	-

Genus SALDA Fabricius

S. anthracina Uhler	AK	YK	NT	-	BC	-	-	-	-	-	-	-	-	-	-
S. buenoi (McDunnough)	-	YK	-	-	BC	AB	SK	MB	ON	-	-	-	-	NF	LB
S. littoralis (Linnaeus)	AK	YK	NT	NU	-	-	-	MB	ON	QC	-	-	-	NF	LB
S. lugubris (Say)	-	YK	NT	-	BC	AB	SK	MB	ON	QC	-	PE	-	NF	-
S. obscura Provancher	AK	YK	NT	-	BC	AB	SK	MB	ON	QC	-	-	-	NF	-
S. provancheri Kelton & Lattin	AK	YK	NT	-	BC	AB	SK	MB	ON	QC	NB	PE	NS	NF	-

Genus TELOLEUCA Reuter

T. bifasciata (Thomson)	AK	YK	-	-	BC	AB	-	-	ON	QC	-	-	-	-	-
T. pellucens (Fabricius)	AK	YK	NT	-	BC	AB	SK	MB	ON	QC	-	-	-	NF	LB

Tribe Saldoidini

Genus CALACANTHIA Reuter

C. trybomi (Sahlberg)	AK	YK	NT	NU	-	-	-	MB	-	-	-	-	-	-	-

Genus IOSCYTUS Reuter

Key to species: McKinnon & Polhemus 1986.

I. politus politus (Uhler)	-	-	-	-	BC	-	-	-	-	-	-	-	-	-	-	[r]

Genus MACROSALDULA Leston & Southwood

M. monae (Drake)	AK	YK	-	-	-	-	-	-	-	-	-	-	-	-	-

Genus MICRACANTHIA Reuter

M. bergrothi (Jakovlev)	-	YK	NT	-	BC	AB	SK	MB	ON	QC	-	PE	-	NF	-
M. fennica (Reuter)	AK	YK	-	-	BC	AB	SK	-	-	-	-	-	-	-	-
M. humilis (Say)	-	-	NT	-	BC	AB	SK	MB	ON	QC	-	PE	-	NF	-
M. quadrimaculata (Champion)	-	YK	-	-	BC	-	-	-	-	-	-	-	-	-	-

Genus SALDULA Van Duzee

S. ablusa Drake & Hottes	-	-	-	-	-	-	-	-	ON	-	-	-	-	-	-
S. andrei andrei Drake	-	-	-	-	BC	-	-	-	-	-	-	-	-	-	-
S. andrei azteca Drake & Hottes	-	-	-	-	BC	AB	-	-	-	-	-	-	-	-	-
S. balli Drake	-	-	-	-	BC	-	-	-	-	-	-	-	-	-	-
S. bouchervillei (Provancher)	AK	-	-	-	BC	AB	SK	MB	ON	QC	-	PE	-	NF	-
S. comatula Parshley	-	-	-	-	BC	AB	SK	-	-	-	-	-	-	-	-
S. confluenta (Say)	-	-	-	-	-	-	-	MB	ON	QC	-	PE	-	-	-
S. dispersa (Uhler)	-	-	-	-	BC	-	-	-	-	-	-	-	-	-	-
S. explanata (Uhler)	-	-	-	-	BC	AB	SK	MB	-	-	-	-	-	-	-
S. laticollis (Reuter)	AK	-	-	-	BC	AB	SK	MB	-	-	-	-	-	NF	-
S. nigrita Parshley	-	YK	-	-	BC	AB	SK	MB	-	QC	-	-	NS	NF	-
S. opacula (Zetterstedt)	AK	YK	NT	-	BC	AB	SK	MB	ON	QC	-	-	NF	-	
S. opiparia Drake & Hottes	-	-	NT	-	BC	AB	SK	MB	ON?	-	-	PE	-	-	-
S. orbiculata (Uhler)	-	-	-	-	-	AB	-	MB	ON	QC	-	-	NS	NF	-
S. pallipes dimidiata (Curtis)	-	-	-	-	BC	-	-	-	-	-	-	-	-	-	-
S. pallipes pallipes (Fabricius)	AK	YK	-	-	BC	AB	-	MB	ON	QC	-	-	NS	NF	-
S. palustris (Douglas)	-	-	-	-	BC	AB	SK	-	-	-	-	-	-	-	-
S. saltatoria (Linnaeus)	AK	YK	NT	-	BC	AB	SK	MB	ON	QC	NB	-	-	NF	-

* introduced; [] indoors; { } extirpated; ? uncertain record; @ unestablished adventive; $ port-of-entry interception

Infraorder Cimicomorpha

Superfamily Cimicoidea

Family ANTHOCORIDAE

Key to genera: Poppius 1909 [world].

Key to genera and species: Kelton 1978 [Canada].

Minute pirate bugs. Mostly predators, occurring on flowers, conifers, deciduous trees and shrubs or under bark. Body rather flattened with somewhat pointed head with ocelli; antennae four-segmented; labium three-segmented, inserted anteriorly on head. Metathorax with distinct scent-gland openings laterally. Forewings, when present; with distinct costal fracture and cuneus. All abdominal terga with laterotergites; female lacking an apophysis on anterior margin of the abdominal sternum; male with asymmetrical genitalia, right paramere greatly reduced or absent. Reproduction involving either traumatic insemination with left paramere modified into an organ to penetrate body wall of female during copulation, or sperm injected into special copulatory tubes that facilitate ovarial fertilization.

Tribe Anthocorini

Genus ACOMPOCORIS Reuter

A. lepidus (Van Duzee)	-	-	-	NU	BC	AB	-	-	-	-	-	-	-	-	-
A. pygmaeus (Fallén)	-	-	-	-	-	-	-	-	ON*	-	NB*	PE*	NS*	-	-

Genus ANTHOCORIS Fallén

A. antevolens White	AK	YK	NT	-	BC	AB	SK	-	ON	QC	-	-	NS	NF	-	
A. confusus Reuter	-	-	-	-	BC*	-	-	-	ON*	-	-	PE*	NS*	-	-	
A. dimorphicus Anderson & Kelton	-	YK	NT	-	-	AB	SK	-	ON	-	-	-	-	-	-	
A. musculus (Say)	AK	YK	NT	-	BC	AB	SK	MB	ON	QC	NB	PE	NS	NF	LB	
A. nemoralis (Fabricius)	-	-	-	-	BC*	-	-	-	ON*	-	-	-	-	-	-	
A. tomentosus Péricart	AK	YK	NT	-	BC	AB	-	MB	-	-	-	-	-	-	-	
A. whitei Reuter	-	-	-	-	BC	-	-	-	-	-	-	-	-	-	-	[r]

Genus ELATOPHILUS Reuter
Key to species: Kelton 1976.

E. brimleyi Kelton	-	-	-	-	-	-	-	-	ON	-	-	-	-	-	-
E. inimicus (Drake & Harris)	-	-	-	-	-	-	-	MB	ON	QC	-	-	-	-	-
E. minutus Kelton	-	-	-	-	-	AB	SK	MB	ON	QC	-	-	-	-	-
E. pullus Kelton & Anderson	-	YK	-	-	BC	-	-	-	-	-	-	-	-	-	-

Genus MELANOCORIS Champion
Key to species: Lattin & Stanton 1993.

M. longirostris Kelton	-	-	-	-	BC	-	-	-	-	-	-	-	-	-	-
M. nigricornis Van Duzee	-	-	-	-	BC	-	-	-	-	-	-	-	-	-	-

Genus TEMNOSTETHUS Fieber

T. gracilis Horváth	-	-	-	-	-	-	-	-	-	-	-	-	NS*	-	-

Genus TETRAPHLEPS Fieber
Key to species: Kelton 1966a.

T. abdulghani Ghauri Note: not established	-	-	-	-	BC*	-	-	-	-	-	NB*	-	NS*	-	-	[b]
T. canadensis Provancher	AK	YK	-	-	BC	AB	SK	MB	ON	QC	NB	PE	NS	NF	-	
T. feratis (Drake & Harris)	-	YK	-	-	BC	AB	-	-	-	-	-	-	-	-	-	
T. latipennis Van Duzee	-	YK	-	-	BC	AB	SK	MB	-	-	NB	-	NS	NF	-	
T. pilosipes Kelton & Anderson	AK	YK	-	-	BC	AB	-	MB	-	QC	-	-	-	NF	-	
T. raoi Ghauri Note: not established	-	-	-	-	-	-	-	-	-	-	NB*	-	NS*	-	-	[b]
T. uniformis Parshley	-	YK	-	-	BC	AB	SK	MB	ON	QC	NB	-	NS	NF	-	

Tribe Dufouriellini

Genus BRACHYSTELES Fieber

B. parvicornis (Costa)	-	-	-	-	-	-	-	-	-	-	-	-	NS*	-	-

Genus CARDIASTETHUS Fieber

C. borealis Kelton	-	-	-	-	BC	-	SK	MB	ON	-	-	-	NS	-	-

Genus DUFOURIELLUS Kirkaldy

D. ater (Dufour)	-	-	-	-	BC	-	-	-	ON	-	-	-	-	-	-

Tribe Oriini

Genus MACROTRACHELIELLA Champion

M. nigra Parshley	-	-	-	-	BC	-	-	MB	ON	QC	NB	-	NS	-	-

Genus ORIUS Wolff

Key to species: Herring 1966 [Western Hemisphere], Kelton 1963 [North America].

O. diespeter Herring	-	YK	-	-	BC	AB	-	-	-	-	-	-	-	-	-
O. insidiosus (Say)	-	-	-	-	BC*	-	-	MB	ON	QC	-	-	-	-	-
O. minutus (Linnaeus)	-	-	-	-	BC*	-	-	-	-	-	-	-	-	-	-
O. tristicolor (White)	AK	YK	NT	-	BC	AB	SK	MB	ON	QC	NB	PE	NS	NF	-

Tribe Scolopini

Genus CALLIODIS Reuter

C. temnostethoides (Reuter)	-	-	-	-	-	-	-	-	ON	QC	NB	PE	NS	-	-

Genus SCOLOPOSCELIS Fieber

S. flavicornis Reuter	-	YK	NT	-	BC	AB	SK	MB	ON	QC	NB	-	NS	NF	-
S. pulchella pulchella (Zetterstedt)	-	-	-	-	-	-	-	-	-	$	-	-	-	-	-

Tribe Xylocorini

Genus XYLOCORIS Dufour

(Subgenus PROXYLOCORIS Carayon)

X. galactinus (Fieber)	-	-	-	-	BC	AB	SK	MB	ON	QC	-	-	-	-	-

(Subgenus XYLOCORIS Dufour)

X. cursitans (Fallén)	-	-	-	-	BC	AB	-	-	ON	QC	-	-	NS	-	-
X. hirtus Kelton	-	-	-	-	-	-	SK	-	ON	QC	-	-	-	-	-
X. umbrinus Van Duzee	-	-	-	-	BC	AB	SK	MB	-	-	-	-	-	-	-

* introduced; [] indoors; { } extirpated; ? uncertain record; @ unestablished adventive; $ port-of-entry interception

Family CIMICIDAE

Key to taxa: Usinger 1966 [world].

Bed bugs. Flattened, oval or rounded, brownish temporary ectoparasites of warm blooded vertebrates. After feeding, bed bugs typically retreat from the host. Apterous, with hemelytra in the form of small pads. Head without ocelli; antennae four-segmented; labium three-segmented, inserted anteriorly on head. Legs with three-segmented tarsi.

Subfamily CIMICINAE

Genus CIMEX Linnaeus

C. adjunctus Barber - - - - - - - MB ON QC - - - NF -
C. brevis Usinger & Ueshima - - - - - - - - - QC - - - - -
C. latipennis Usinger & Ueshima - - - - BC - - - - - - - - - -
C. lectularius Linnaeus - - - - BC AB SK MB ON QC - PE NS NF -
bed bug; punaise des lits
C. pilosellus (Horváth) - - - - BC AB SK - - - - - - - -
bat bug; punaise de la chauve-souris

Genus OECIACUS Stål

O. vicarius Horváth - - - - BC AB - MB ON QC - - NS - -
swallow bug; punaise de l'hirondelle

Subfamily HAEMATOSIPHONINAE

Genus HESPEROCIMEX List

H. coloradensis List - - - - BC - - - - - - - - - - [r]

Family LASIOCHILIDAE

Key to genus and species: Kelton 1978 [Canada].

Lasiochilids. Small, generally ground-dwelling predators. Body rather flattened, the head with ocelli; antennae four-segmented, with terminal two segments thinner than first two segments and with long pilosities; labium three-segmented, inserted anteriorly on head. Metathorax with scent-gland canal curved posteriorly. Forewings with deep costal fracture; membrane with stub located anteriorly near corium-membrane boundary, with four or five free veins. Abdominal segments one and two only; terga three to eight forming simple plates; abdominal spiracle 1 absent. Male genitalia asymmetrical; left paramere more or less sickle-shaped, not grooved for receiving phallus. Reproduction not involving traumatic insemination.

Genus LASIOCHILUS Reuter

(Subgenus DILASIA Reuter)

L. fusculus (Reuter) - - - - - - - - ON QC - - - - -

Family LYCTOCORIDAE

Key to genus and species: Kelton 1967, 1978 [Canada].

Lyctocorids. Predatory on insects and mites in decaying vegetable matter, under bark, in birds' nests and rodent burrows, sometimes sucking blood of humans and domestic animals. Body rather flattened; head porrect with ocelli; antennae four-segmented, terminal two segments thinner than base of second segment; labium with first segment very short. Metathorax with scent-gland grooves present and ostiolar canal angular and bent anteriorly; mesosternum with longitudinal median groove. Forewing with distinct stub anteriorly near corium-membrane boundary, and with one a few free veins. All abdominal terga with laterotergites; female with internal apophysis on anterior margin of seventh abdominal sternum; male genitalia asymmetrical; left paramere not grooved for receiving phallus; apex of aedeagus in form of acus. Reproduction involves traumatic insemination via puncture of intersegmental membrane between terga six and seven on right side of abdomen. Fertilization takes place in the vitellarium, with some embryonic development prior to oviposition.

Genus LYCTOCORIS Hahn
Key to species: Kelton 1967, 1978.

(Subgenus DOLICHOMERIUM Kirkaldy)

L. stalii (Reuter) - - - - BC - - MB - - - - - - -

(Subgenus LYCTOCORIS Hahn)

L. campestris (Fabricius) - - - - BC - - MB ON QC - - - - -
L. canadensis Kelton - - - - - - - MB - QC - - - - -
L. okanaganus Kelton & Anderson - - - - BC - - - - - - - - - - [r]
L. rostratus Kelton & Anderson - - - - BC - - - - - - - - - - [r]
L. tuberosus Kelton & Anderson - - - - BC - - - - - - - - - -

Family NABIDAE

Key to taxa: Harris 1928 [North America].

Damsel bugs. Active predators. Mostly dull brownish, occurring on vegetation in both short- and long-winged forms. Some species shiny black and ground dwelling. Head with ocelli; antennae with prepedicellite of variable length between segments 1 and 2; labium four-segmented. Fore legs with a dense pad of hairs (fossula spongiosa) at distal end of tibia. Hemelytra with costal fracture present or absent; membrane with two or three elongate cells, and usually with emanating veins and a stub. Abdomen with three dorsal scent gland openings between terga 4/5, 5/6 and 6/7; male genitalia usually symmetrical.

Subfamily NABINAE

Tribe Nabini

Genus ANAPTUS Kerzhner

A. major (Costa) - - - - BC* - - - - - - - - - -

Genus HOPLISTOSCELIS Reuter

H. heidemanni (Reuter) - - - - BC - - - - - - - - - - [r]

Genus LASIOMERUS Reuter

L. annulatus (Reuter) - - - - - - - - ON* QC* - - - - -

Genus HIMACERUS Wolff

H. apterus (Fabricius) - - - - - - - - - - - - NS* - -

Genus NABICULA Kirby
Key to species: Larivière 1994a [Canada].

(Subgenus DOLICHONABIS Reuter)

N. americolimbata (Carayon) AK YK NT - BC AB SK MB ON QC NB PE NS NF LB
N. limbata (Dahlbom) - - - - - - - - - - - - - NF* -
N. nigrovittata nearctica Kerzhner AK YK NT - BC AB SK MB ON QC NB PE NS NF LB

(Subgenus LIMNONABIS Kerzhner)

N. propinqua (Reuter) - - - - BC AB SK MB ON QC NB PE NS - -

(Subgenus NABICULA Kirby)

N. flavomarginata (Scholtz) AK YK NT - BC AB - - - QC NB PE NS NF LB
N. subcoleoptrata Kirby - - - - BC AB SK MB ON QC NB PE NS - -
N. vanduzeei (Kirkaldy) - - - - BC AB SK - - - - - - - -

* introduced; [] indoors; { } extirpated; ? uncertain record; @ unestablished adventive; $ port-of-entry interception

Genus NABIS Latreille

(Subgenus NABIS Latreille)

N. roseipennis Reuter	-	-	-	-	BC	AB	SK	MB	ON	QC	NB	-	NS	-	-
N. rufusculus Reuter	-	-	-	-	BC	AB	SK	MB	ON	QC	NB	PE	NS	NF	-

(Subgenus REDUVIOLUS Kirby)

N. alternatus alternatus Parshley	-	-	-	-	BC	AB	-	-	-	-	-	-	-	-	-
N. alternatus uniformis Harris	-	-	-	-	BC	-	-	-	-	-	-	-	-	-	-
N. americoferus Carayon	-	YK	NT	-	BC	AB	SK	MB	ON	QC	NB	PE	NS	NF	-
N. inscriptus (Kirby)	AK	YK	NT	-	BC	AB	-	-	-	QC	-	-	-	NF	-
N. kalmii Reuter	-	-	-	-	-	-	-	-	-	-	-	-	NS	NF	-

Genus OMANONABIS Asquith & Lattin

O. lovetti (Harris)	-	-	-	-	BC	-	-	-	-	-	-	-	-	-	-	[r]

Subfamily PROSTEMMATINAE

Tribe Prostemmatini

Genus PAGASA Stål

(Subgenus LAMPROPAGASA Reuter)

P. fusca (Stein)	-	YK	NT	-	BC	AB	SK	MB	ON	QC	-	-	NS	NF	LB

Superfamily Miroidea

Family MICROPHYSIDAE

Microphysids. Minute predators, usually found associated with mosses and lichens growing on ground or on vegetation. Body oval or oblong and flattened. Macropterous forms with large ocelli on head; labium appearing three-segmented. Metapleura without either ostiolar groove or evaporatorium. Hemelytra with costal margin expanded; costal fracture and cuneus present; membrane with single closed cell, and with a distinct stub in postero-lateral angle; hemelytra often staphylinoid. Tarsi two-segmented. Male genitalia symmetrical; ovipositor laciniate.

Genus LORICULA Curtis

L. bipunctata (Perris)	-	-	-	-	BC*	-	-	-	-	-	-	-	-	-	-
L. elegantula (Baerensprung)	-	-	-	-	$	-	-	-	-	-	-	-	-	-	-
L. pselaphiformis Curtis	-	-	-	-	-	-	-	-	-	-	-	-	NS*	-	-

Genus MYRMEDOBIA Baerensprung

M. exilis (Fallén)	-	-	-	-	-	-	-	-	-	-	-	-	-	NF*	-

Family MIRIDAE

Catalogue of species: Schuh 1995 [world].

Key to world subfamilies, tribes and genera: Carvalho 1955 [world].

Key to genera and species: Kelton 1980 [prairie provinces].

Plant bugs. Usually plant sucking, but with some species obligate or facultative predators. Body generally fragile, often cryptically coloured green or brown; sometimes bright red or black. Head with or without ocelli; antennae and labium both four-segmented. Metathorax usually with distinct auriculate scent gland evaporative area. Hemelytra usually with distinct costal fracture and cuneus; membrane with one or two closed cells. Trochanters two-segmented. Male parameres asymmetrical.

Subfamily BRYOCORINAE

Tribe Bryocorini

Genus MONALOCORIS Dahlbom

M. americanus Wagner & Slater	AK	-	-	-	BC	AB	SK	MB	ON	QC	NB	PE	NS	NF	LB

Tribe Dicyphini

Key to genera: Cassis 1984.

Genus CAMPYLONEURA Fieber

C. virgula (Herrich-Schaeffer)	-	-	-	-	BC*	-	-	-	-	-	-	-	-	-	-

Genus DICYPHUS Fieber

Key to species: Cassis 1984.

D. discrepans Knight	AK	YK	-	-	BC	AB	SK	MB	ON	QC	NB	-	NS	-	-
D. famelicus (Uhler)	-	-	-	-	-	-	-	-	ON	QC	NB	-	NS	-	-
D. hesperus Knight	-	-	-	-	BC	AB	SK	MB	ON	QC	NB	-	-	-	-
D. pallicornis (Fieber)	-	-	-	-	BC*	-	-	-	-	-	-	-	-	-	-
D. vestitus Uhler	-	-	-	-	BC	-	-	-	-	-	-	-	-	-	-

Genus MACROLOPHUS Fieber

Key to species: Knight 1926 [North America].

M. rivalis (Knight)	-	-	-	-	BC	-	-	-	-	-	-	-	-	-	-	[r]
M. separatus (Uhler)	-	-	-	-	-	-	-	-	ON	-	-	-	-	-	-	
M. tenuicornis Blatchley	-	-	-	-	-	-	-	-	ON	QC	NB	PE	NS	-	-	

Genus TUPIOCORIS China & Carvalho

T. agilis (Uhler)	-	-	-	-	BC	-	-	-	-	-	-	-	-	-	-	
T. californicus (Stål)	-	-	-	-	BC	-	-	-	-	-	-	-	-	-	-	[r]
T. confusus (Kelton)	-	YK	-	-	BC	AB	-	-	-	-	-	-	-	-	-	
T. elongatus (Van Duzee)	-	-	-	-	BC	-	-	-	-	-	-	-	-	-	-	
T. rubi (Knight)	-	YK	-	-	BC	AB	SK	MB	ON	QC	NB	PE	NS	NF	LB	
T. similis (Kelton)	AK	YK	NT	-	BC	AB	SK	-	ON	QC	NB	-	NS	NF	LB	
T. tibialis (Kelton)	-	-	-	-	BC	AB	-	-	-	-	-	-	-	-	-	
T. tinctus (Knight)	-	-	-	-	BC	-	-	-	-	-	-	-	-	-	-	

Genus USINGERELLA China & Carvalho

U. bakeri (Knight)	-	-	-	-	BC	-	-	-	-	-	-	-	-	-	-

Tribe Eccritotarsini

Genus SIXEONOTUS Reuter

S. insignis Reuter	-	-	-	-	-	-	-	-	-	QC	-	-	-	-	-
S. rostratus Knight	-	-	-	-	-	AB	SK	-	-	-	-	-	-	-	-

* introduced; [] indoors; { } extirpated; ? uncertain record; @ unestablished adventive; $ port-of-entry interception

Subfamily CYLAPINAE

Tribe Cylapini

Genus CYLAPUS Say

C. tenuicornis (Say)	-	-	-	-	-	-	-	-	ON	-	-	-	-	-	-	

Genus FULVIUS Stål

Key to species: Kelton 1985 [Canada].

F. imbecilis (Say)	-	-	-	-	-	AB	-	MB	ON	-	-	-	-	-	-	
F. slateri Wheeler	-	-	-	-	-	-	-	MB	ON	QC	-	PE	-	-	-	

Subfamily DERAEOCORINAE

Tribe Clivinematini

Genus BOTHYNOTUS Fieber

B. pilosus (Boheman)	AK	YK	-	-	BC	-	-	-	-	-	-	-	-	-	-	

Genus CLIVINEMA Reuter

C. fuscum Downes	-	-	-	-	BC	-	-	-	-	-	-	-	-	-	-	[r]
C. mediale Knight	-	-	-	-	BC	-	-	-	-	-	-	-	-	-	-	
C. villosum Reuter	-	-	-	-	-	AB	-	-	-	-	-	-	-	-	-	[r]

Genus LARGIDEA Van Duzee

Key to species: Knight 1968b.

L. grossa Van Duzee	-	-	-	-	BC	-	-	-	-	-	-	-	-	-	-	
L. shoshonea Knight	-	YK	-	-	BC	AB	SK	MB	ON	-	NB	-	-	-	-	

Tribe Deraeocorini

Genus DERAEOCAPSUS Knight

D. fraternus (Van Duzee)	-	-	-	-	BC	-	-	-	-	-	-	-	-	-	-	

Genus DERAEOCORIS Kirschbaum

Key to species: Knight 1921 [North America], Razafimahatratra 1981 [western].

D. albigulus Knight	-	-	-	-	BC	AB	SK	MB	ON	QC	NB	-	NS	-	-	
D. alnicola Knight	-	-	-	-	-	-	-	-	ON	QC	-	-	-	-	-	
D. aphidiphagus Knight	-	-	-	-	-	AB	SK	MB	ON	QC	NB	PE	NS	-	-	
D. bakeri Knight	-	-	-	-	BC	-	-	-	-	-	-	-	-	-	-	[r]
D. barberi Knight	-	-	-	-	BC	-	-	-	-	-	-	-	-	-	-	
D. betulae Knight	-	-	-	-	-	-	-	-	ON	QC	-	-	NS	-	-	
D. borealis (Van Duzee)	-	-	-	-	-	-	-	-	ON	QC	-	-	NS	-	-	
D. brevis (Uhler)	AK	YK	NT	-	BC	AB	SK	MB	-	-	-	-	-	-	-	
D. diveni Knight	-	YK	-	-	BC	AB	-	-	-	-	-	-	-	-	-	
D. fasciolus castus Knight	-	-	-	-	-	-	-	-	-	QC	-	-	-	-	-	
D. fasciolus fasciolus Knight	-	-	-	-	BC	AB	SK	MB	ON	QC	NB	PE	NS	-	-	
D. fulgidus (Van Duzee)	-	-	-	-	BC	-	-	-	-	-	-	-	-	-	-	[r]
D. fulvescens (Reuter)	-	-	-	-	BC	-	-	-	-	-	-	-	-	-	-	
D. grandis (Uhler)	-	-	-	-	-	-	-	-	ON	QC	-	-	-	-	-	
D. histrio (Reuter)	-	-	-	-	-	AB	SK	MB	ON	QC	-	-	-	-	-	
D. incertus Knight	-	-	-	-	BC	-	-	-	-	-	-	-	-	-	-	
D. kennicotti Knight	-	YK	NT	-	BC	AB	SK	MB	ON	QC	NB	-	NS	-	-	
D. laricicola Knight	-	YK	-	-	-	-	SK	MB	ON	QC	NB	-	NS	NF	LB	
D. mutatus Knight	-	-	-	-	BC	-	-	-	-	-	-	-	-	-	-	
D. nebulosus (Uhler)	-	-	-	-	-	-	-	-	ON	QC	NB	-	NS	NF	-	
D. nitenatus Knight	-	-	-	-	-	-	SK	MB	ON	QC	-	-	-	-	-	

Deraeocoris nubilus Knight	-	-	-	-	-	-	-	-	ON	QC	NB	PE	NS	NF	-	
D. ornatus Knight	-	-	-	-	BC	AB	SK	MB	-	-	-	-	-	-	-	
D. piceicola Knight	-	YK	-	-	BC	AB	-	-	-	-	-	-	-	-	-	
D. pinicola Knight	-	-	-	-	-	-	-	-	ON	QC	-	-	NS	-	-	
D. poecilus McAtee	-	-	-	-	-	-	-	-	ON	QC	-	-	-	-	-	
D. punctulatus (Fallén)	AK	YK	NT	-	BC	-	-	-	-	-	-	-	-	-	-	
D. quercicola Knight	-	-	-	-	-	-	SK	MB	ON	QC	-	-	-	-	-	
D. rubripes Kelton	-	-	-	-	-	-	-	MB	ON	QC	NB	PE	NS	-	-	
D. rubroclarus Knight	-	-	-	-	BC	-	-	-	-	-	-	-	-	-	-	
D. sayi (Reuter)	-	-	-	-	BC	-	-	MB	ON	-	-	-	-	-	-	
D. schwarzii (Uhler)	-	-	-	-	BC	-	-	-	-	-	-	-	-	-	-	
D. triannulipes Knight	-	-	-	-	BC	AB	SK	MB	ON	QC	-	-	-	-	-	
D. validus (Reuter)	-	-	-	-	BC	-	-	-	-	-	-	-	-	-	-	

Genus EURYCHILOPTERELLA Reuter

E. luridula Reuter	-	-	-	-	-	-	-	-	ON	-	-	PE	NS	-	-	
E. pacifica Stonedahl	-	-	-	-	BC	-	-	-	-	-	-	-	-	-	-	[r]

Genus HESPEROPHYLUM Reuter & Poppius

H. heidemanni Reuter & Poppius	-	-	-	-	-	-	-	-	-	-	-	-	NS	-	-	[r]

Tribe Hyaliodini

Genus HYALIODES Reuter

H. harti Knight	-	-	-	-	BC	AB	SK	MB	ON	QC	-	PE	NS	-	-	
H. vitripennis (Say)	-	-	-	-	-	-	-	-	ON	QC	-	-	NS	-	-	

Tribe Surinamellini

Genus EUSTICTUS Reuter

E. necopinus Knight	-	-	-	-	BC	-	-	MB	ON	QC	-	-	-	-	-	

Subfamily ISOMETOPINAE

Tribe Isometopini

Genus CORTICORIS McAtee & Malloch

Key to species: Henry & Herring 1979.

C. signatus (Heidemann)	-	-	-	-	-	-	-	-	ON	-	-	-	-	-	-	[r]

Genus DIPHLEPS Bergoth

D. unica Bergroth	-	-	-	-	-	-	-	-	ON	-	-	-	-	-	-	

Genus MYIOMMA Puton

Key to species: Henry 1979.

M. cixiiforme (Uhler)	-	-	-	-	-	-	-	-	ON	QC	-	-	-	-	-	[r]

Subfamily MIRINAE

Tribe Herdoniini

Genus PARADACERLA Carvalho & Usinger

Key to species: Kelton & Knight 1959.

P. downesi (Knight)	-	-	-	-	BC	-	-	-	-	-	-	-	-	-	-	
P. formicina (Parshley)	-	-	-	-	BC	-	-	-	-	-	-	-	-	-	-	[r]

* introduced; [] indoors; { } extirpated; ? uncertain record; @ unestablished adventive; $ port-of-entry interception

Tribe Mirini

Genus ADELPHOCORIS Reuter

A. lineolatus (Goeze) alfalfa plant bug; capside de la luzerne	-	-	-	-	BC*	AB*	SK*	MB*	ON*	QC*	NB*	PE*	NS*	NF*	-	[A]
A. rapidus (Say) rapid plant bug; capside du bouleau	-	-	-	-	BC	AB	SK	MB	ON	QC	NB	PE	NS	-	-	[A]
A. superbus (Uhler) superb plant bug; capside des prés	-	-	-	-	BC	AB	SK	-	-	-	-	-	-	-	-	[A]

Genus AGNOCORIS Reuter

Key to species: Moore 1955, 1956.

A. pulverulentus (Uhler)	AK	-	-	-	-	AB	SK	MB	ON	QC	NB	-	NS	-	-	
A. rubicundus (Fallén)	AK	YK	NT	-	BC	AB	SK	MB	ON	QC	NB	-	-	-	-	
A. utahensis Moore	-	-	-	-	BC	AB	SK	MB	ON	-	-	-	-	-	-	

Genus BOLTERIA Uhler

Key to species: Kelton 1972a.

B. luteifrons Knight	-	-	-	-	-	-	-	-	ON	QC	-	-	-	-	-	

Genus CAMPTOZYGUM Reuter

C. aequale (Villers)	-	-	-	-	-	-	-	-	ON*	QC*	-	-	NS*	-	-	

Genus CAPSUS Fabricius

Key to species: Vinokurov 1977.

C. ater (Linnaeus)	-	-	-	-	BC*	AB*	-	-	ON*	QC*	NB*	PE*	NS*	NF*	-	[A]
C. cinctus (Kolenati)	AK	YK	NT	-	BC	AB	SK	MB	ON	QC	-	-	-	-	-	[A]

Genus CLOSTEROTOMUS Fieber

C. fulvomaculatus (DeGeer)	AK	YK	NT	-	BC	-	-	MB	-	-	-	-	-	-	-	
C. norvegicus (Gmelin) strawberry bug; capside du frasier	-	-	-	-	BC*	-	-	-	-	QC*	NB*	PE*	NS*	NF*	-	[A]

Genus COCCOBAPHES Uhler

C. frontifer (Walker)	-	-	-	-	-	-	-	-	ON	QC	-	-	-	-	-	

Genus DICHROOSCYTUS Fieber

Key to species: Kelton 1972b.

D. alpinus Kelton	-	-	-	-	-	AB	-	-	-	-	-	-	-	-	-	
D. convexifrons Knight	-	-	-	-	BC	AB	-	-	-	-	-	-	-	-	-	
D. elegans Heidemann	-	-	-	-	-	-	SK	MB	ON	QC	NB	PE	NS	-	-	
D. flagellatus Kelton	-	-	-	-	BC	-	-	-	-	-	-	-	-	-	-	[r]
D. flavescens Knight	-	-	-	-	BC	-	-	-	-	-	-	-	-	-	-	
D. irroratus Van Duzee	-	-	-	-	BC	-	-	-	-	-	-	-	-	-	-	
D. latifrons Knight	AK	YK	-	-	BC	AB	SK	MB	ON	QC	NB	PE	NS	NF	-	
D. rainieri Knight	-	-	-	-	BC	-	-	-	-	-	-	-	-	-	-	
D. repletus (Heidemann)	-	-	-	-	-	-	-	-	ON	QC	NB	PE	NS	-	-	
D. rostratus Kelton	-	YK	-	-	BC	AB	-	-	-	-	-	-	-	-	-	
D. ruberellus Knight	-	-	-	-	BC	AB	SK	MB	-	-	-	-	-	-	-	
D. rubidus Kelton	-	-	-	-	BC	AB	-	-	-	-	-	-	-	-	-	
D. rufipennis (Fallén)	-	-	-	-	-	-	-	-	ON*	-	-	-	-	-	-	
D. rufusculus Kelton	-	-	-	-	BC	-	-	-	-	-	-	-	-	-	-	[r]
D. suspectus Reuter	-	-	-	-	-	AB	SK	MB	ON	QC	NB	-	NS	-	-	

Genus GARGANUS Stål

G. fusiformis (Say)	-	-	-	-	-	-	-	-	ON	-	-	-	-	-	-	

Genus HENRYLYGUS Schwartz

Key to species: Schwartz & Foottit 1998.

H. nubilus (Van Duzee)	-	-	-	-	BC	AB	-	-	-	-	-	-	-	-	-	
H. ultranubilus (Knight)	-	-	-	-	BC	AB	-	-	-	QC	-	-	-	-	-	

Genus IRBISIA Reuter

Key to species: Schwartz 1984.

I. brachycera (Uhler)	-	-	-	-	BC	AB	SK	-	-	-	-	-	-	-	-	
I. elongata Knight	-	-	-	-	BC	AB	-	-	-	-	-	-	-	-	-	
I. fuscipubescens Knight	-	-	-	-	BC	AB	SK	-	-	-	-	-	-	-	-	
I. knighti Schwartz & Lattin	-	-	-	-	BC	-	-	-	-	-	-	-	-	-	-	[r]
I. nigripes Knight	-	-	-	-	BC	AB	-	-	-	-	-	-	-	-	-	
I. pacifica (Uhler)	-	-	-	-	BC	-	-	-	-	-	-	-	-	-	-	
I. sericans (Stål)	AK	-	-	-	BC	-	-	-	-	QC	-	-	-	-	-	[A]
I. serrata Bliven	-	-	-	-	BC	-	-	-	-	-	-	-	-	-	-	
I. shulli Knight	-	-	-	-	BC	-	-	-	-	-	-	-	-	-	-	[r]

Genus LYGIDEA Reuter

L. annexa (Uhler)	AK	YK	NT	-	BC	AB	-	-	-	-	-	-	-	-	-	
L. mendax Reuter apple red bug; lygide du pommier	-	-	-	-	-	-	-	-	ON	QC	NB	-	NS	-	-	[A]
L. obscura Reuter	-	-	-	-	BC	-	-	-	ON	-	-	-	NS	-	-	
L. rosacea Reuter	-	-	-	-	BC	AB	SK	MB	ON	QC	-	-	-	NF	-	
L. rubecula (Uhler)	-	-	-	-	-	-	-	MB	ON	-	-	-	-	-	-	
L. salicis Knight	-	-	-	-	-	AB	SK	MB	ON	QC	-	-	-	NF	-	
L. viburni Knight	-	-	-	-	-	-	-	-	ON	QC	-	PE	NS	-	-	

Genus LYGOCORIS Reuter

Key to species: Kelton 1971b [Canada].

(Subgenus APOLYGUS China)

L. lucorum (Meyer-Dür)	-	-	-	-	-	-	-	-	-	QC*	-	-	-	-	-	

(Subgenus LYGOCORIS Reuter)

L. pabulinus (Linnaeus)	AK	YK	-	-	BC	AB	SK	MB	ON	QC	NB	-	NS	NF	-	
L. rugicollis (Fallén)	AK	YK	NT	-	BC	AB	-	MB	-	-	-	-	-	NF	LB	

(Subgenus NEOLYGUS Knight)

L. alni (Knight)	-	-	-	-	BC	-	-	-	ON	QC	-	-	NS	-	-	
L. atricallus Kelton	-	-	-	-	-	-	-	-	ON	QC	-	-	-	-	-	
L. atritylus (Knight)	-	-	-	-	BC	AB	SK	MB	ON	QC	NB	PE	NS	-	-	
L. belfragii (Reuter)	-	-	-	-	-	-	-	MB	ON	QC	NB	PE	NS	NF	-	[A]
L. canadensis (Knight)	-	-	-	-	-	-	-	MB	ON	QC	-	-	-	NF	-	
L. caryae (Knight) hickory plant bug; punaise du caryer	-	-	-	-	-	-	-	-	ON	QC	-	-	-	-	-	[A]
L. clavigenitalis (Knight)	-	-	-	-	-	-	-	-	ON	QC	-	-	NS	-	-	
L. communis (Knight) pear (or green apple) plant bug; punaise du poirier	AK	YK	NT	-	BC	AB	SK	MB	ON	QC	NB	PE	NS	NF	LB	[A]
L. contaminatus (Fallén)	AK	YK	-	-	BC	AB	-	-	ON	QC	NB	PE	NS	NF	-	
L. fagi (Knight)	-	-	-	-	-	-	-	-	ON	QC	-	-	NS	-	-	
L. geneseensis (Knight)	-	-	-	-	-	-	-	-	ON	-	-	-	-	-	-	
L. hirticulus (Van Duzee)	-	-	-	-	-	-	-	-	ON	QC	NB	PE	NS	-	-	
L. inconspicuus (Knight)	-	-	-	-	-	-	-	-	ON	QC	-	-	-	-	-	[A]
L. invitus (Say)	-	-	-	-	-	-	-	-	ON	QC	-	-	NS	-	-	
L. johnsoni (Knight)	-	-	-	-	-	-	-	-	ON	-	-	-	-	-	-	
L. knighti Kelton	-	-	-	-	-	-	-	MB	ON	-	-	-	-	-	-	[A]
L. laureae (Knight)	-	-	-	-	-	-	-	-	ON	QC	-	-	-	-	-	
L. omnivagus (Knight)	-	-	-	-	-	-	-	MB	ON	QC	NB	-	NS	-	-	[A]
L. ostryae (Knight)	-	-	-	-	-	-	-	-	ON	QC	-	-	-	-	-	
L. parrotti (Knight)	-	-	-	-	-	-	-	-	ON	QC	-	-	-	-	-	

Lygocoris parshleyi (Knight)	-	-	-	-	-	-	-	-	-	-	NB	PE	NS	NF	-	
L. piceicola (Kelton)	-	-	-	-	-	-	-	MB	ON	QC	NB	-	-	-	-	
L. quercalbae (Knight) oak plant bug; punaise du chêne	-	-	-	-	-	-	-	MB	ON	QC	-	-	NS	-	-	[A]
L. semivittatus (Knight)	-	-	-	-	-	-	-	-	ON	QC	-	-	-	-	-	
L. tiliae (Knight)	-	-	-	-	-	-	-	-	ON	QC	-	-	-	-	-	
L. univittatus (Knight)	-	-	-	-	-	-	-	-	ON	QC	-	-	-	-	-	
L. viburni (Knight)	-	-	-	-	-	-	-	-	ON	QC	NB	PE	NS	-	-	[A]
L. vitticollis (Reuter)	-	-	-	-	-	-	-	-	ON	QC	NB	-	NS	-	-	
L. walleyi Kelton	-	-	-	-	-	-	-	-	ON	-	-	-	-	-	-	

Genus **LYGUS** Hahn

Key to species: Schwartz & Foottit 1998.

L. atriflavus Knight	-	-	-	-	BC	-	-	-	-	-	-	-	-	-	-	
L. atritibialis Knight	-	-	-	-	BC	-	SK	MB	-	-	-	-	-	-	-	
L. borealis (Kelton)	AK	YK	NT	NU	BC	AB	SK	MB	ON	QC	-	-	-	NF	LB	[A]
L. bradleyi Knight	-	-	-	-	BC	-	-	-	-	-	-	-	-	-	-	
L. ceanothi Knight	-	-	-	-	BC	AB	-	-	-	-	-	-	-	-	-	
L. convexicollis Reuter	-	-	-	-	BC	AB	SK	-	-	-	-	-	-	-	-	
L. elisus Van Duzee lucerne plant (pale legume) bug; punaise de la luzerne	AK	YK	NT	-	BC	AB	SK	MB	ON	-	-	-	-	-	-	[A]
L. hesperus Knight western tarnished plant bug; punaise occidentale	-	-	-	-	BC	AB	-	-	-	-	-	-	-	-	-	[A]
L. humeralis Knight	-	YK	-	-	BC	AB	-	-	-	-	-	-	-	-	-	
L. keltoni Schwartz	-	-	-	-	BC	AB	SK	-	-	-	-	-	-	-	-	[A]
L. lineolaris (Palisot) tarnished plant bug; punaise terne	AK	YK	NT	NU	BC	AB	SK	MB	ON	QC	NB	PE	NS	NF	-	[A]
L. lupini Schwartz	AK	YK	NT	-	BC	-	-	-	-	-	-	-	-	-	-	
L. oregonae Knight	-	-	-	-	BC	-	-	-	-	-	-	-	-	-	-	[A]
L. plagiatus Uhler	-	-	-	-	-	AB	SK	MB	ON	QC	-	-	-	-	-	[A]
L. potentillae Kelton	AK	YK	NT	-	BC	AB	SK	MB	ON	QC	NB	-	-	NF	LB	
L. punctatus (Zetterstedt)	AK	YK	NT	-	BC	AB	SK	MB	ON	QC	-	-	-	NF	LB	[A]
L. robustus (Uhler)	-	-	-	-	BC	AB	-	-	-	-	-	-	-	-	-	
L. rubroclarus Knight	-	-	-	-	BC	AB	SK	MB	ON	QC	NB	PE	NS	NF	-	[A]
L. rubrosignatus Knight	-	YK	NT	-	BC	AB	SK	MB	-	QC	-	-	-	-	-	[A]
L. rufidorsus (Kelton)	-	-	NT	-	BC	AB	SK	MB	ON	QC	NB	PE	NS	-	-	[A]
L. rugulipennis Poppius	AK	YK	NT	-	BC	AB	-	-	-	-	-	-	-	-	-	
L. scudderi Schwartz	-	YK	-	-	BC	-	-	-	-	-	-	-	-	-	-	
L. shulli Knight	AK	YK	NT	-	BC	AB	SK	MB	ON	QC	NB	-	-	NF	LB	[A]
L. solidaginis (Kelton)	-	-	-	-	BC	AB	SK	MB	-	-	-	-	-	-	-	[A]
L. striatus Knight	-	-	-	-	BC	AB	SK	-	-	-	-	-	-	-	-	
L. unctuosus (Kelton)	-	-	NT	-	BC	AB	SK	MB	ON	QC	NB	-	-	-	-	[A]
L. vanduzeei Knight	-	-	-	-	-	-	SK	MB	ON	QC	NB	PE	NS	-	-	

Genus **METRIORRHYNCHOMIRIS** Kirkaldy

M. dislocatus (Say)	-	-	-	-	-	AB	SK	MB	ON	QC	-	-	-	-	-
M. illini (Knight)	-	-	-	-	-	-	-	-	ON	-	-	-	-	-	-

Genus **NEOBORELLA** Knight

Key to species: Kelton & Herring 1978 [North America].

N. canadensis Kelton & Herring	-	-	-	-	-	AB	SK	-	-	-	-	-	-	-	-	
N. xanthenes Herring	-	-	-	-	BC	-	-	-	-	-	-	-	-	-	-	[r]

Genus **NEUROCOLPUS** Reuter

Key to species: Henry & Kim 1984.

N. jessiae Knight	-	-	-	-	-	-	-	-	ON	QC	-	-	-	-	-	
N. longirostris Knight	-	-	-	-	BC	-	-	-	-	-	-	-	-	-	-	
N. nubilus (Say)	-	-	-	-	BC	-	-	MB	ON	QC	-	-	-	-	-	[A]

Neurocolpus pumilus Henry	-	-	-	-	-	-	-	-	ON	QC	-	-	-	-	-	
N. tiliae Knight	-	-	-	-	-	-	-	-	ON	QC	-	-	-	-	-	

Genus NONLYGUS Schwartz

N. nubilatus (Knight)	-	-	-	-	BC	-	-	-	-	-	-	-	-	-	-	

Genus ORTHOPS Fieber

O. scutellatus Uhler	AK	YK	-	-	BC	AB	SK	MB	ON	QC	NB	PE	NS	NF	LB	[A]

Genus PHYTOCORIS Fallén

Key to species: Stonedahl 1988 [western North America].

P. americanus Carvalho	-	-	-	-	-	-	-	-	ON	QC	NB	PE	NS	-	-	
P. antennalis Reuter	-	-	-	-	-	-	-	-	-	QC	-	-	-	-	-	
P. borealis Knight	-	-	-	-	-	-	-	MB	ON	-	-	-	-	-	-	
P. brevifurcatus Knight	-	-	-	-	-	-	-	-	ON	QC	-	-	-	-	-	
P. brimleyi Knight	-	-	-	-	-	AB	-	MB	ON	-	-	-	-	-	-	
P. brooksi Kelton	-	-	-	-	-	-	SK	MB	-	-	-	-	-	-	-	
P. buenoi Knight	-	-	-	-	-	-	-	-	ON	QC	-	-	-	-	-	
P. californicus Knight	-	-	-	-	BC	-	-	-	-	-	-	-	-	-	-	
P. canadensis Van Duzee	-	-	-	-	-	-	-	-	ON	QC	-	PE	NS	-	-	
P. caryae Knight	-	-	-	-	-	-	-	-	ON	-	-	-	-	-	-	
P. conspersipes Reuter	-	-	-	-	-	-	SK	MB	ON	QC	NB	-	NS	-	-	
P. conspurcatus Knight	-	-	-	-	BC	AB	SK	MB	ON	QC	-	-	NS	-	-	
P. corticevivens Knight	-	-	-	-	-	-	-	-	ON	QC	-	-	NS	-	-	
P. cortitectus Knight	-	-	-	-	-	-	-	-	ON	-	-	-	NS	-	-	
P. crawfordi Knight	-	-	-	-	-	-	-	-	ON	QC	NB	PE	NS	-	-	
P. dentatus Knight	-	-	-	-	BC	-	-	-	-	-	-	-	-	-	-	
P. depictus Knight	-	-	-	-	-	-	-	-	ON	QC	-	-	-	-	-	
P. dimidiatus Kirschbaum	-	-	-	-	BC*	-	-	-	-	-	-	-	NS*	-	-	
P. diversus Knight	-	-	-	-	-	-	-	-	ON	QC	-	-	NS	-	-	
P. driesbachi Knight	-	-	-	-	BC	AB	SK	MB	ON	QC	NB	-	-	-	-	
P. dumicola Stonedahl	-	-	-	-	BC	-	-	-	-	-	-	-	-	-	-	
P. erectus Van Duzee	-	-	-	-	-	-	SK	-	ON	QC	-	-	NS	-	-	
P. eurekae Bliven	-	-	-	-	BC	-	-	-	-	-	-	-	-	-	-	
P. eximius Reuter	-	-	-	-	-	-	-	-	ON	QC	-	-	-	-	-	
P. fenestratus Reuter	-	-	-	-	-	-	-	-	ON	QC	NB	-	-	-	-	
P. fraterculus Van Duzee	-	-	-	-	BC	-	-	-	-	-	-	-	-	-	-	
P. fumatus Reuter	-	-	-	-	-	-	-	-	-	-	-	-	NS	-	-	
P. fulvus Knight	-	-	-	-	-	-	-	-	ON	QC	NB	-	NS	-	-	
P. husseyi Knight	-	-	-	-	-	-	-	-	ON	QC	-	-	NS	-	-	
P. infuscatus Reuter	-	-	-	-	-	-	-	-	ON	QC	-	-	-	-	-	
P. inops Uhler	-	-	-	-	BC	-	-	-	-	QC	-	-	-	-	-	
P. jucundus Van Duzee	-	-	-	-	BC	-	-	-	-	-	-	-	-	-	-	
P. juliae Stonedahl	-	-	-	-	BC	-	-	-	-	-	-	-	-	-	-	
P. junceus Knight	-	-	-	-	-	AB	-	-	-	-	-	-	-	-	-	
P. keltoni Kerzhner & Schuh	-	-	-	-	BC	AB	-	-	-	-	-	-	-	-	-	
P. knowltoni Knight	-	YK	-	-	BC	AB	-	-	-	-	-	-	-	-	-	
P. lacunosus Knight	-	-	-	-	-	-	-	-	-	QC	-	-	-	-	-	
P. laevis (Uhler)	-	-	-	-	-	AB	SK	-	-	-	-	-	-	-	-	
P. lasiomerus Reuter	-	-	-	-	BC	AB	SK	MB	ON	QC	NB	-	NS	-	-	
P. listi Knight	-	-	-	-	-	-	SK	-	-	-	-	-	-	-	-	
P. michiganae Knight	-	-	-	-	-	-	-	MB	-	QC	-	-	-	-	-	
P. minutulus Reuter	-	-	-	-	-	-	-	-	ON	QC	-	-	-	-	-	
P. neglectus Knight	-	-	-	-	BC	AB	SK	MB	ON	QC	NB	PE	NS	-	-	
P. nigricollis Knight	-	-	-	-	-	-	-	-	-	-	-	-	NS	-	-	
P. occidentalis Stonedahl	-	-	-	-	BC	-	-	-	-	-	-	-	-	-	-	[r]
P. onustus Van Duzee	-	-	-	-	-	-	-	MB	ON	QC	-	PE	NS	-	-	

* introduced; [] indoors; { } extirpated; ? uncertain record; @ unestablished adventive; $ port-of-entry interception

Phytocoris pallidicornis Reuter	-	-	-	-	BC	AB	SK	MB	ON	QC	NB	PE	NS	-	-	
P. pinicola Knight	-	-	-	-	-	-	SK	MB	-	QC	NB	-	NS	-	-	
P. plenus Van Duzee	-	-	-	-	BC	-	-	-	-	-	-	-	-	-	-	
P. politus (Reuter)	-	-	-	-	BC	-	-	-	-	-	-	-	-	-	-	
P. populi (Linnaeus)	-	-	-	-	BC*	-	-	-	-	-	-	-	NS*	-	-	
P. proctori Knight	-	-	-	-	-	-	-	-	-	QC	-	-	-	-	-	
P. puella Reuter	-	-	-	-	-	-	-	MB	ON	QC	-	-	-	-	-	
P. purshiae Stonedahl	-	-	-	-	BC	-	-	-	-	-	-	-	-	-	-	[r]
P. purvus Knight	-	-	-	-	-	-	-	-	ON	QC	-	-	-	-	-	
P. quercicola Knight	-	-	-	-	-	-	-	-	ON	-	-	-	-	-	-	
P. rostratus Knight	-	-	-	-	BC	-	-	-	-	-	-	-	-	-	-	
P. rubellus Knight	-	-	-	-	-	-	-	-	ON	-	-	-	-	-	-	
P. salicis Knight	-	-	-	-	-	-	SK	MB	ON	QC	NB	-	NS	-	-	
P. schotti Knight	-	-	-	-	-	-	-	-	ON	-	-	-	-	-	-	
P. spicatus Knight	-	-	-	-	-	-	-	-	ON	QC	-	-	NS	-	-	
P. stellatus Van Duzee	-	-	-	-	BC	AB	-	-	-	-	-	-	-	-	-	
P. strigosus Knight	-	-	-	-	BC	-	-	-	-	-	-	-	-	-	-	
P. sulcatus Knight	-	-	-	-	-	-	-	-	ON	QC	-	-	-	-	-	
P. tenuis Van Duzee	-	-	-	-	BC	-	-	-	-	-	-	-	-	-	-	
P. tibialis Reuter	-	-	-	-	-	-	-	-	ON	QC	-	-	-	-	-	
P. tiliae (Fabricius)	-	-	-	-	BC*	-	-	-	-	-	-	-	NS*	-	-	
P. ulmi (Linnaeus)	-	-	-	-	-	-	-	-	-	-	-	-	NS*	-	-	
P. validus Reuter	-	-	-	-	-	AB	SK	MB	-	-	-	-	-	-	-	
P. yollabollae Bliven	-	-	-	-	BC	-	-	-	-	-	-	-	-	-	-	

Genus **PINALITUS** Kelton

Key to species: Kelton 1977 [North America].

P. approximatus (Stål)	AK	YK	NT	-	BC	AB	SK	MB	ON	QC	NB	PE	NS	NF	LB	
P. cervinus (Herrich-Schaeffer)	-	-	-	-	-	-	-	-	-	-	-	-	NS*	-	-	
P. rostratus Kelton	-	YK	-	-	BC	AB	-	-	-	QC	NB	PE	NS	NF	-	
P. rubricatus (Fallén)	-	-	-	-	-	-	-	-	-	-	-	-	NS*	-	-	
P. solivagus (Van Duzee)	-	-	-	-	BC	-	-	-	-	-	-	-	-	-	-	

Genus **PLATYLYGUS** Van Duzee

Key to species: Kelton & Knight 1970.

P. luridus (Reuter)	-	YK	NT	-	BC	AB	SK	MB	ON	QC	NB	-	NS	-	-	
P. piceicola Kelton	-	YK	-	-	BC	AB	-	-	-	-	-	-	-	-	-	
P. pseudotsugae Kelton	-	-	-	-	BC	-	-	-	-	-	-	-	-	-	-	
P. rolfsi Knight	-	-	-	-	BC	-	-	-	-	-	-	-	-	-	-	
P. rubripes Knight	-	-	-	-	BC	AB	-	-	-	-	-	-	-	-	-	

Genus **POECILOCAPSUS** Reuter

P. lineatus (Fabricius) fourlined plant bug; punaise quadrilignée	-	-	-	-	-	-	SK	MB	ON	QC	NB	PE	NS	-	-	[A]

Genus **POLYMERUS** Hahn

Key to species: Schwartz *et al.* 1991 [supplement to Kelton 1980].

P. americanus (Reuter)	-	-	-	-	-	-	-	-	ON	-	-	-	-	-	-	
P. balli Knight	-	-	-	-	-	AB	SK	MB	-	-	-	-	-	-	-	
P. basalis (Reuter)	-	-	-	-	-	-	-	-	ON	QC	-	-	-	-	-	
P. basivittis basivittis (Reuter)	-	-	-	-	-	AB	SK	MB	ON	-	-	-	-	-	-	
P. basivittis pallidulus Knight	-	-	-	-	BC	AB	-	-	-	-	-	-	-	-	-	
P. brevirostris Knight	-	-	-	-	-	AB	SK	MB	ON	-	-	-	-	-	-	
P. chrysopsis Knight	-	-	-	-	BC	-	SK	MB	-	-	-	-	-	-	-	
P. cognatus (Fieber)	-	-	-	-	BC	AB	SK	MB	ON	QC	NB	-	-	-	-	
P. diffusus (Uhler)	-	-	-	-	BC	AB	SK	-	-	-	-	-	-	-	-	
P. flavocostatus Knight	-	-	-	-	-	-	-	MB	ON	QC	-	-	-	-	-	
P. hirtus Knight	-	-	-	-	-	AB	SK	-	-	-	-	-	-	-	-	

Polymerus opacus Knight	-	-	-	-	-	-	-	-	ON	QC	NB	-	NS	-	-	
P. pallescens (Walker)	-	-	-	-	-	-	-	MB	-	-	-	-	-	-	-	
P. proximus Knight	-	-	-	-	-	-	-	-	ON	QC	-	-	-	-	-	
P. punctipes Knight	-	-	-	-	-	-	-	-	ON	QC	NB	-	NS	-	-	
P. rubrocuneatus Knight	-	-	-	-	-	AB	SK	MB	-	-	-	-	-	-	-	
P. rufipes Knight	-	-	-	-	BC	AB	-	-	-	-	-	-	-	-	-	
P. sculleni Knight	-	-	-	-	-	-	SK	-	-	-	-	-	-	-	-	
P. severini Knight	-	-	-	-	-	AB	-	-	-	-	-	-	-	-	-	
P. tumidifrons Knight	-	YK	-	-	BC	AB	-	-	-	-	-	-	-	-	-	
P. unifasciatus (Fabricius)	AK	YK	NT	-	BC	AB	SK	MB	ON	QC	-	-	NS	NF	-	
P. venaticus (Uhler)	-	-	-	-	BC	AB	SK	MB	ON	QC	-	-	NS	-	-	
P. vulneratus (Panzer)	AK	YK	NT	-	BC	-	-	-	-	-	-	-	-	-	-	

Genus SALIGNUS Kelton

S. tahoensis (Knight)	AK	YK	NT	-	BC	AB	-	MB	ON	QC	NB	-	-	NF	LB	

Genus STENOTUS Jakovlev

S. binotatus (Fabricius)	-	-	-	-	BC*	-	-	MB*	ON*	QC*	-	PE*	NS*	NF*	-	[A]

Genus TAEDIA Distant

Key to species: Knight 1930.

T. casta (McAtee)	-	-	-	-	-	-	-	-	ON	QC	-	-	-	-	-	
T. colon (Say)	-	-	-	-	-	-	-	-	ON	QC	-	-	-	-	-	
T. gleditsiae (Knight)	-	-	-	-	-	-	-	-	ON	-	-	-	-	-	-	
T. pallidula (McAtee)	-	-	-	-	-	-	SK	MB	ON	QC	-	-	NS	-	-	[A]
T. salicis (Knight)	-	-	-	-	-	-	-	-	ON	-	-	-	-	-	-	
T. scrupea (Say)	-	-	-	-	-	-	-	-	ON	QC	-	-	-	-	-	[A]

Genus TROPIDOSTEPTES Uhler

T. adustus (Knight)	-	-	-	-	-	-	-	-	ON	-	-	-	-	-	-	
T. amoenus amoenus Reuter ash plant bug; punaise du frêne	-	-	-	-	-	AB	SK	MB	ON	QC	-	-	NS	-	-	
T. amoenus atriscutis Knight	-	-	-	-	-	-	-	-	ON	-	-	-	-	-	-	
T. amoenus signatus Reuter	-	-	-	-	-	-	-	-	-	QC	-	-	-	-	-	
T. brooksi Kelton	-	-	-	-	-	-	SK	MB	ON	QC	-	-	NS	-	-	
T. canadensis Van Duzee	-	-	-	-	-	-	-	MB	ON	QC	-	-	NS	-	-	
T. cardinalis Uhler	-	-	-	-	-	-	-	-	ON	QC	-	-	-	-	-	
T. commissuralis (Reuter)	-	-	-	-	-	-	-	MB?	ON	QC	-	-	NS	-	-	
T. geminus (Say)	-	-	-	-	-	-	-	-	ON	QC	-	-	-	-	-	
T. glaber (Knight)	-	-	-	-	-	-	-	-	ON	QC	-	-	-	-	-	
T. neglectus (Knight)	-	-	-	-	-	-	-	-	ON	QC	-	-	-	-	-	
T. pacificus (Van Duzee)	-	-	-	-	BC	-	-	-	-	-	-	-	-	-	-	
T. palmeri (Reuter)	-	-	-	-	-	-	SK	MB	ON	QC	-	-	NS	-	-	
T. pettiti Reuter	-	-	-	-	-	-	-	MB	ON	QC	-	-	-	-	-	
T. plagifer Reuter	-	-	-	-	-	-	-	MB	ON	QC	-	-	-	-	-	
T. pubescens (Knight)	-	-	-	-	-	-	-	MB	ON	QC	-	-	NS	-	-	
T. rufusculus (Knight)	-	-	-	-	-	-	-	-	ON	-	-	-	-	-	-	

Tribe Restheniini

Genus OPISTHEURISTA Carvalho

O. clandestina (Van Duzee)	-	-	-	-	-	-	-	MB	ON	-	-	-	-	-	-	

Genus PREPOPS Reuter

P. bivittis (Stål)	-	YK	NT	-	BC	AB	SK	MB	ON	-	-	-	-	NF	-	
P. borealis (Knight)	-	-	-	-	BC	AB	SK	MB	ON	QC	NB	PE	NS	-	-	
P. circumcinctus (Say)	-	-	-	-	-	-	-	-	-	QC	NB	PE	NS	-	-	

* introduced; [] indoors; { } extirpated; ? uncertain record; @ unestablished adventive; $ port-of-entry interception

Prepops eremicola (Knight)	-	-	-	-	-	AB	SK	MB	-	-	-	-	-	-	-	
P. fraterculus (Knight)	-	-	-	-	-	-	-	-	ON	QC	-	-	-	-	-	
P. fraternus (Knight)	-	-	-	-	-	-	-	MB	ON	QC	-	-	-	-	-	
P. insignis (Say)	-	-	-	-	-	-	-	-	ON	QC	-	-	-	-	-	
P. insitivus (Say)	-	-	-	-	-	-	-	-	ON	-	-	-	-	-	-	
P. nigricollis (Reuter)	-	-	-	-	-	-	-	-	ON	QC	NB	-	-	-	-	
P. nigripilus (Knight)	-	YK	NT	-	BC	AB	-	MB	-	QC	-	-	-	-	-	
P. nigroscutellatus (Knight)	-	-	-	-	-	-	-	-	ON	-	-	-	-	-	-	
P. rubellicollis (Knight)	-	-	-	-	BC	AB	SK	MB	ON	QC	-	-	NS	NF	-	
P. rubrovittatus (Stål)	-	-	-	-	-	-	-	-	ON	QC	NB	PE	NS	-	-	
P. zonatus (Knight)	-	-	-	-	-	AB	SK	MB	ON	QC	-	-	-	-	-	

Tribe Stenodemini

Key to genera: Eyles & Carvalho 1975 [world].

Genus ACTINOCORIS Reuter

A. signatus Reuter	-	YK	NT	NU	-	AB	SK	-	-	-	-	-	-	-	-

Genus COLLARIA Provancher

Key to species: Carvalho & Fontes 1981 [USA].

C. meilleurii Provancher	-	-	-	-	-	AB	SK	MB	ON	QC	NB	PE	NS	-	-
C. oculata (Reuter)	-	-	-	-	-	AB?	-	-	ON	QC	NB	-	-	-	-

Genus LEPTOPTERNA Fieber

Key to species: Vinokurov 1982.

L. amoena Uhler	-	-	NT	-	BC	AB	SK	-	ON	QC	NB	PE	-	NF	LB	
L. dolabrata (Linnaeus) meadow plant bug; punaise des prés	-	-	-	-	BC*	-	-	MB*	ON*	QC*	NB*	PE*	NS*	NF*	-	[A]
L. ferrugata (Fallén)	AK	YK	NT	-	-	-	-	-	-	-	-	-	-	-	-	

Genus LITOMIRIS Slater

Key to species: Schwartz 1989.

L. curtus (Knight)	-	-	-	-	BC	-	-	-	-	-	-	-	-	-	-
L. debilis (Uhler)	AK	YK	NT	-	BC	AB	SK	MB	ON	QC	-	-	-	NF	-

Genus MEGALOCEROEA Fieber

M. recticornis (Geoffroy)	-	-	-	-	BC*	-	-	-	ON*	QC*	NB*	PE*	NS*	NF*	-

Genus MIMOCEPS Uhler

M. insignis Uhler	AK	YK	NT	-	BC	AB	SK	MB	ON	QC	-	-	-	-	-

Genus PITHANUS Fieber

P. hrabei Stehlik	-	-	-	-	-	-	SK	MB	-	-	-	-	-	-	-
P. maerkelii (Herrich-Schaeffer)	-	-	-	-	BC*	AB*	-	-	ON*	QC*	NB*	PE*	NS*	NF*	-

Genus STENODEMA Laporte

Key to species: Kelton 1961 [North America].

(Subgenus BRACHYSTIRA Fieber)

S. trispinosa Reuter	AK	YK	NT	-	BC	AB	SK	MB	ON	QC	NB	PE	NS	NF	LB

(Subgenus STENODEMA Laporte)

S. pilosipes Kelton	AK	YK	NT	-	BC	AB	SK	MB	-	-	-	-	-	-	-
S. vicinum (Provancher)	-	-	-	-	BC	AB	SK	MB	ON	QC	NB	PE	NS	NF	LB

Genus TERATOCORIS Fieber

Key to species: Kelton 1966b [North America].

T. borealis Kelton	-	-	NT	NU	-	-	-	-	-	-	-	-	-	-	-
T. caricis Kirkaldy	AK	YK	NT	-	BC	AB	SK	MB	-	QC	-	-	-	NF	LB

Teratocoris discolor Uhler	-	-	-	-	BC	AB	SK	MB	ON	QC	-	-	-	-	-
T. paludum Sahlberg	AK	YK	NT	-	BC	AB	SK	MB	ON	QC	NB	-	-	NF	LB
T. saundersi Douglas & Scott	AK	YK	NT	NU	BC	-	SK	MB	-	QC	-	-	-	NF	LB
T. viridis Douglas & Scott	AK	YK	NT	NU	-	-	-	-	-	-	-	-	-	-	-

Genus TRIGONOTYLUS Fieber

Key to species: Kelton 1971a [North America].

T. americanus Carvalho	AK	YK	-	-	BC	AB	SK	MB	-	-	-	-	-	-	-
T. antennatus Kelton	-	-	-	-	BC	-	SK	MB	-	-	-	-	-	-	-
T. brooksi Kelton	-	-	-	-	BC	-	SK	-	-	-	-	-	-	-	-
T. caelestialium (Kirkaldy)	AK	-	-	-	BC	AB	SK	MB	ON	QC	NB	PE	NS	NF	-
T. canadensis Kelton	-	-	-	-	-	AB	SK	MB	-	-	-	-	-	-	-
T. confusus Reuter	-	-	-	-	-	-	-	-	-	-	-	-	NS	-	-
T. flavicornis Kelton	-	-	-	-	-	-	SK	MB	-	-	-	-	-	-	-
T. longipes Slater & Wagner	-	-	-	-	BC	-	-	-	-	-	NB	PE	NS	-	-
T. saileri Carvalho	-	-	-	-	-	-	-	-	-	-	NB	PE	NS	-	-
T. tarsalis (Reuter)	-	-	-	-	-	AB	SK	MB	ON	QC	NB	PE	NS	NF	-
T. tenuis Reuter	-	-	-	-	-	AB	SK	MB	-	-	-	-	-	-	-
T. uhleri (Reuter)	-	-	-	-	-	-	-	-	ON	QC	NB	PE	-	-	-
T. viridis (Provancher)	AK	YK	NT	-	BC	AB	SK	MB	ON	QC	NB	PE	NS	NF	LB

Subfamily ORTHOTYLINAE

Tribe Halticini

Genus HALTICUS Hahn

Key to species: Henry 1983 [Western Hemisphere].

H. apterus (Linnaeus)	-	-	-	-	-	-	-	-	ON*	QC*	NB*	PE*	NS*	NF*	-
H. bractatus (Say)	-	-	-	-	-	-	-	-	ON	QC	NB	-	-	-	-
H. intermedius Uhler	-	-	-	-	-	-	SK	MB	ON	QC	-	-	NS	-	-

Genus LABOPS Burmeister

Key to species: Slater 1954 [North America].

L. brooksi Slater	-	-	-	-	BC	AB	SK	MB	ON	-	-	-	-	-	-	
L. burmeisteri Stål	AK	YK	NT	-	BC	AB	-	-	ON	QC	NB	-	-	-	-	
L. chelifer Slater	AK	YK	NT	-	-	-	-	-	-	-	-	-	-	-	-	
L. hesperius Uhler black grass bug	-	YK	NT	-	BC	AB	SK	MB	ON	QC	-	-	-	-	-	[A]
L. hirtus Knight	-	-	NT	-	BC	AB	SK	-	ON	QC	-	-	-	-	-	
L. tumidifrons Knight	-	YK	-	-	BC	AB	SK	-	-	-	-	-	-	-	-	
L. verae Knight	AK	YK	NT	NU	BC	AB	-	MB	-	-	-	-	-	-	-	

Genus ORTHOCEPHALUS Fieber

O. coriaceus (Fabricius)	-	-	-	-	BC*	-	-	-	ON*	QC*	NB*	PE*	NS*	NF*	-
O. saltator (Hahn)	-	-	-	-	-	-	-	-	ON*	QC*	NB*	PE*	NS*	NF*	-

Tribe Orthotylini

Genus AOPLONEMA Knight

Key to species: Knight 1928.

A. princeps (Uhler)	-	-	-	-	BC	AB	SK	-	-	-	-	-	-	-	-	
A. uhleri (Van Duzee)	-	-	-	-	BC	AB	-	-	-	-	-	-	-	-	-	[r]

Genus BLEPHARIDOPTERUS Kolenati

Key to species: Kelton 1965a [partial key].

B. angulatus (Fallén)	-	-	-	-	BC*	AB*	-	-	-	-	NB*	PE*	NS*	NF*	-
B. chlorionis (Say)	-	-	-	-	-	-	-	-	ON	QC	-	-	-	-	-
B. diaphanus (Kirschbaum)	-	-	-	-	BC	AB	SK	MB	ON	QC	-	-	NS	-	-
B. provancheri (Burque)	AK	YK	-	-	BC	AB	SK	MB	ON	QC	NB	PE	NS	NF	-

* introduced; [] indoors; { } extirpated; ? uncertain record; @ unestablished adventive; $ port-of-entry interception

Genus BROOKSETTA Kelton

Key to species: Kelton 1979b.

B. chelifer (Knight)	-	-	-	-	-	AB	SK	MB	-	-	-	-	-	-	-
B. inconspicua (Uhler)	-	-	-	-	BC	AB	SK	MB	-	-	-	-	-	-	-
B. incurva (Knight)	-	-	-	-	BC	AB	SK	MB	-	-	-	-	-	-	-
B. viridicata (Uhler)	-	YK	-	-	BC	AB	SK	MB	-	-	-	-	-	-	-

Genus CERATOCAPSUS Reuter

Key to species: Knight 1941; Henry 1979a [eastern North America, partial keys].

C. apicatus Van Duzee	-	-	-	-	BC	-	-	-	-	-	-	-	-	-	-
C. cunealis Henry	-	-	-	-	BC	-	-	-	-	-	-	-	-	-	-
C. denticulatus Knight	-	-	-	-	BC	-	-	-	-	-	-	-	-	-	-
C. digitulus Knight	-	-	-	-	-	-	-	MB	ON	QC	-	-	-	-	-
C. downesi Knight	-	-	-	-	BC	-	-	-	-	-	-	-	-	-	-
C. drakei Knight	-	-	-	-	BC	AB	SK	-	ON	QC	-	-	-	-	-
C. fasciatus (Uhler)	-	-	-	-	-	-	-	-	ON	-	-	-	-	-	-
C. fuscinus Knight	-	-	-	-	-	-	-	-	ON	QC	-	-	-	-	-
C. geminatus Knight	-	-	-	-	-	AB	SK	MB	-	-	-	-	-	-	-
C. incisus Knight	-	-	-	-	-	-	-	-	ON	-	-	-	-	-	-
C. keltoni Henry	-	-	-	-	-	-	-	-	ON	-	-	-	-	-	-
C. modestus (Uhler)	-	-	-	-	-	-	SK	MB	ON	QC	-	-	-	-	-
C. nigellus Knight	-	-	-	-	-	-	-	-	ON	QC	-	-	-	-	-
C. nigrocephalus Knight	-	-	-	-	BC	-	SK	MB	ON	QC	-	-	NS	-	-
C. pilosulus Knight	-	-	-	-	-	-	-	MB	ON	QC	-	-	-	-	-
C. pumilus (Uhler)	-	-	-	-	-	-	-	-	ON	QC	-	PE	NS	-	-
C. uniformis Knight	-	-	-	-	-	-	-	-	ON	-	-	-	-	-	-
C. vicinus Knight	-	-	-	-	-	-	-	-	ON	-	-	-	-	-	-

Genus CYRTORHINUS Fieber

Key to species: Carvalho & Southwood 1955.

C. caricis (Fallén)	AK	YK	NT	-	BC	AB	SK	MB	ON	QC	-	-	-	NF	-

Genus DICHAETOCORIS Knight

Key to species: Kelton 1965a.

D. piceicola (Knight)	-	YK	-	-	BC	AB	-	-	-	-	-	-	-	-	-

Genus FIEBEROCAPSUS Carvalho & Southwood

F. flaveolus (Reuter)	-	YK	-	-	BC	AB	SK	MB	ON	QC	-	-	-	-	-

Genus HADRONEMA Uhler

Key to species: Knight 1928.

H. bispinosum Knight	-	-	-	-	-	AB	SK	-	-	-	-	-	-	-	-
H. militare Uhler	-	-	-	-	BC	AB	SK	MB	ON	-	-	-	-	-	-
H. pictum Uhler	-	-	-	-	-	AB?	-	-	-	-	-	-	-	-	-
H. simplex Knight	-	-	-	-	BC	AB	SK	MB	-	-	-	-	-	-	-

Genus HETEROCORDYLUS Fieber

H. malinus Slingerland	-	-	-	-	-	-	-	-	ON	QC	-	-	-	-	-	[A]

Genus HETEROTOMA Lepeletier & Serville

H. planicornis (Pallas)	-	-	-	-	BC*	-	-	-	-	-	-	-	NS*	-	-

Genus ILNACORA Reuter

Key to species: Knight 1963, Knight & Schaffner 1976.

I. albifrons Knight	-	-	-	-	BC	AB	SK	MB	ON	-	-	-	-	-	-
I. chloris (Uhler)	-	-	-	-	-	-	-	-	-	QC	-	-	-	-	-
I. malina (Uhler)	-	-	-	-	-	-	-	-	ON	QC	NB	-	NS	-	-

A agricultural, F forestry, G greenhouse, O ornamental pest; r rare; b biocontrol agent

Ilnacora stalii Reuter	-	-	-	-	-	AB	SK	MB	ON	QC	-	-	-	-	-	
I. vittifrons Knight	-	-	-	-	-	-	SK	MB	-	-	-	-	-	-	-	

Genus ILNACORELLA Knight

I. argentata Knight	-	-	-	-	BC	-	-	-	-	-	-	-	-	-	-	
I. sulcata Knight	-	-	-	-	BC	AB	-	-	-	-	-	-	-	-	-	

Genus LABOPIDEA Uhler

Key to species: Kelton 1979a, Henry 1982b.

L. artemisiae (Sahlberg)	AK	-	NT	-	-	-	-	-	-	-	-	-	-	-	-	
L. bermani Kerzhner	AK	YK	NT	NU	-	-	-	-	-	-	-	-	-	-	-	
L. brooksi Kelton	-	-	-	-	-	AB	SK	-	-	-	-	-	-	-	-	
L. discolor (Sahlberg)	-	-	NT	NU	-	-	-	MB	-	-	-	-	-	-	-	
L. lenensis (Lindberg)	-	-	-	-	BC	AB	SK	MB	-	-	-	-	-	-	-	
L. nigrisetosa Knight	-	YK	-	-	BC	AB	-	-	-	-	-	-	-	-	-	
L. pallida Knight	-	-	-	-	-	AB	SK	-	-	-	-	-	-	-	-	
L. planifrons Knight	-	-	-	-	-	-	SK	MB	-	-	-	-	-	-	-	
L. simplex (Uhler)	-	-	-	-	BC	AB	SK	MB	-	-	-	-	-	-	-	

Genus LOPIDEA Uhler

Key to species: Asquith 1991 [North America].

L. ampla Van Duzee	-	-	-	-	BC	-	-	-	-	-	-	-	-	-	-	
L. balli Knight	-	-	-	-	-	AB	SK	-	-	-	-	-	-	-	-	
L. caesar (Reuter)	-	-	-	-	-	-	-	-	ON	-	-	-	-	-	-	
L. confluenta (Say)	-	-	-	-	-	AB	-	MB	ON	QC	-	-	-	-	-	
L. confraterna (Gibson)	-	-	-	-	-	AB	-	-	-	-	-	-	-	-	-	
L. dakota Knight	AK	YK	NT	-	BC	AB	SK	MB	ON	QC	NB	-	-	-	-	[A]
L. heidemanni Knight	-	-	-	-	-	-	-	-	ON	QC	-	-	-	-	-	
L. incurva Knight	-	-	-	-	-	-	-	-	ON	-	-	-	-	-	-	
L. instabilis (Reuter)	-	-	-	-	-	-	-	-	ON	QC	-	-	-	-	-	
L. lathyri Knight	-	-	-	-	-	AB	SK	MB	ON	-	-	-	-	-	-	
L. media (Say)	-	-	-	-	-	AB	SK	MB	ON	QC	NB	-	NS	-	-	
L. minor Knight	-	-	-	-	BC	AB	SK	MB	-	-	-	-	-	-	-	
L. nigridea aculeata Van Duzee	-	-	-	-	BC	-	-	-	-	-	-	-	-	-	-	
L. nigridea nigridea Uhler	-	-	-	-	BC	-	-	-	-	-	-	-	-	-	-	
L. nigridea serica Knight	AK	YK	-	-	-	AB	SK	MB	-	-	-	-	-	-	-	
L. picta Knight	-	-	-	-	BC	AB	-	-	-	-	-	-	-	-	-	[r]
L. robiniae (Uhler)	-	-	-	-	-	-	-	-	ON	QC	-	-	-	-	-	
L. salicis Knight	-	-	-	-	-	-	-	-	ON	-	-	-	-	-	-	
L. taurina Van Duzee	-	-	-	-	BC	-	-	-	-	-	-	-	-	-	-	
L. teton Knight	-	-	-	-	-	AB	SK	MB	-	-	-	-	-	-	-	

Genus MALACOCORIS Fieber

M. chlorizans (Panzer)	-	-	-	-	BC*	-	-	-	-	-	-	-	-	-	-	

Genus MECOMMA Fieber

Key to species: Kelton & Knight 1962.

M. angustatum (Uhler)	AK	YK	NT	-	BC	AB	SK	MB	ON	QC	NB	PE	-	NF	-	
M. bradora Kelton	-	-	-	-	-	-	-	-	-	QC	-	-	-	-	-	
M. gilvipes (Stål)	AK	YK	NT	-	BC	AB	SK	MB	ON	QC	NB	PE	NS	NF	LB	

Genus MELANOTRICHUS Reuter

Key to species: Henry 1991 [eastern North America], Knight 1968b [western North America].

M. albocostatus (Van Duzee)	-	YK	-	-	BC	AB	SK	MB	-	-	-	-	-	-	-	
M. atriplicis Knight	-	-	-	-	-	AB	-	-	-	-	-	-	-	-	-	
M. brindleyi Knight	-	-	-	-	BC	AB	SK	MB	ON	QC	-	PE	NS	NF	-	
M. catulus (Van Duzee)	-	-	-	-	-	-	-	-	ON	QC	-	-	-	-	-	

* introduced; [] indoors; { } extirpated; ? uncertain record; @ unestablished adventive; $ port-of-entry interception

Melanotrichus coagulatus (Uhler)	-	YK	-	-	BC	AB	SK	MB	-	-	-	-	-	-	-	
M. concolor (Kirschbaum)	-	-	-	-	BC*	-	-	-	-	QC*	-	-	NS*	-	-	
M. elongatus Kelton	-	-	-	-	-	AB	SK	-	-	-	-	-	-	-	-	
M. flavosparsus (Sahlberg)	-	-	NT*	-	BC*	AB*	SK*	MB*	ON*	QC*	NB*	PE*	NS*	-	-	[A]
M. leviculus Knight	-	-	-	-	BC	AB	SK	MB	-	-	-	-	-	-	-	
M. mistus (Knight)	AK	YK	-	-	BC	AB	-	-	-	-	-	-	-	-	-	
M. vestitus (Uhler)	-	-	-	-	BC	AB	-	-	-	-	-	-	-	-	-	
M. virescens (Douglas & Scott)	-	-	-	-	BC*	-	-	-	-	-	-	-	NS*	-	-	
M. wallisi Kelton	-	-	-	-	-	AB	SK	-	-	-	-	-	-	-	-	

Genus NOCTUOCORIS Knight

N. fumidus (Van Duzee)	-	-	-	-	BC	-	-	MB	ON	QC	-	-	-	-	-

Genus ORTHOTYLUS Fieber

(Subgenus NEOMECOMMA Southwood)

O. candidatus Van Duzee	-	YK	-	-	BC	AB	SK	-	ON	QC	-	-	-	NF	-

(Subgenus ORTHOTYLUS Fieber)

O. alni Knight	-	YK	NT	-	BC	AB	-	MB	ON	QC	NB	-	-	NF	LB
O. angulatus (Uhler)	-	-	-	-	-	AB	-	-	-	-	-	-	-	-	-
O. basicornis Knight	-	-	-	-	-	-	SK	-	ON	QC	-	-	-	-	-
O. contrastus Van Duzee	-	-	-	-	BC	-	-	-	-	-	-	-	-	-	-
O. cruciatus Van Duzee	-	-	-	-	-	-	-	-	-	QC	NB	-	NS	-	-
O. dorsalis (Provancher)	-	-	-	-	BC	AB	SK	MB	ON	QC	NB	-	NS	NF	-
O. fuscicornis Knight	-	-	-	-	-	AB	SK	MB	-	-	-	-	-	-	-
O. katmai (Knight)	AK	YK	NT	NU	BC	AB	-	MB	ON	QC	-	-	-	-	-
O. marginatus (Uhler)	-	-	-	-	-	-	-	-	ON?	QC?	-	-	-	-	-
O. modestus Van Duzee	-	-	-	-	-	-	-	-	ON	QC	-	-	-	-	-
O. nassatus (Fabricius)	-	-	-	-	-	-	-	-	ON*	-	-	PE*	NS*	-	-
O. necopinus Van Duzee	-	-	-	-	-	-	-	-	ON	QC	-	-	-	NF	-
O. neglectus Knight	-	YK	-	-	BC	AB	SK	MB	ON	QC	NB	-	NS	NF	-
O. notabilis Knight	-	-	-	-	-	AB	SK	MB	ON	-	-	-	-	-	-
O. ochrotrichus Fieber	-	-	-	-	BC*	-	-	-	-	-	-	-	-	-	-
O. ornatus Van Duzee	-	-	-	-	-	AB	-	-	ON	QC	NB	-	-	-	-
O. pacificus Van Duzee	-	-	-	-	BC	AB	-	-	-	-	-	-	-	-	-
O. pennsylvanicus Henry	-	-	-	-	-	AB	-	MB	ON	QC	NB	-	NS	-	-
O. robiniae Johnston	-	-	-	-	-	-	-	-	ON	-	-	-	-	-	-
O. serus Van Duzee	-	-	-	-	-	-	-	-	ON	-	-	-	-	-	-
O. submarginatus (Say)	-	-	-	-	-	-	-	-	ON	QC	-	-	-	-	-
O. uniformis Van Duzee	-	-	-	-	-	-	-	-	-	QC?	-	-	-	-	-
O. ute Knight	AK	-	-	-	BC	AB	-	-	-	-	-	-	-	-	-
O. viridinervis (Kirschbaum)	-	-	-	-	-	-	-	-	ON*	-	-	-	NS*	-	-
O. viridis Van Duzee	-	-	-	-	-	AB?	-	-	-	QC	-	-	-	-	-

Genus PARAPROBA Distant

P. capitata (Van Duzee)	-	-	-	-	-	-	-	-	ON	QC	NB	-	NS	-	-
P. cincta Van Duzee	-	-	-	-	BC	-	-	-	-	-	-	-	-	-	-
P. hamata Van Duzee	-	-	-	-	BC	-	-	-	-	-	-	-	-	-	-

Genus PARTHENICUS Reuter

Key to species: Henry 1982a [eastern North America].

P. basicornis Knight	-	-	-	-	BC	-	-	-	-	-	-	-	-	-	-
P. brindleyi Knight	-	-	-	-	BC	-	-	-	-	-	-	-	-	-	-
P. brooksi Kelton	-	-	-	-	-	-	SK	-	-	-	-	-	-	-	-
P. giffardi Van Duzee	-	-	-	-	BC	-	-	-	-	-	-	-	-	-	-
P. grex Van Duzee	-	-	-	-	BC	-	-	-	-	-	-	-	-	-	-
P. juniperi (Heidemann)	-	-	-	-	-	-	-	-	ON	-	-	-	-	-	-

Parthenicus pallidicollis Van Duzee	-	-	-	-	BC	-	-	-	-	-	-	-	-	-	-	
P. sabulosus Van Duzee	-	-	-	-	BC	-	-	-	-	-	-	-	-	-	-	

Genus PSEUDOLOXOPS Kirkaldy

P. coccineus (Meyer-Dür)	-	-	-	-	-	-	-	-	ON*	-	-	-	-	-	-	

Genus PSEUDOPSALLUS Van Duzee

Key to species: Stonedahl & Schwartz 1986.

P. anograe Knight	-	-	-	-	-	AB	-	MB	-	-	-	-	-	-	-	
P. demensus (Van Duzee)	-	-	-	-	-	AB	SK	MB	-	-	-	-	-	-	-	
P. occidentalis Stonedahl & Schwartz	-	-	-	-	BC	-	-	-	-	-	-	-	-	-	-	
P. sericatus (Uhler)	-	-	-	-	-	AB	SK	-	-	-	-	-	-	-	-	

Genus PSEUDOXENETUS Reuter

P. regalis (Uhler)	-	-	-	-	-	-	-	-	ON	QC	-	-	-	-	-	

Genus REUTERIA Puton

Key to species: Henry 1976.

R. bifurcata Knight	-	-	-	-	-	-	-	-	ON	-	-	-	-	-	-	
R. fuscicornis Knight	-	-	-	-	-	-	-	-	ON	QC	-	-	-	-	-	
R. irrorata (Say)	-	-	-	-	-	-	-	-	ON	QC	-	-	-	-	-	
R. platani Knight	-	-	-	-	-	-	-	-	ON	-	-	-	-	-	-	
R. querci Knight	-	-	-	-	-	-	-	MB	ON	QC	-	-	-	-	-	

Genus SCHAFFNERIA Knight

Key to species: Henry 1994.

S. davisi (Knight)	-	-	-	-	-	-	-	MB	ON	-	-	-	-	-	-	
S. pilophoroides (Knight)	-	-	-	-	-	AB	SK	-	-	-	-	-	-	-	-	

Genus SERICOPHANES Reuter

S. heidemanni Poppius	-	-	-	-	BC	AB	SK	-	ON	QC	-	-	-	-	-	

Genus SLATEROCORIS Wagner

Key to species: Kelton 1968, Knight 1970.

S. ambrosiae (Knight)	-	-	-	-	-	-	-	-	ON	-	-	-	-	-	-	
S. atritibialis (Knight)	-	-	-	-	BC	AB	SK	MB	ON	QC	NB	-	NS	-	-	
S. breviatus (Knight)	-	-	-	-	BC	AB	SK	MB	ON	QC	NB	-	NS	-	-	
S. pallidicornis (Knight)	-	-	-	-	-	AB	SK	MB	-	-	-	-	NS	-	-	
S. pilosus Kelton	-	-	-	-	BC	-	-	-	-	-	-	-	-	-	-	[r]
S. robustus (Uhler)	-	-	-	-	BC	AB	-	-	-	-	-	-	-	-	-	[r]
S. stygicus (Say)	-	-	-	-	BC	AB	SK	MB	ON	QC	NB	-	NS	-	-	

Subfamily PHYLINAE

Tribe Hallodapini

Genus COQUILLETTIA Uhler

C. albertae Kelton	-	-	-	-	-	AB	-	-	-	-	-	-	-	-	-	
C. insignis Uhler	-	-	-	-	BC	AB	SK	-	-	-	-	-	-	-	-	

Genus ORECTODERUS Uhler

Key to species: Knight 1968a.

O. arcuatus Knight	-	-	-	-	BC	-	-	-	-	-	-	-	-	-	-	[r]
O. montanus Knight	-	-	-	-	-	AB	-	-	-	-	-	-	-	-	-	[r]
O. obliquus Uhler	AK	YK	NT	-	BC	AB	SK	MB	ON	QC	NB	-	NS	-	-	
O. salicis Knight	-	-	-	-	-	AB	-	-	-	-	-	-	-	-	-	[r]

* introduced; [] indoors; { } extirpated; ? uncertain record; @ unestablished adventive; $ port-of-entry interception

Genus TELEORHINUS Uhler
Key to species: Knight 1968b.

T. brindleyi Knight	-	-	-	-	BC	-	-	-	-	-	-	-	-	-	-	[r]

Tribe Leucophoropterini

Genus TYTTHUS Fieber
Key to species: Carvalho & Southwood 1955.

T. pubescens (Knight)	AK	YK	-	-	BC	AB	SK	-	-	QC	-	-	-	NF	-
T. pygmaeus (Zetterstedt)	AK	YK	-	-	BC	AB	SK	-	ON	-	-	-	-	NF	LB
T. vagus (Knight)	-	-	-	-	-	-	-	-	-	-	NB	-	NS	-	-

Tribe Phylini

Genus AMBLYTYLUS Fieber

A. nasutus (Kirschbaum)	-	-	-	-	BC*	-	-	-	ON*	-	-	-	-	-	-

Genus ASCIODEMA Reuter

A. obsoletum (Fieber)	-	-	-	-	BC*	-	-	-	-	-	-	-	-	-	-

Genus ATOMOSCELIS Reuter

A. onustus (Fieber)	-	-	-	-	BC	AB	SK	MB	-	-	-	-	-	-	-

Genus ATRACTOTOMUS Fieber
Key to species: Stonedahl 1990 [Holarctic].
Note: *A. mali* is not properly placed here.

A. atricolor (Knight)	-	-	-	-	BC	AB	SK	MB	ON	-	-	-	-	-	-
A. cerocarpi Knight	-	-	-	-	BC	-	-	-	-	-	-	-	-	-	-
A. cooperi Stonedahl	-	-	-	-	BC	-	-	-	-	-	-	-	-	-	-
A. kolenatii (Flor)	-	YK	-	-	BC	-	-	-	-	-	-	-	-	-	-
A. magnicornis (Fallén)	-	-	-	-	-	-	-	-	ON*	QC*	-	-	NS*	NF*	-
A. mali (Meyer-Dür) [incert. sed.]	-	-	-	-	BC*	-	-	-	-	-	NB*	PE*	NS*	-	-
A. ramentum Stonedahl	-	-	-	-	-	-	-	-	ON	-	-	-	-	-	-
A. rubidus (Uhler)	-	-	-	-	BC	-	-	-	ON	QC	-	-	-	-	-

Genus CAMPYLOMMA Reuter

C. verbasci (Meyer-Dür) mullein bug; punaise de la molène	-	-	-	-	BC*	-	-	-	ON*	QC*	NB*	-	NS*	-	-

Genus CHLAMYDATUS Curtis
Key to species: Kelton 1965b [North America].

C. artemisiae Kelton	-	-	-	-	-	AB	-	-	-	-	-	-	-	-	-	
C. associatus (Uhler)	-	-	-	-	BC	AB	SK	MB	ON	QC	-	-	-	-	-	[A]
C. brevicornis Knight	-	-	-	-	BC	-	-	-	-	-	-	-	-	-	-	[r]
C. montanus Knight	-	YK	-	-	BC	AB	SK	MB	-	-	-	-	-	-	-	
C. obliquus (Uhler)	-	-	-	-	BC	AB	-	-	-	-	-	-	-	-	-	
C. opacus (Zetterstedt)	-	-	-	NU	-	-	-	MB	-	QC	-	-	-	-	-	
C. pallidicornis Knight	-	-	-	-	BC	AB	-	-	-	-	-	-	-	-	-	
C. pulicarius (Fallén)	AK	YK	NT	-	BC	AB	SK	MB	ON	QC	NB	PE	NS	NF	-	
C. pullus (Reuter)	AK	YK	NT	NU	BC	AB	SK	MB	ON	QC	-	-	-	NF	LB	
C. ruficornis Knight	-	-	-	-	-	AB	SK	-	-	-	-	-	-	-	-	
C. schuhi Knight	-	-	-	-	BC	-	-	-	-	-	-	-	-	-	-	[r]
C. suavis (Reuter)	-	-	-	-	-	-	-	-	ON	-	-	-	-	-	-	
C. wilkinsoni (Douglas & Scott)	AK	YK	NT	NU	-	-	-	MB	-	QC	-	-	-	NF	-	

Genus COMPSIDOLON Reuter

C. pumilum (Jakovlev)	-	YK	NT	-	-	-	-	-	-	-	-	-	-	-	-
C. salicellum (Herrich-Schaeffer)	-	-	-	-	BC*	-	-	-	ON*	-	-	PE*	NS*	-	-

A agricultural, F forestry, G greenhouse, O ornamental pest; r rare; b biocontrol agent

Genus CONIFEROCORIS Schwartz & Schuh
Key to species: Schwartz & Schuh 1999.

C. pinicolus Schwartz & Schuh	-	-	-	-	BC	-	-	-	-	-	-	-	-	-	-

Genus CONOSTETHUS Fieber

C. americanus Knight	-	-	NT	-	-	AB	SK	-	-	-	-	-	-	-	-

Genus CRIOCORIS Fieber

C. saliens (Reuter)	-	-	-	-	BC	AB	SK	MB	ON	QC	-	-	NS	-	-

Genus DACOTA Uhler

D. hesperia Uhler	-	-	-	-	BC	AB	SK	MB	-	-	-	-	-	-	-

Genus EUROPIELLA Reuter

E. angulata (Uhler)	-	-	-	-	-	AB	SK	-	-	-	-	-	-	-	-
E. artemisiae (Becker)	AK	YK	NT	-	BC	AB	SK	-	-	-	-	-	-	-	-
E. consors (Uhler)	-	-	-	-	BC	AB	SK	MB	-	-	-	-	-	-	-
E. decolor (Uhler)	AK	YK	-	-	BC	AB	SK	MB	ON	QC	-	-	-	NF	-
E. pilosula (Uhler)	-	-	-	-	BC	-	-	-	-	-	-	-	-	-	-

Genus HOPLOMACHUS Fieber

H. affiguratus (Uhler)	-	-	-	-	BC	AB	-	-	-	-	-	-	-	-	-

Genus ICODEMA Reuter

I. nigrolineatum (Knight)	-	-	-	-	-	-	-	MB	ON	-	-	-	-	-	-

Genus KNIGHTOMIROIDES Stonedahl & Schwartz

K. contortae Stonedahl & Schwartz	-	-	-	-	BC	-	-	-	-	-	-	-	-	-	-
K. ponderosae Stonedahl & Schwartz	-	-	-	-	BC	-	-	-	-	-	-	-	-	-	-

Genus LEPIDARGYRUS Mumonov

L. ancorifer (Fieber)	-	-	-	-	BC*	-	-	-	ON*	-	-	-	-	-	-

Genus LOPUS Hahn

L. decolor (Fallén)	-	-	-	-	BC*	-	-	-	ON*	QC*	NB*	PE*	NS*	NF*	-

Genus MACROTYLUS Fieber
Key to species: Van Duzee 1916.

M. multipunctatus Van Duzee	-	-	-	-	BC	-	-	-	-	-	-	-	-	-	-
M. sexguttatus (Provancher)	-	-	-	-	-	-	-	-	ON	QC	-	-	-	-	-

Genus MEGALOCOLEUS Reuter

M. molliculus (Fallén)	-	-	-	-	-	-	-	-	ON*	QC*	NB*	PE*	NS*	NF*	-

Genus MEGALOPSALLUS Knight

M. femoralis Kelton	-	-	-	-	-	AB	SK	-	-	-	-	-	-	-	-
M. lycii (Knight)	-	-	-	-	-	AB	SK	-	-	-	-	-	-	-	-
M. montanae (Knight)	-	-	-	-	-	-	SK	-	-	-	-	-	-	-	-
M. rubropictipes Knight	-	-	-	-	-	AB	SK	-	-	-	-	-	-	-	-

Genus MICROPHYLELLUS Reuter

M. adustus binotatus Knight	-	-	-	-	BC	-	-	-	-	-	-	-	-	-	-
M. elongatus Knight	-	-	-	-	-	-	-	-	ON	QC	-	-	-	-	-
M. flavipes (Provancher)	-	-	-	-	-	-	-	-	ON	QC	-	-	-	-	-
M. fuscicornis Knight	-	-	-	-	-	-	SK	MB	-	-	-	-	-	-	-
M. longirostris Knight	-	-	-	-	-	-	-	MB	ON	QC	-	-	-	-	-
M. maculipennis Knight	-	-	-	-	-	-	-	MB	-	-	-	-	-	-	-

* introduced; [] indoors; { } extirpated; ? uncertain record; @ unestablished adventive; $ port-of-entry interception

Microphylellus modestus Reuter	-	-	-	-	-	-	SK	MB	ON	QC	-	-	-	-	-
M. tsugae Knight	-	-	-	-	-	-	-	-	ON	QC	NB	-	NS	-	-
M. tumidifrons Knight	-	-	-	-	-	-	SK	MB	ON	QC	-	-	NS	-	-

Genus MONOSYNAMMA Scott

M. bohemanni (Fallén)	AK	YK	NT	-	BC	AB	SK	MB	ON	QC	-	-	NS	NF	LB

Genus PARAPSALLUS Wagner

P. vitellinus (Scholtz)	-	-	-	-	-	-	-	-	ON*	QC*	-	-	-	-	-

Genus PHOENICOCORIS Reuter

P. longirostris (Knight)	-	YK	-	-	BC	AB	SK	-	-	-	-	-	-	-	-
P. minusculus (Knight)	-	-	-	-	-	-	-	-	ON	QC	-	-	-	-	-
P. rostratus (Knight)	-	YK	-	-	BC	AB	SK	MB	ON	QC	NB	-	NS	-	-
P. strobicola (Knight)	-	-	-	-	-	-	-	-	ON	QC	NB	-	NS	-	-

Genus PHYLLOPIDEA Knight
Key to species: Knight 1968b.

P. montana Knight	-	-	-	-	BC	AB	-	-	-	-	-	-	-	-	-

Genus PHYLUS Hahn

P. coryli (Linnaeus)	-	-	-	-	BC*	-	-	-	-	-	-	-	-	-	-

Genus PICEOPHYLUS Schwartz & Schuh

P. keltoni Schwartz & Schuh	-	-	-	-	-	AB	-	-	ON	QC	NB	PE	NS	-	-

Genus PINOPHYLUS Schwartz & Schuh

P. rolfsi (Knight)	-	YK	-	-	BC	AB	-	-	-	-	-	-	-	-	-

Genus PLAGIOGNATHUS Fieber

P. albatus albatus (Van Duzee)	-	-	-	-	-	-	-	-	ON	QC	-	-	-	-	-	
P. albatus vittiscutis Knight	-	-	-	-	-	-	-	-	-	QC	-	-	-	-	-	
P. albonotatus Knight	-	-	-	-	-	AB	SK	MB	ON	QC	NB	-	NS	-	-	
P. alboradialis Knight	-	YK	NT	-	BC	AB	SK	MB	ON	QC	NB	-	-	NF	-	
P. annulatus Uhler	-	-	-	-	BC	AB	SK	-	ON	QC	-	-	-	-	-	
P. arbustorum (Fabricius)	-	-	-	-	BC*	-	-	-	-	-	-	-	-	NF*	-	
P. blatchleyi Reuter	-	-	-	-	-	-	-	MB	ON	QC	-	-	-	-	-	
P. brevirostris Knight	-	-	-	-	-	-	-	-	ON	QC	-	-	-	NF	-	
P. brunneus (Provancher)	AK	YK	NT	-	BC	AB	SK	MB	-	-	-	-	-	-	-	[A]
P. caryae Knight	-	-	-	-	-	-	-	-	ON	QC	-	-	-	-	-	
P. chrysanthemi (Wolff)	-	-	-	-	BC*	-	-	MB*	ON*	QC*	NB*	PE*	NS*	NF*	-	[A]
P. cornicola Knight	-	-	-	-	-	-	-	-	ON	QC	-	-	-	-	-	
P. cuneatus Knight	-	-	-	-	-	-	-	-	ON	QC	-	-	-	-	-	
P. davisi Knight	-	-	-	-	BC	AB	SK	MB	-	-	-	-	-	-	-	
P. delicatus (Uhler)	-	-	-	-	-	-	-	-	ON	-	-	-	-	-	-	
P. dispar Knight	-	-	-	-	-	-	-	MB	ON	QC	-	-	-	-	-	
P. flavicornis Knight	-	-	-	-	-	-	-	-	ON	QC	NB	-	NS	-	-	
P. flavidus Knight	-	-	-	-	-	AB	SK	MB	-	QC	-	-	-	-	-	
P. flavoscutellatus Knight	-	-	-	-	-	-	-	-	ON	QC	-	-	NS	NF	-	
P. fumidus Uhler	-	-	-	-	BC	-	-	-	-	-	-	-	-	-	-	
P. fusciflavus Knight	-	-	-	-	BC	-	-	-	-	-	-	-	-	-	-	
P. fuscipes Knight	-	-	-	-	-	AB	-	-	ON	-	-	-	-	NF	-	
P. fuscosus (Provancher)	-	-	-	-	BC	AB	SK	MB	ON	QC	-	-	-	-	-	
P. guttatipes (Uhler)	-	-	-	-	-	AB	SK	MB	-	-	-	-	-	-	-	
P. laricicola Knight	-	YK	-	-	BC	AB	SK	MB	ON	QC	-	-	-	-	-	
P. moerens (Reuter)	-	-	-	-	BC	AB	-	-	-	-	-	-	-	-	-	

Plagiognathus mundus Van Duzee	-	-	-	-	BC	-	-	-	-	-	-	-	-	-	-	
P. negundinis Knight	-	-	-	-	-	-	SK	-	ON	QC	-	-	-	-	-	
P. nigritus Knight	-	-	-	-	-	AB	SK	MB	ON	QC	-	-	-	-	-	
P. nigronitens Knight	-	-	-	-	-	AB	SK	MB	ON	QC	-	-	-	-	-	
P. obscurus albocuneatus Knight	-	-	-	-	-	-	-	-	ON	QC	-	PE	NS	NF	-	
P. obscurus obscurus Uhler obscure plant bug; punaise sombre	-	-	-	-	BC	AB	SK	MB	ON	QC	NB	PE	NS	NF	-	
P. politus flaveolus Knight	-	-	-	-	-	-	-	-	ON	QC	-	-	-	-	-	
P. politus politus Uhler	-	-	-	-	-	-	-	MB	ON	QC	-	-	NS	NF	-	[A]
P. punctatipes Knight	-	-	-	-	-	-	-	-	ON	QC	NB	-	NS	-	-	
P. repetitus Knight	-	-	-	-	-	-	-	-	ON	QC	-	-	NS	-	-	
P. repletus Knight	-	-	-	-	-	-	-	-	-	QC	-	-	-	-	-	
P. ribesi Kelton	-	-	-	-	BC	-	-	-	-	-	-	-	-	-	-	
P. rosicola Knight	-	-	-	-	-	-	-	-	ON	QC	-	-	-	-	-	
P. salicicola Knight	-	-	-	-	-	-	-	-	ON	QC	-	-	-	-	-	
P. shoshonea Knight	-	-	-	-	BC	AB	-	-	-	-	-	-	-	-	-	
P. similis Knight	-	-	-	-	-	-	-	-	ON	-	-	-	-	-	-	
P. subovatus Knight	-	-	-	-	-	-	-	-	ON	-	-	-	-	-	-	
P. suffuscipennis Knight	AK	YK	-	-	BC	AB	SK	MB	ON	QC	-	-	-	-	-	
P. tinctus Knight	-	-	-	-	-	-	-	-	ON	-	-	-	-	-	-	

Genus PLESIODEMA Reuter

P. sericeum (Heidemann)	-	-	-	-	-	-	-	-	ON	-	-	-	-	-	-	

Genus PRONOTOCREPIS Knight

P. clavicornis Knight	-	-	-	-	BC	-	-	-	-	-	-	-	-	-	-	[r]

Genus PSALLOVIUS Henry

Key to species: Schwartz & Schuh 1999.

P. nigroantennatus Schwartz & Schuh	-	-	-	-	BC	-	-	-	-	-	-	-	-	-	-	
P. piceicola (Knight)	-	-	-	-	BC	AB	SK	MB	ON	QC	NB	PE	NS	NF	-	

Genus PSALLUS Fieber

P. aethiops (Zetterstedt)	AK	YK	-	-	BC	AB	SK	MB	ON	QC	NB	-	NS	NF	LB	
P. alnicenatus Knight	-	-	-	-	-	-	-	-	ON	QC	-	-	NS	-	-	
P. betuleti (Fallén)	AK	YK	NT	-	BC	-	-	-	ON	QC	NB	-	NS	-	-	
P. falleni Reuter	AK	YK	NT	-	BC	AB	-	-	ON	QC	NB	PE	NS	-	-	
P. flavellus Stichel	-	-	-	-	-	-	-	-	-	-	-	-	NS*	-	-	
P. haematodes (Gmelin)	-	-	-	-	BC*	-	-	-	-	-	-	-	-	-	-	
P. kimi Josifov	AK	YK	NT	-	BC	-	-	-	-	QC	-	-	-	-	-	
P. lepidus Fieber	-	-	-	-	-	-	-	-	-	-	-	PE*	NS*	-	-	
P. morrisoni Knight	-	-	-	-	-	-	-	-	ON	QC	-	-	NS	-	-	
P. parshleyi Knight	AK	YK	NT	-	BC	AB	SK	MB	-	QC	NB	-	NS	NF	-	
P. physocarpi Henry	-	-	-	-	-	-	-	-	ON	-	-	-	-	-	-	
P. variabilis (Fallén)	-	-	-	-	-	-	-	-	ON*	-	-	-	-	-	-	

Genus PSEUDATOMOSCELIS Poppius

P. seriatus (Reuter)	-	-	-	-	-	-	SK	-	-	-	-	-	-	-	-	

Genus REUTEROSCOPUS Kirkaldy

Key to species: Kelton 1964, Knight 1965.

R. ornatus (Reuter)	-	-	-	-	-	-	-	-	ON	-	-	-	-	-	-	

Genus RHINOCAPSUS Uhler

R. rubricans (Provancher)	-	-	-	-	-	-	SK	MB	ON	QC	-	-	-	-	-	
R. vanduzeei Uhler	-	-	-	-	-	-	-	MB	ON	QC	NB	PE	NS	-	-	

* introduced; [] indoors; { } extirpated; ? uncertain record; @ unestablished adventive; $ port-of-entry interception

Genus **SEMIUM** Reuter
Key to species: Kelton 1973.

S. hirtum Reuter	-	-	-	-	-	-	-	-	ON	-	-	-	-	-	-

Genus **STHENARUS** Fieber

S. cuneotinctus (Van Duzee) [incert. sed.]	-	-	-	-	BC	-	-	-	-	-	-	-	-	-	-
S. rotermundi (Scholtz)	-	-	-	-	-	-	-	-	ON*	-	-	-	-	-	-

Tribe Pilophorini

Genus **PILOPHORUS** Hahn
Key to species: Schuh & Schwartz 1988 [New World].

P. americanus Poppius	-	YK	-	-	BC	AB	SK	MB	-	-	-	-	-	-	-	
P. amoenus Uhler	-	-	-	-	-	-	-	MB	ON	QC	NB	-	-	-	-	
P. brunneus Poppius	-	-	-	-	-	-	-	-	ON	-	-	-	-	-	-	
P. clavatus (Linnaeus)	-	-	-	-	BC	AB	SK	MB	ON	QC	-	-	NS	-	-	
P. confusus Kirschbaum	-	-	-	-	-	-	-	-	-	QC*	-	-	NS*	-	-	
P. crassipes Heidemann	-	-	-	-	-	AB	-	MB	ON	QC	-	-	-	-	-	
P. diffusus Knight	-	YK	-	-	BC	AB	-	-	-	-	-	-	-	-	-	
P. furvus Knight	-	-	-	-	-	-	-	MB	-	-	-	-	-	-	-	
P. geminus Knight	-	-	-	-	-	-	-	MB	ON	QC	-	-	-	-	-	
P. gracilis Uhler	-	-	-	-	-	-	-	MB	ON	QC	-	-	-	-	-	
P. laetus Heidemann	-	-	-	-	-	-	-	-	ON	-	-	-	-	-	-	
P. neoclavatus Schuh & Schwartz	-	-	-	-	-	AB	SK	MB	ON	QC	-	-	-	-	-	
P. perplexus Douglas & Scott	-	-	-	-	BC*	-	-	-	ON*	QC*	-	PE*	NS*	-	-	
P. piceicola Knight	-	-	-	-	-	AB	SK	MB	ON	QC	NB	PE	NS	-	-	
P. salicis Knight	-	-	-	-	-	AB	SK	MB	-	-	-	-	-	-	-	
P. setiger Knight	-	-	-	-	-	-	-	MB	-	-	-	-	-	-	-	
P. stonedahli Schuh & Schwartz	-	-	-	-	BC	-	-	-	-	-	-	-	-	-	-	[r]
P. strobicola Knight	-	-	-	-	-	-	-	-	ON	QC	NB	-	NS	-	-	
P. tibialis Van Duzee	-	-	-	-	BC	-	-	-	-	-	-	-	-	-	-	
P. uhleri Knight	-	-	-	-	BC	AB	SK	MB	ON	QC	NB	PE	NS	-	-	
P. vicarius Poppius	AK	YK	-	-	BC	AB	SK	MB	-	-	-	-	-	-	-	
P. walshii Uhler	-	-	-	-	-	-	-	-	ON	-	-	-	-	-	-	

Superfamily Tingoidea

Family TINGIDAE

Key to subfamilies:Drake & Davis 1960.

Catalogue to world species: Drake & Ruhoff 1965.

Lacebugs. Head without ocelli; antennae 4-segmented with first two segments short and stout; forewings lacking division into clavus, corium and membrane, but with numerous small, lace-like cells; tarsi with 2 segments. Plant feeders, often quite host specific.

Subfamily TINGINAE

Tribe Tingini

Genus **ACALYPTA** Westwood
Key to species: Drake & Lattin 1963.

A. cooleyi Drake	-	YK	-	-	BC	-	-	-	-	-	-	-	-	-	-
A. elegans Horváth	AK	YK	NT	-	BC	AB	-	MB	-	-	-	-	-	NF	-
A. lillianis Torre Bueno	AK	YK	NT	-	BC	AB	-	-	ON	QC	NB	-	-	NF	-
A. parvula (Fallén)	-	-	-	-	BC	-	-	-	ON	-	NB	-	-	-	-
A. saundersi (Downes)	-	-	-	-	BC	-	-	-	-	-	-	-	-	-	-

Genus ALVEOTINGIS Osborn & Drake

A. grossocerata Osborne & Drake - - - - - - - - ON QC - - - - -

Genus CORYTHAICA Stål

C. venusta (Champion) - - - - BC AB - - - - - - - - -

Genus CORYTHUCHA Stål

Key to species (partial): Gibson 1918, Bailey 1951.

C. arcuata (Say) - - - - - - - MB ON QC - - - - -
oak lace bug; punaise reticulé du chêne

C. bellula Gibson - - - - - - - MB ON - NB - - - -

C. ciliata (Say) - - - - - - - - ON QC - - - - -
sycamore lace bug; punaise reticulé du platane

C. cydoniae (Fitch) - - - - BC AB SK MB ON - - - - - -

C. distincta Osborn & Drake - - - - BC - - - - - - - - - - [A]

C. elegans Drake - - - - BC AB SK MB ON - - - NS - -
willow lace bug; punaise reticulé des peupliers

C. heidemanni Drake - - - - - - - - ON QC NB - NS NF -
alder lace bug; punaise reticulé de l'aulne

C. hewitti Drake - - - - BC - - MB ON - - - - - -

C. immaculata Osborn & Drake - - - - BC AB - - - - - - - - -

C. juglandis (Fitch) - - - - BC - - - - QC - - - - -
walnut lace bug; punaise reticulé du noyer

C. marmorata (Uhler) AK YK - - BC AB SK MB ON QC - - - - -

C. mollicula Osborn & Drake - - - - BC - - MB ON QC - - - - -

C. morrilli Osborn & Drake - - - - BC - - - - - - - - - -

C. obliqua Osborn & Drake - - - - BC - - - - - - - - - -

C. padi Drake - - - - BC - - - - - - - - - -

C. pallipes Parshley - - - - - - - MB ON QC NB - - NF -
birch lace bug; punaise reticulé du bouleau

C. pergandei Heidemann - - - - BC - - - ON QC - - NS - -

C. salicata Gibson - - NT - BC - SK MB ON - - - - - -

C. ulmi Osborn & Drake - - - - - - - - ON QC - - - - -

Genus DICTYLA Stål

D. echii (Schrank) - - - - - - - - ON* QC* - - - - -

D. labeculata (Uhler) - YK - - BC AB SK - - - - - - - -

Genus GALEATUS Curtis

G. spinifrons (Fallén) AK - - - - AB SK MB ON QC - - - - -

Genus GARGAPHIA Stål

Key to species: Drake 1917.

G. angulata Heidemann - - - - - - - - ON - - - - - -

G. opacula Uhler - - - - BC - - - - - - - - - -

G. solani Heidemann - - - - BC - - - - - - - - - -

G. tiliae (Walsh) - - - - - - - - ON QC NB - - - -
basswood lace bug; punaise reticulé du tilleul

Genus HESPEROTINGIS Parshley

H. antennata Parshley - - - - BC - SK MB - - - - - - -

H. fuscata Parshley - - - - BC AB - - - - - - - - -

H. occidentalis Drake - - - - BC AB - - - - - - - -

Genus LEPTOPHARSA Stål

L. heidemanni (Osborne & Drake) - - - - - - - - ON - - - - - -

Genus LEPTOYPHA Stål

L. mutica (Say) - - - - - - SK MB ON QC - - - - -

* introduced; [] indoors; { } extirpated; ? uncertain record; @ unestablished adventive; $ port-of-entry interception

Genus MELANORHOPALA Stål
Key to species: Henry & Wheeler 1986.

M. clavata (Stål)	-	-	-	-	BC	AB	SK	MB	ON	QC	-	-	NS	-	-

Genus PHYSATOCHEILA Fieber

P. brevirostris Osborn & Drake	-	-	-	-	-	-	-	-	ON	QC	-	-	-	-	-
P. plexa (Say)	-	-	-	-	BC	AB	SK	MB	ON	QC	-	-	NS	-	-
P. variegata Parshley	-	-	-	-	BC	AB	SK	MB	ON	QC	NB	-	-	-	-

Genus STEPHANITIS Stål
(Subgenus STEPHANITIS **Stål**)

S. rhododendri Horváth	-	-	-	-	BC*	-	-	-	ON*	QC*	NB*	-	NS*	NF*	-

Genus TELEONEMIA Costa

T. nigrina Champion	-	-	-	-	BC	-	-	-	-	-	-	-	-	-	-

Tribe Ypsotingini

Genus DICTYONOTA Curtis
(Subgenus DICTYONOTA **Curtis**)

D. fuliginosa Costa	-	-	-	-	BC*	-	-	-	-	-	-	-	-	-	-

(Subgenus ALCLETHA **Kirkaldy**)

D. tricornis (Schrank)	-	-	-	-	-	-	-	-	-	QC	NB	PE	NS	-	-

Superfamily Reduvioidea

Key to families and subfamilies: Usinger 1944.

Family REDUVIIDAE

Catalogue of taxa: Maldonado-Capriles 1990 [world].

Key to taxa: Fracker 1912 [North America].

Assassin bugs. Mostly active predators, with some blood feeding on vertebrate hosts. Head usually with ocelli, but lacking trichobothria; rostrum usually a stout, three-segmented and with first segment curving away from head in side view. Prosternum with a characteristic median longitudinal groove that is minutely transversely striated. Forewing lacking costal fracture; membrane usually with two complete cells and a few veins emanating posteriorly. First abdominal segment with paired Brindley's gland opening dorso-laterally on thoracico-abdominal junction. Male genitalia usually symmetrical; female ovipositor usually plate-like.

Subfamily APIOMERINAE

Key to genera: Costa Lima *et al.* 1948 [world].

Genus APIOMERUS Hahn
Key to species: Szerlip 1980.

A. crassipes crassipes (Fabricius)	-	-	-	-	BC?	-	-	-	-	-	-	-	-	-	-

Subfamily EMESINAE

Key to tribes, genera and species: Wygodzinsky 1966.

Tribe Metapterini

Genus BARCE Stål

B. fraterna (Say)	-	-	-	-	BC	AB	SK	MB	ON	QC	NB	-	NS	-	-
B. uhleri Banks	-	-	-	-	-	-	-	-	ON	-	-	-	-	-	-

Genus EMESAYA McAtee & Malloch

E. brevipennis brevipennis (Say)	-	-	-	-	-	-	-	-	ON	-	-	-	-	-	-

Tribe Ploiariolini

Genus EMPICORIS Wolff

E. culiciformis (DeGeer)	-	-	-	-	BC*	-	-	-	ON*	-	-	-	-	-	-
E. errabundus (Say)	-	-	-	-	BC	-	-	MB	ON	QC	-	-	NS	-	-
E. orthoneuron McAtee & Malloch	-	-	-	-	BC	-	-	-	-	-	-	-	-	-	-
E. parshleyi (Bergroth)	-	-	-	-	-	-	-	-	-	QC	-	-	-	-	-
E. pilosus (Fieber)	-	-	-	-	BC*	-	-	-	ON*	QC*	-	-	NS*	-	-
E. rubromaculatus (Blackburn)	-	-	-	-	BC	-	-	-	-	-	-	-	-	-	-
E. vagabundus (Linnaeus)	-	-	-	-	BC*	-	-	-	ON*	QC*	-	PE*	NS*	NF*	-

Subfamily HARPACTORINAE

Genus ACHOLLA Stål

A. multispinosa (DeGeer)	-	-	-	-	-	-	-	-	ON	-	-	-	-	-	-

Genus ARILUS Hahn

A. cristatus (Linnaeus)	-	-	-	-	-	-	-	-	ON	-	-	-	-	-	-

Genus FITCHIA Stål

Key to species: McPherson *et al.* 1992.

F. aptera Stål	-	-	-	-	-	-	-	-	ON	QC	NB	-	-	-	-
F. spinosula Stål	-	-	-	-	BC	AB	SK	-	-	-	-	-	-	-	-

Genus PSELLIOPUS Bergroth

Key to species: Barber 1924 [USA].

P. spinicollis (Champion)	-	-	-	-	BC	-	-	-	-	-	-	-	-	-	-

Genus RHYNOCORIS Hahn

R. leucospilus (Stål)	AK	-	-	-	-	-	-	-	-	-	-	-	-	-	-
R. ventralis ventralis (Say)	-	-	-	-	BC	AB	SK	MB	-	-	-	-	-	-	-

Genus SINEA Amyot & Serville

Key to species: Caudell 1901.

S. diadema (Fabricius) spined assassin bug	-	-	NT	-	BC	AB	SK	MB	ON	QC	NB	-	NS	-	-

Genus ZELUS Fabricius

Key to species: Hart 1986 [North America].

(Subgenus PINDUS Stål)

Z. tetracanthus Stål	-	-	-	-	BC	AB	SK	MB	ON	QC	NB	-	NS	-	-

(Subgenus ZELUS Fabricius)

Z. luridus Stål	-	-	-	-	-	-	-	-	ON	QC	-	-	-	-	-

* introduced; [] indoors; {} extirpated; ? uncertain record; @ unestablished adventive; $ port-of-entry interception

Subfamily PEIRATINAE

Genus MELANOLESTES Stål

M. picipes (Herrich-Schaeffer) - - - - - - - - - QC - - - - -

Subfamily REDUVIINAE

Key to genera: Lent & Wygodzinsky 1947 [New World].

Genus REDUVIUS Fabricius

Key to species: Wygodzinsky & Usinger 1964 [North America].

R. personatus (Linnaeus) - - - - BC - - - ON QC NB - - - -
masked hunter; chasseur masqué

Subfamily STENOPODAINAE

Key to genera and species: Barber 1930 [New World].

Genus PYGOLAMPIS Germar

P. pectoralis (Say) - - - - - AB - MB ON QC NB - - - -
P. sericea Stål - - - - BC - - - ON QC - - - - -

Genus STENOPODA Laporte

S. cinerea Laporte - - - - - - - - - - NB - - - -

Family PHYMATIDAE

Key to taxa: Kormilev 1962 [world].

Ambush bugs. Cryptically coloured ambush predators, usually found hidden in flower heads. Stout bodied with greatly enlarged fore femora; fore tibiae and tarsus fused, often lying in scrobe on femur. Fourth antennal segment clavate. Prosternum with a median longitudinal groove that is minutely transversely striate.

Subfamily PHYMATINAE

Genus PHYMATA Latreille

(Subgenus PHYMATA Latreille)

P. americana americana Melin - - - - - AB SK MB ON QC - - - - -
P. americana metcalfi Evans - - - - BC - - - - - - - - - -
P. pennsylvanica Handlirsch - - - - BC - - - ON QC - - - - -
P. vicina vicina Handlirsch - - - - BC AB SK MB - - - - - - -

Infraorder Pentatomomorpha

Key to superfamilies: Henry 1997a.

Superfamily Aradoidea

Family ARADIDAE

Key to subfamilies, genera and species: Usinger & Matsuda 1959 [world].

List of species: Kormilev & Froeschner 1987 [world].

Flatbugs. Usually found under bark of dead trees, feeding in fungi. Dorso-ventrally flattened, brownish or blackish, elongate oval in shape, with granular surface; ocelli lacking; antennae stout.

Subfamily ANEURINAE

Genus ANEURUS Curtis

Key to species: Picchi 1977 [North America].

A. borealis Picchi	AK	YK	-	-	BC	AB	SK	-	ON	QC	-	-	-	-	-
A. inconstans Uhler	-	-	-	-	BC	AB	SK	MB	ON	QC	-	-	NS	-	-
A. simplex Uhler	AK	YK	NT	-	BC	AB	-	MB	ON	QC	-	-	NS	-	LB

Subfamily ARADINAE

Genus ARADUS Fabricius

Key to species: Matsuda 1977 [Canada].

(Subgenus ARADUS Fabricius)

A. abbas Bergroth	AK	YK	NT	-	BC	AB	SK	MB	ON	QC	NB	-	NS	NF	-
A. acutus Say	AK	-	-	-	-	-	-	MB	-	-	-	-	-	-	-
A. aequalis Say	-	-	-	-	-	-	-	-	ON	QC	-	-	-	-	-
A. alaskanus Kormilev & Heiss	AK	-	-	-	-	-	-	-	-	-	-	-	-	-	-
A. antennalis Parshley	-	-	-	-	BC	-	-	-	-	-	-	-	-	-	-
A. approximatus Parshley	-	-	-	-	BC	-	-	MB	-	QC	-	-	-	-	-
A. basalis Parshley	-	-	-	-	-	-	-	-	-	QC	NB	-	-	-	-
A. behrensi Bergroth	-	-	-	-	BC	-	-	-	-	-	-	-	-	-	-
A. blaisdelli Van Duzee	-	-	-	-	BC	-	SK	-	-	-	-	-	-	-	-
A. borealis Heidemann	-	YK	-	-	-	-	SK	MB	ON	QC	-	-	-	-	-
A. coarctatus Heidemann	-	-	-	-	BC	-	-	-	-	-	-	-	-	-	-
A. compressus Heidemann	-	-	-	-	BC	AB	-	-	-	-	-	-	-	-	-
A. crenatus Say	-	-	-	-	-	-	-	-	ON	QC	-	-	-	-	-
A. debilis Uhler	-	-	NT	-	BC	-	-	-	-	-	-	-	-	-	-
A. depictus Van Duzee	-	-	-	-	BC	-	-	-	-	-	-	-	-	-	-
A. duzeei Bergroth	-	-	-	-	-	-	-	-	ON	QC	-	-	-	-	-
A. falleni Stål	-	-	-	-	BC?	-	-	-	-	-	-	-	-	-	-
A. funestus Bergroth	AK	YK	-	-	BC	AB	-	-	ON	QC	-	-	-	-	-
A. fuscomaculatus Stål	-	-	-	-	BC	-	-	-	-	-	-	-	-	-	-
A. gracilis Parshley	-	-	-	-	-	AB	-	-	-	-	-	-	-	-	-
A. implanus Parshley	-	-	-	-	-	-	-	-	ON	QC	-	-	-	-	-
A. inornatus Uhler	-	-	-	-	-	-	SK	MB	ON	QC	-	-	-	-	-
A. insolitus Van Duzee	-	-	-	-	BC	AB	-	-	ON	-	-	-	-	-	-
A. intectus Parshley	-	YK	-	-	BC	AB	SK	MB	-	-	-	-	-	-	-
A. kormilevi Heiss pine flat bug	-	-	-	-	BC	AB	SK	MB	ON	QC	NB	-	NS	-	-
A. lugubris lugubris Fallén	AK	YK	NT	-	BC	AB	SK	MB	ON	QC	-	-	NS	NF	LB

* introduced; [] indoors; { } extirpated; ? uncertain record; @ unestablished adventive; $ port-of-entry interception

Aradus lugubris nigricornis Reuter	AK	-	NT	-	BC	-	-	-	ON	-	-	-	-	-	-
A. medioximus Parshley	-	-	-	-	BC	-	-	-	-	-	-	-	-	-	-
A. montanus Bergroth	-	-	-	-	-	-	-	-	-	QC	-	-	-	-	-
A. paganicus Parshley	-	-	-	-	BC	-	-	-	ON	-	-	-	-	-	-
A. parshleyi Van Duzee	-	-	NT	-	BC	AB	-	-	-	-	-	-	-	-	-
A. parvicornis Parshley	-	-	-	-	BC	-	-	-	-	-	-	-	-	-	-
A. persimilis Van Duzee	-	-	-	-	BC	AB	-	-	-	-	-	-	-	NF	-
A. proboscideus Walker	AK	-	-	-	BC	AB	-	MB	ON	QC	NB	-	NS	-	LB
A. quadrilineatus Say	-	YK	NT	-	BC	AB	SK	MB	ON	QC	NB	-	NS	-	-
A. robustus Uhler	-	-	-	-	-	-	-	MB	ON	QC	-	-	-	-	-
A. saileri Kormilev	AK	-	-	-	-	-	-	-	-	-	-	-	-	-	-
A. saskatchewanensis Matsuda	-	-	-	-	-	-	SK	-	-	-	-	-	-	-	-
A. serratus Usinger	-	-	-	-	-	AB	-	-	-	-	-	-	-	-	-
A. shermani Heidemann	-	YK	-	-	-	-	SK	-	ON	QC	-	-	-	-	-
A. signaticornis Sahlberg	AK	YK	-	-	-	-	-	-	-	-	-	-	-	-	-
A. similis Say	AK	-	-	-	BC	-	-	-	ON	QC	-	-	NS	-	-
A. taylori Van Duzee	-	-	-	-	BC	-	-	-	-	-	-	-	-	-	-
A. tuberculifer Kirby	AK	YK	NT	-	BC	AB	-	MB	ON	QC	-	-	-	NF	-
A. uniannulatus Parshley	-	YK	NT	-	-	AB	-	MB	ON	QC	-	-	-	-	-
A. uniformis Heidemann	-	-	-	-	-	-	-	-	ON	-	-	-	-	-	-
A. vadosus Van Duzee	-	-	-	-	BC	AB	-	-	-	-	-	-	-	-	-

(Subgenus QUILNUS Stål)

A. heidemanni Bergroth	-	-	-	-	BC	AB	-	-	-	-	-	-	-	-	-
A. niger Stål	-	-	-	-	-	-	-	-	ON	QC	NB	-	NS	-	-
A. nigrinus canadensis Parshley	-	-	-	-	-	AB	-	-	-	-	-	-	-	-	-

Subfamily MEZIRINAE

Genus MEZIRA Amyot & Serville

Key to species: Usinger 1936.

M. pacifica Usinger	-	-	-	-	BC	-	-	-	-	-	-	-	-	-	-	[r]

Superfamily Coreoidea

Family ALYDIDAE

Key to genera: Stål 1868 [New World].

Broad headed bugs. Plant feeders. Elongate, with slender body; broad and prominent head with ocelli; bucculae not extending beyond base of antennae; membrane of forewing with numerous veins; metathoracic scent-glands usually with conspicuous lateral opening. Larvae are ant mimics.

Subfamily ALYDINAE

Key to taxa: Fracker 1918.

Genus ALYDUS Fabricius

A. calcaratus (Linnaeus)	AK	-	-	-	BC	AB	SK	MB	-	QC	-	-	-	-	-
A. conspersus conspersus Montandon	-	-	-	-	BC	AB	SK	MB	ON	QC	NB	-	-	-	-
A. conspersus infuscatus Fracker	-	-	-	-	BC	AB	-	-	-	-	-	-	-	-	-
A. eurinus eurinus (Say)	AK	YK	NT	-	BC	AB	SK	MB	ON	QC	NB	PE	NS	-	-
A. eurinus obesus Fraker	-	-	-	-	-	AB	-	-	-	-	-	-	-	-	-
A. scutellatus Van Duzee	-	-	-	-	BC	AB	-	-	-	-	-	-	-	-	-

Genus MEGALOTOMUS Fieber

M. quinquespinosus (Say) - - - - BC AB SK MB ON QC - - - - -

Genus STACHYOCNEMUS Stål

S. apicalis (Dallas) - - - - - AB - - - - - - - - -

Genus TOLLIUS Stål

T. curtulus (Stål) - - - - BC AB SK - - - - - - - -
T. setosus Van Duzee - - - - BC - - - - - - - - - - [r]

Subfamily MICROELYTRINAE

Genus PROTENOR Stål

P. belfragei Haglund - - - - - - SK MB ON QC NB PE - - -

Family COREIDAE

Key to subfamilies and tribes: Packauskas 1994 [New World].

Key to genera: Stål 1868 [New World].

Squash bugs and allies. Phytophagous, often pests. Body broad, robust, brownish; head with ocelli; bucculae extending posteriorly beyond base of antennae; membrane of forewing with many veins; metathoracic scent-glands with large openings laterally; hind femur often with spines; hind tibia sometimes with lateral expansions.

Subfamily COREINAE

Tribe Acanthocephalini

Genus ACANTHOCEPHALA Laporte

(Subgenus METAPODIESSA Kirkaldy)

A. terminalis (Dallas) - - - - - - - - ON - - - - - - [r]

Tribe Acanthocerini

Generic revision: O'Shea 1980.

Genus EUTHOCHTHA Mayr

E. galeator (Fabricius) - - - - - - - - ON - - - - - -

Tribe Anisoscelidini

Key to genera: Hussey 1953, Osuna 1984.

Genus LEPTOGLOSSUS Guérin-Méneville

Key to species: Allen 1969 [North America].

L. occidentalis Heidemann - - - - BC AB SK - ON* - NB* - - - - [F]
western conifer-seed bug

Tribe Chelinideini

Genus CHELINIDEA Uhler

Key to species: Herring 1980.

C. vittigera Uhler - - - - - AB SK - - @ - - - - -

* introduced; [] indoors; { } extirpated; ? uncertain record; @ unestablished adventive; $ port-of-entry interception

Tribe Coreini

Genus ANASA Amyot & Serville
Key to species: Brailovsky 1985.

A. armigera (Say) - - - - - - - - ON QC - - - - - [A]
A. tristis (DeGeer) - - - - BC - - - ON QC NB PE - - - [A]
squash bug; punaise de la courge

Genus CATORHINTHA Stål

C. mendica Stål - - - - - - - MB ON - - - - - -

Genus NISOSCOLOPOCERUS Barber

N. apiculatus Barber - - - - - AB SK MB - - - - - - - [r]

Subfamily MEROPACHYDINAE

Key to tribes, genera and species: Kormilev 1954.

Tribe Merocorini

Genus MEROCORIS Perty

M. distinctus Dallas - - - - - AB SK MB ON QC - - - - -

Subfamily PSEUDOPHLOEINAE

Genus CERALEPTUS Costa
Key to species: Froeschner 1963 [Western Hemisphere].

C. denticulatus Froeschner - - - - BC - - - - - - - - - -
C. pacificus Barber - - - - BC - - - - - - - - - -

Genus CORIOMERIS Westwood
Key to species: Dolling & Yonke 1976.

C. humilis (Uhler) AK YK NT - BC AB SK MB ON QC - - - - -
C. insularis Dolling & Yonke - - - - BC - - - - - - - - - - [r]

Family RHOPALIDAE

Catalogue of species: Göllner-Scheiding 1983.
Key to subfamilies, tribes and genera: Chopra 1967.
Key to genera and species: Hoebeke & Wheeler 1982 [eastern North America].

Scentless plant bugs. Phytophagous. Head with large raised ocelli; corium with closed cells; membrane with numerous veins; metathoracic scent-gland openings absent or very small.

Subfamily RHOPALINAE

Tribe Harmostini

Genus AUFEIUS Stål

A. impressicollis Stål - - - - BC - - - - - - - - - - [r]

Genus HARMOSTES Burmeister
Key to species: Göllner-Scheiding 1978.
(Subgenus HARMOSTES Burmeister)

H. dorsalis Burmeister - - - - BC - - - - - - - - - - [r]
H. reflexulus (Say) - - - - BC AB SK - ON QC - - - - -

Tribe Neisthreini

Genus **ARHYSSUS** Stål

Key to species: Chopra 1968.

A. barberi Harris	-	-	-	-	BC	-	-	-	-	-	-	-	-	-	-	[r]
A. crassus Harris	-	-	-	-	BC	-	-	-	-	-	-	-	-	-	-	
A. distinctus Chopra	-	-	-	-	BC	-	-	-	-	-	-	-	-	-	-	[r]
A. lateralis (Say)	-	-	-	-	-	-	SK	MB	ON	QC	NB	-	-	-	-	
A. nigristernum (Signoret)	-	-	-	-	-	-	-	-	ON	QC	-	-	-	-	-	
A. scutatus (Stål)	-	-	-	-	BC	AB	-	-	-	-	-	-	-	-	-	
A. validus (Uhler)	-	-	-	-	BC	-	-	-	-	-	-	-	-	-	-	[r]

Tribe Rhopalini

Genus **LIORHYSSUS** Stål

Key to species: Göllner-Scheiding 1976.

L. hyalinus (Fabricius)	-	-	-	-	BC	-	-	MB	ON	-	-	-	-	-	-

Genus **STICTOPLEURUS** Stål

Key to species: Göllner-Scheiding 1975.

S. plutonius (Baker)	-	-	-	-	BC	-	-	-	-	-	-	-	-	-	-
S. punctiventris (Dallas)	AK	YK	NT	-	BC	AB	SK	MB	ON	QC	NB	PE	NS	NF	-
S. viridicatus (Uhler)	-	YK	-	-	BC	AB	SK	MB	-	-	-	-	-	-	-

Subfamily SERINETHINAE

Genus **BOISEA** Kirkaldy

B. rubrolineata (Barber) western boxelder bug	-	-	-	-	BC	-	-	-	-	-	-	-	-	-	-
B. trivittata (Say) boxelder bug; punaise de l'érable négondo	-	-	-	-	-	AB	SK	MB	ON	-	-	-	-	-	-

Superfamily Lygaeoidea

Revised classification of superfamily and key to families: Henry 1997a.

Catalogue of taxa: Slater 1964 [as Lygaeidae sensu lato, excluding Berytidae; world].

Regional guide: Béique & Robert 1963, 1964 [as Lygaeidae *sensu lato*, Québec].

Family ARTHENEIDAE

Head with ocelli, but these not circled by groove; pronotum with lateral margins explanate or with a wide flattened carina; hemelytral membrane with 4-5 simple veins; tarsi 3-segmented; abdominal spiracles on segment 2 dorsal, on segments 3 and 4 ventral; trichobothria present on abdominal sterna 4 and 5; male abdominal sternum 7 without clusters or comb of setae ventrally.

Genus **CHILACIS** Fieber

C. typhae (Perris)	-	-	-	-	BC*	-	-	-	ON*	-	-	-	-	-	-

* introduced; [] indoors; { } extirpated; ? uncertain record; @ unestablished adventive; $ port-of-entry interception

Family BERYTIDAE

Key to genera: Henry 1997b [world].

Key to genera and species: Scudder 1991 [Canada].

Stilt bugs. Phytophagous, but sometimes predaceous. Body very slender and elongate; head with ocelli nearly encircled by distinct groove; antennae long, 4-segmented, with first segment elongate, slender, often clavate, subequal in thickness to and nearly always longer than segments 2 and 3; rostrum 4-segmented; legs elongate; abdominal spiracles dorsal.

Subfamily BERYTINAE

Genus BERYTINUS Kirkaldy

B. minor (Herrich-Schaeffer) - - - - - - - - ON* QC* NB* PE* NS* NF* -

Genus NEONEIDES Stusak

N. muticus (Say) - - - - BC AB SK MB ON QC NB - NS - -

Subfamily GAMPSOCORINAE

Genus HOPLINUS Stål

H. echinata (Uhler) - - - - BC - - - - - - - - - - [r]

Subfamily METACANTHINAE

Genus JALYSUS Stål

J. spinosus (Say) - - - - - - - - ON - - - - - -

J. wickhami Van Duzee - - - - BC - - MB ON QC - - - - -

Family BLISSIDAE

Key to genera: Slater 1979 [world].

Head with ocelli not encircled by a groove; pronotal calli without impressed transverse grooves; scutellum without a cross-shaped carina; membrane with 4-5 veins; spiracles on abdominal segment 8 ventral; abdominal terga transverse, sutures never curved forward.

Genus BLISSUS Burmeister

Key to species: Leonard 1968 [eastern North America].

B. arenarius Barber - - - - - - - - - - NB PE NS - -

B. canadensis Leonard - - - - - AB SK MB - - - - - - -

B. leucopterus hirtus Montandon - - - - - - - - ON QC NB PE NS NF - [A]
hairy cinch bug; punaise velue

B. occiduus Barber - - - - BC - - - - - - - - - - [A]
western cinch bug; punaise de l'ouest

Genus ISCHNODEMUS Fieber

Key to species: Slater 1979.

I. falicus (Say) - - - - - - - - ON QC NB - NS - -

I. hesperius Parshley - - - - - - - MB - - - - - - -

Family CYMIDAE

Key to taxa: Hamid 1975 [world].

Head with ocelli nearly encircled by distinct furrow; bucculae short; first antennal segment short, stout and barrel-shaped; hemelytra distinctly punctate on clavus and corium; corium opaque throughout, never notched; hemelytral membrane with 4 or 5 simple veins; trichobothria present on abdominal sterna 2 to 7; abdominal spiracles dorsal; tarsi with 3 segments.

Tribe Cymini

Genus CYMUS Hahn

Key to species: Hamid 1975.

C. angustatus Stål	-	-	-	-	-	-	-	MB	ON	QC	NB	PE	NS	-	-
C. coriacipennis (Stål)	-	-	-	-	BC	-	-	-	-	-	-	-	-	-	-
C. discors Horváth	-	-	-	-	-	-	-	MB	ON	QC	NB	-	NS	-	-
C. luridus Stål	-	-	NT	-	BC	AB	SK	MB	ON	QC	NB	PE	NS	NF	-
C. robustus Barber	-	-	-	-	-	-	-	-	ON	-	-	-	-	-	-

Family GEOCORIDAE

Key to genera: McAttee 1914, Readio & Sweet 1982 [Nearctic].

Head with prominent eyes; ocelli not encircled by a groove; hemelytral membrane with 4 or 5 simple veins; spiracles of abdominal segments 3 and 4 dorsal, those of segments 5 and 6 ventral; sutures between terga 4/5 and 5/6 curved forward in middle.

Genus GEOCORIS Fallén

Key to species: Readio & Sweet 1982 [eastern North America].

G. atricolor Montandon	-	-	-	-	BC	AB	-	-	-	-	-	-	-	-	-
G. bullatus (Say)	AK	YK	NT	-	BC	AB	SK	MB	ON	QC	NB	-	NS	NF	LB
G. discopterus Stål	-	YK	-	-	BC	AB	SK	-	-	-	-	-	-	-	-
G. howardi Montandon	AK	YK	NT	-	BC	AB	-	-	-	-	-	-	-	-	-
G. limbatus Stål	-	-	-	-	BC	AB	SK	MB	ON	QC	-	-	-	-	-
G. lividipennis Stål	-	-	-	-	BC?	-	-	-	-	-	-	-	-	-	-
G. pallens Stål	-	-	-	-	BC	-	-	-	-	-	-	-	-	-	-
G. punctipes (Say)	-	-	-	-	@	-	-	-	-	-	-	-	-	-	-
G. uliginosus (Say)	-	-	-	-	-	-	-	-	ON	QC	-	-	-	NF	-

Genus ISTHMOCORIS McAtee

Key to species: McAtee 1914, Readio & Sweet 1982.

I. piceus (Say)	-	-	-	-	-	-	-	-	ON	-	-	-	-	-	-	[r]

Family HETEROGASTRIDAE

Key to genera: Scudder 1962b [world].

Head without trichobothria; fore femur weakly incrassate; hemelytral membrane with 4 or 5 simple veins, with a distinct closed cell; all abdominal spiracles ventral.

Genus HETEROGASTER Schilling

H. behrensii (Uhler)	-	-	-	-	BC	-	-	-	-	-	-	-	-	-	-	[r]

* introduced; [] indoors; { } extirpated; ? uncertain record; @ unestablished adventive; $ port-of-entry interception

Family LYGAEIDAE

Catalogue: Slater 1964, Slater & O'Donnell 1995 [Lygaeidae sensu lato; world].

Head with each ocelli not encircled by groove; each pronotal callus with an impressed, transverse, usually shiny groove; scutellum usually with a cross-shaped carina; spiracles of abdominal segments 2 to 7 dorsal; all abdominal terga transverse, sutures never curved forward; trichobothria present on sterna 4 and 5.

Subfamily ISCHNORHYNCHINAE

Key to genera: Scudder 1962c [world].

Genus KLEIDOCERYS Stephens

Key to species: Barber 1953 [North America], Scudder 1962b [world].

K. franciscanus (Stål)	-	-	-	-	BC	-	-	-	-	-	-	-	-	-	-	
K. modestus Barber	-	-	-	-	BC	-	-	-	-	-	-	-	-	-	-	
K. ovalis Barber	-	-	-	-	BC	-	-	-	ON	-	-	-	-	-	-	
K. resedae fuscomaculatus Barber	-	-	-	-	BC	-	-	-	-	-	-	-	-	-	-	
K. resedae geminatus (Say) birch catkin bug; punaise des chatons du bouleau	-	-	-	-	-	-	-	-	ON	QC	-	-	NS	-	-	
K. resedae resedae (Panzer)	AK	YK	NT	-	BC	AB	SK	MB	ON	QC	NB	-	NS	NF	LB	

Subfamily LYGAEINAE

Key to genera: Ashlock 1975, Brailovsky 1978a [North America].

Genus LYGAEOSPILUS Barber

Key to species: Scudder 1981.

L. brevipilus Scudder	-	-	-	-	BC	-	-	-	-	-	-	-	-	-	-	[r]
L. tripunctatus (Dallas)	-	-	-	-	-	-	SK	MB	-	QC	-	-	-	-	-	

Genus LYGAEUS Fabricius

Key to species: Brailovsky 1978b [Western Hemisphere], Torre-Bueno 1946 [USA].

L. kalmii angustomarginatus Parshley	-	-	-	-	-	-	-	MB	ON	QC	-	-	NS	-	-	
L. kalmii kalmii Stål	-	-	-	-	BC	AB	SK	MB	-	-	-	-	-	-	-	
L. turcicus Fabricius	-	-	-	-	-	-	-	-	ON	QC	-	-	-	-	-	

Genus MELACORYPHUS Slater

M. lateralis (Dallas)	-	-	-	-	BC	-	SK	-	-	-	-	-	-	-	-	

Genus MELANOPLEURUS Stål

Key to species: Scudder 1981.

M. perplexus Scudder	-	-	-	-	BC	AB	SK	MB	-	-	-	-	-	-	-	
M. pyrropterus (Stål)	-	-	-	-	-	AB	-	-	-	-	-	-	-	-	-	[r]

Genus NEACORYPHUS Scudder

N. bicrucis (Say)	-	-	-	-	BC	-	-	-	ON	-	-	-	-	-	-	

Genus ONCOPELTUS Stål

Key to species: Torre-Bueno 1946 [North America].

(Subgenus ERYTHRISCHIUS Stål)

O. fasciatus (Dallas) large milkweed bug; punaise de l'asclépiade	-	-	-	-	-	-	-	-	ON	QC	-	-	-	-	-	

Subfamily ORSILLINAE

Key to tribes and genera: Ashlock 1967 [world].

Tribe Metrargini

Genus XYONYSIUS Ashlock & Lattin

Key to species: Barber 1947.

X. californicus alabamensis (Baker)	-	-	-	-	-	AB	SK	MB	ON	-	-	-	-	-	-	
X. californicus californicus (Stål)	-	-	-	-	BC	-	-	-	-	-	-	-	-	-	-	

Tribe Nysiini

Genus NYSIUS Dallas

Key to species: Barber 1947 [North America].

N. angustatus Uhler	-	-	NT	-	BC	AB	SK	MB	ON	QC	NB	-	-	-	-	
N. fuscovittatus Barber	AK	YK	-	-	BC	AB	-	-	-	-	-	-	-	-	-	
N. grandis Baker	-	-	-	-	-	AB	-	MB	-	-	-	-	-	-	-	
N. groenlandicus (Zetterstedt)	AK	YK	-	-	-	-	-	MB	ON	QC	-	PE	-	NF	LB	
N. niger Baker northern false cinch bug	AK	YK	NT	-	BC	AB	SK	MB	ON	QC	NB	-	-	NF	-	
N. paludicola Barber	-	-	-	-	BC	-	-	-	-	-	-	-	-	-	-	[r]
N. raphanus Howard	-	-	-	-	BC	AB	SK	-	ON	-	-	-	-	-	-	
N. tenellus Barber	-	-	-	-	BC	-	-	-	-	-	-	-	-	-	-	
N. thymi (Wolff)	AK	YK	-	-	BC	AB	SK	MB	ON	QC	-	-	-	-	-	

Tribe Orsillini

Genus BELONOCHILUS Uhler

B. numenius (Say)	-	-	-	-	-	-	-	-	ON	-	-	-	-	-	-	[r]

Genus NEORTHOLOMUS Hamilton

Key to species: S.W. Hamilton 1983.

N. scolopax (Say)	-	-	NT	-	BC	AB	SK	MB	ON	QC	NB	PE	NS	-	-	

Family OXYCARENIDAE

Key to genera: Barber 1917, Torre Bueno 1946 [North America].

Head with each ocellus not encircled by groove; pronotum with lateral margins rounded or at most weakly carinate; membrane with 4 or 5 veins; female with abdomen truncate caudally; male with transverse cones or clusters of setae on sternum 7; spiracles of abdominal segments 3 and 4 ventral, that on segment 2 dorsal.

Genus CROPHIUS Stål

Key to species: Barber 1938.

C. angustatus Van Duzee	-	-	-	-	BC	-	-	-	-	-	-	-	-	-	-
C. bohemani (Stål)	-	-	-	-	BC	-	-	-	-	-	-	-	-	-	-
C. disconotus (Say)	-	YK	NT	-	BC	AB	SK	MB	ON	QC	NB	-	-	-	LB
C. ramosus Barber	AK	YK	-	-	BC	AB	SK	-	-	-	-	-	-	-	-

* introduced; [] indoors; { } extirpated; ? uncertain record; @ unestablished adventive; $ port-of-entry interception

Family PACHYGRONTHIDAE

Key to taxa: Slater 1955 [world].

Head without trichobothria; fore femur strongly incrassate; membrane with 4 or 5 veins, without closed basal cell; abdominal spiracles ventral; female with ovipositor dividing sterna 6 and 7.

Tribe Pachygronthini

Genus OEDANCALA Amyot & Serville

O. dorsalis (Say)	-	-	-	-	-	-	-	-	ON	QC	-	-	-	-	-

Tribe Teracriini

Genus PHLEGYAS Stål

P. abbreviatus (Uhler)	-	-	-	-	-	-	-	MB	ON	QC	NB	-	NS	-	-
P. annulicrus Stål	-	-	-	-	BC	-	-	-	-	-	-	-	-	-	-

Family RHYPAROCHROMIDAE

Key to tribes: Sweet 1967.

Head with trichobothria; membrane with 4 or 5 veins; spiracle of abdominal segment 2 ventral; female with ovipositor dividing sternum 7 only; suture between sterna 4 and 5 usually curving anteriorly and ending before lateral abdominal margin.

Subfamily PLINTHISINAE

Tribe Plinthisini

Genus PLINTHISUS Stephens
Key to species: Barber 1918.

P. americanus Van Duzee	-	-	-	-	-	AB	SK	MB	ON	QC	NB	-	-	-	-	
P. indentatus Barber	-	-	-	-	-	-	SK	-	-	-	-	-	-	-	-	[r]

Subfamily RHYPAROCHROMINAE

Tribe Antillocorini

Key to genera: Slater 1980 [New World].

Genus ANTILLOCORIS Kirkaldy
Key to species: Barber 1952a.

A. minutus (Bergroth)	-	-	-	-	-	-	-	-	ON	QC	NB	-	NS	NF	-
A. pilosulus (Stål)	-	-	-	-	-	-	-	-	ON	QC	-	-	-	-	-

Genus BOTOCUDO Kirkaldy

B. modestus (Barber)	-	-	-	-	BC	-	-	-	-	-	-	-	-	-	-	[r]

Tribe Drymini

Key to genera: Ashlock 1979 [North America].

Genus DRYMUS Fieber

D. crassus Van Duzee	-	-	-	-	-	-	-	-	ON	QC	-	-	-	-	-
D. unus (Say)	-	-	-	-	-	-	SK	MB	ON	QC	NB	PE	NS	-	-

Genus **EREMOCORIS** Fieber

Key to species: Sweet 1977 [eastern North America]; Walley 1929 [North America].

E. borealis (Dallas)	AK	YK	-	-	BC	AB	SK	MB	ON	QC	NB	-	NS	NF	LB	
E. canadensis Walley	-	-	-	-	BC	-	-	-	-	-	-	-	-	-	-	[r]
E. ferus (Say)	-	-	-	-	BC	-	-	-	ON	QC	-	-	NS	-	-	
E. melanotus Walley	-	-	-	-	BC	-	-	-	-	-	-	-	-	-	-	
E. obscurus Van Duzee	-	-	-	-	BC	-	-	-	-	-	-	-	-	-	-	
E. setosus Blatchley	-	-	-	-	-	-	-	-	-	QC	-	-	-	-	-	[r]

Genus **GASTRODES** Westwood

Key to species: Usinger 1938.

G. grossipes (DeGeer)	-	-	-	-	-	-	-	-	$	-	-	-	-	-	-	
G. intermedius Usinger	-	-	-	-	BC	-	-	-	-	-	-	-	-	-	-	[r]
G. pacificus (Provancher)	-	-	-	-	BC	-	-	-	-	-	-	-	-	-	-	
G. walleyi Usinger	-	-	-	-	-	-	-	-	ON	-	-	-	-	-	-	[r]

Genus **SCOLOPOSTETHUS** Fieber

S. atlanticus Horváth	-	-	-	-	BC	-	-	MB	ON	QC	-	-	-	NF	-	
S. diffidens Horváth	-	-	-	-	BC	-	-	-	ON	QC	-	-	NS	-	-	
S. pacificus Barber	-	-	-	-	BC	-	-	-	-	-	-	-	-	-	-	
S. thomsoni Reuter	AK	YK	NT	-	BC	AB	SK	MB	ON	QC	NB	PE	NS	NF	LB	
S. tropicus (Distant)	-	-	-	-	BC	-	-	-	-	-	-	-	-	-	-	[r]

Tribe Gonianotini

Key to genera: Kiritshenko & Scudder 1973 [world].

Genus **ATRAZONOTUS** Slater & Ashlock

A. umbrosus (Distant)	-	-	NT	-	-	-	SK	MB	ON	-	-	-	-	-	-	

Genus **EMBLETHIS** Fieber

E. vicarius Horváth	AK	YK	NT	-	BC	AB	SK	MB	ON	QC	-	-	-	-	-	

Genus **MALEZONOTUS** Barber

Key to species: Ashlock 1958.

M. angustatus (Van Duzee)	-	-	-	-	BC	-	-	-	-	-	-	-	-	-	-	
M. arcuatus Ashlock	-	-	-	-	BC	-	-	-	-	-	-	-	-	-	-	[r]
M. grossus Van Duzee	-	-	-	-	BC	-	-	-	-	-	-	-	-	-	-	[r]
M. sodalicius (Uhler)	-	-	-	-	BC	-	-	-	-	-	-	-	-	-	-	

Genus **TRAPEZONOTUS** Fieber

T. arenarius (Linnaeus)	-	YK	NT	-	BC	AB	SK	MB	ON	QC	-	-	-	-	-	

Tribe Lethaeini

Key to taxa: O'Donnell 1986 [western hemisphere].

Genus **CRYPHULA** Stål

Key to species: Scudder 1962a.

C. trimaculata (Distant)	-	-	-	-	-	-	-	-	ON	-	-	-	-	-	-	[r]

Genus **XESTOCORIS** Van Duzee

X. nitens Van Duzee	-	-	-	-	-	-	-	-	ON	QC	-	-	-	NF	-	

Tribe Megalonotini

Genus **LAMPRODEMA** Fieber

L. maurum (Fabricius)	-	-	-	-	-	-	-	-	-	-	-	-	NS*	-	-	

* introduced; [] indoors; { } extirpated; ? uncertain record; @ unestablished adventive; $ port-of-entry interception

Genus MEGALONOTUS Fieber

M. sabulicola (Thomson) - - - - BC* - - - ON* - - - - - -

Genus SPHRAGISTICUS Stål

S. nebulosus (Fallén) AK YK NT - BC AB SK MB ON QC - - NS - -

Tribe Myodochini

Key to genera: Harrington 1980 [world].

Genus CARPILIS Stål

C. consimilis Barber - - - - - - - - - QC - - - - - [r]

Genus FROESCHNERIA Harrington

F. piligera (Stål) - - - - - - - - ON - - - - - -

Genus HERAEUS Stål

H. plebejus Stål - - - - - - - - ON QC - - - - -

Genus KOLENETRUS Barber

K. plenus (Distant) - YK - - BC AB SK - ON QC - - - - -

Genus LIGYROCORIS Stål

Key to species: Sweet 1986 [eastern North America].

L. caricis Sweet - - - - - - - - - QC - - - - -
L. depictus Barber - - - - - - - - - QC - - - - -
L. diffusus (Uhler) - YK NT - BC AB SK MB ON QC NB - NS NF -
L. latimarginatus Barber - - - - BC - - - - - - - - - -
L. sylvestris (Linnaeus) AK YK NT - BC AB SK MB ON QC NB PE NS NF LB

Genus MYODOCHA Latreille

Key to species: Blatchley 1926, Torre-Bueno 1946.

M. serripes Olivier - - - - - - - - ON QC - - - - -

Genus NEOPAMERA Harrington

Key to species: Barber 1952b [as *Pachybrachius*].

N. albocincta (Barber) - - - - - - - - ON - - - - - - [r]
N. bilobata bilobata (Say) - - - - - - - - ON - - - - - - [r]

Genus PACHYBRACHIUS Hahn

Key to species: Barber 1952b.

P. fracticollis (Schilling) - - - - - - - - - QC* - - - - -
P. luridus Hahn - - - - - - - - - QC* - - - - -

Genus PERIGENES Distant

Key to species: Torre-Bueno 1946.

P. constrictus (Say) AK - - - - - - - ON QC - - NS - -

Genus PSEUDOCNEMODUS Barber

P. canadensis (Provancher) - - - - BC - - - ON QC - - - - -

Genus PSEUDOPACHYBRACHIUS Malipatil
Key to species: Zheng & Slater 1984 [world].

P. basalis (Dallas)	-	-	-	-	-	-	-	-	ON	QC	-	-	-	-	-

Genus SISAMNES Distant

S. claviger (Uhler)	-	-	-	-	BC	-	-	-	-	-	-	-	-	-	-

Genus SLATEROBIUS Harrington
Key to species: Slater *et al.* 1993.

S. insignis (Uhler)	AK	YK	NT	-	BC	AB	SK	MB	ON	QC	-	-	NS	NF	LB

Genus ZERIDONEUS Barber

Z. costalis (Van Duzee)	-	-	-	-	BC	AB	SK	MB	ON	QC	-	-	-	-	-

Tribe Ozophorini

Key to genera: Ashlock & Slater 1982 [Western Hemisphere].

Genus OZOPHORA Uhler
Key to species: Slater 1988 [western USA].

O. occidentalis Slater	-	-	-	-	BC	-	-	-	-	-	-	-	-	-	-	[r]
O. picturata Uhler	-	-	-	-	-	-	-	-	ON	-	-	-	-	-	-	[r]

Tribe Rhyparochromini

Key to genera: Scudder 1984, Sweet 1991 [North America].

Genus CORDILLONOTUS Scudder

C. stellatus Scudder	-	-	-	-	BC	-	-	-	-	-	-	-	-	-	-	[r]

Genus PERITRECHUS Fieber
Key to species: Scudder 1999 [North America].

P. convivus Stål	AK	YK	NT	-	BC	AB	SK	MB	ON	QC	NB	PE	NS	NF	-	[A]
P. fraternus Uhler	-	-	-	-	BC	AB	SK	MB	ON	QC	-	-	-	-	-	
P. pilosulus Scudder	-	-	-	-	BC	-	-	-	-	-	-	-	-	-	-	
P. tristis Van Duzee	-	-	-	-	BC	-	-	-	-	-	-	-	-	-	-	

Genus UHLERIOLA Horváth

U. floralis (Uhler)	-	-	-	-	-	AB	SK	MB	-	-	-	-	-	-	-

Tribe Stygnocorini

Genus STYGNOCORIS Douglas & Scott
Key to species: Southwood & Leston 1959.

S. rusticus (Fallén)	-	-	-	-	BC*	AB*	-	-	ON*	QC*	-	PE*	NS*	NF*	-
S. sabulosus (Schilling)	-	-	-	-	BC*	AB*	-	-	ON*	QC*	NB*	PE*	NS*	NF*	-

Tribe Udeocorini

Genus NEOSURIS Barber
Key to species: Brailovsky 1978c.

N. castanea (Barber)	-	-	-	-	BC	-	-	-	-	-	-	-	-	-	-	[r]

* introduced; [] indoors; { } extirpated; ? uncertain record; @ unestablished adventive; $ port-of-entry interception

Superfamily Piesmatoidea

Family PIESMATIDAE

Key to genera: Drake & Davis 1958.

Ash-gray leaf bugs. Phytophagous. Head without ocelli; forewing with an irregular network of cells; tarsi with 2 segments; abdominal spiracles dorsal; abdomen with trichobothria lacking on sterna 4 and 5.

Subfamily PIESMATINAE

Genus PIESMA Lepeletier & Serville

Key to species: Drake & Davis 1958.

P. cinereum (Say)	-	YK	-	-	BC	AB	SK	MB	ON	QC	-	-	-	-	-	[A]
P. costatum (Uhler)	-	-	-	-	-	AB	-	-	-	-	-	-	-	-	-	
P. explanatum McAtee	-	-	-	-	BC	AB	SK	MB	-	-	-	-	-	-	-	
P. patruele McAtee	-	-	-	-	-	AB	-	-	-	-	-	-	-	-	-	

Superfamily Pentatomoidea

Keys and family diagnoses: Rolston & McDonald 1979 [Western Hemisphere].

Key to taxa: McPherson 1982 [eastern North America].

Family ACANTHOSOMATIDAE

Key to genera and species: Thomas 1991.

Acanthosomatids. Phytophagous stinkbugs. Head keeled laterally; antennae 5-segmented; antenniferous tubercles not visible from above; scutellum large and triangular, usually about half or less than half as long as abdomen; mesosternum with a prominent median keel; tarsi with 2 segments.

Subfamily ACANTHOSOMATINAE

Genus ELASMOSTETHUS Fieber

E. atricornis (Van Duzee)	-	-	-	-	-	-	-	-	ON	QC	-	-	-	-	-
E. cruciatus (Say) redcrossed stink bug; punaise à croix rouge	-	-	NT	-	BC	AB	SK	MB	ON	QC	NB	PE	NS	NF	LB
E. interstinctus (Linnaeus)	AK	YK	NT	-	BC	-	-	-	-	-	-	-	-	-	-

Genus ELASMUCHA Stål

E. lateralis (Say)	AK	YK	NT	-	BC	AB	SK	MB	ON	QC	NB	PE	NS	NF	LB

Family CYDNIDAE

Key to taxa: Froeschner 1960 [Western Hemisphere].

Burrowing bugs. Phytophagous, usually feeding on roots of plants. Black or reddish with body ovoid and convex; antennae 5-segmented; fore tibia flattened, spiny, adapted for digging; middle and hind tibiae with stout spines; middle and hind coxae with fringe of stiff bristles at tips; ostiolar groove elongate.

Subfamily AMNESTINAE

Genus AMNESTUS Dallas

A. pallidus Zimmer	-	-	-	-	BC	-	-	-	ON	QC	-	-	-	-	-	
A. pusillus Uhler	-	-	-	-	-	-	-	-	ON	QC	-	-	-	-	-	
A. spinifrons (Say)	-	-	-	-	-	-	-	-	ON	QC	-	-	-	-	-	

Subfamily CYDNINAE

Genus DALLASIELLUS Berg

(Subgenus PSEUDOPANGAEUS Froeschner)

D. discrepans (Uhler)	-	-	-	-	BC	-	-	-	-	-	-	-	-	-	-	[r]

Genus MACROPORUS Uhler

M. repetitus Uhler	-	-	-	-	BC	-	-	-	-	-	-	-	-	-	-	[r]

Genus MELANAETHUS Uhler

M. punctatissimus (Signoret)	AK	-	-	-	-	-	-	-	-	-	-	-	-	-	-	
M. robustus Uhler	-	-	-	-	-	-	-	-	ON	-	-	-	-	-	-	[r]

Genus MICROPORUS Uhler

M. obliquus Uhler	-	-	-	-	BC	AB	SK	-	-	-	-	-	-	-	-	[r]

Genus PANGAEUS Stål

(Subgenus HOMALOPORUS Uhler)

P. bilineatus (Say)	-	-	-	-	-	-	-	-	-	QC	-	-	-	-	-	[r]

Subfamily SEHIRINAE

Genus SEHIRUS Amyot & Serville

S. cinctus albonotatus Dallas	-	-	NT	-	BC	AB	SK	MB	ON	QC	NB	PE	NS	NF	-	

* introduced; [] indoors; { } extirpated; ? uncertain record; @ unestablished adventive; $ port-of-entry interception

Family PENTATOMIDAE

Keys and diagnoses for subfamilies: Rolston & McDonald 1979 [Western Hemisphere].

Stink bugs. Phytophagous or predaceous (subfamily Asopinae). Variously coloured; body usually broad and ovoid; antennae 5-segmented; scutellum large, usually triangular or subtriangular; claval commissure reduced or absent; mesosternum usually lacking a median carina; tibiae without rows of thick black spines; tarsi 3-segmented.

Subfamily ASOPINAE

Key to genera: Schouteden 1907 [world].

Key to genera and species: Thomas 1992 [Western Hemisphere].

Genus APATETICUS Dallas

(Subgenus APOECILUS Stål)

A. bracteatus (Fitch)	-	-	-	-	BC	AB	SK	MB	ON	QC	NB	PE	NS	-	-
A. cynicus (Say)	-	-	-	-	-	-	-	MB	ON	QC	-	PE	-	NF	-

Genus PERILLUS Stål

P. bioculatus (Fabricius) twospotted stink bug; punaise bimaculée	-	-	-	-	BC	AB	SK	MB	ON	QC	-	-	-	-	-
P. circumcinctus Stål	-	-	-	-	-	-	SK	MB	ON	QC	-	-	-	-	-
P. exaptus (Say)	-	-	-	-	BC	AB	SK	MB	ON	QC	NB	-	NS	NF	-

Genus PICROMERUS Amyot & Serville

P. bidens (Linnaeus)	-	-	-	-	-	-	-	-	ON*	QC*	NB*	PE*	NS*	-	-

Genus PODISUS Herrich-Schaeffer

(Subgenus PODISUS Herrich-Schaeffer)

P. brevispinus Phillips	-	-	-	-	BC	AB	SK	MB	ON	QC	NB	PE	NS	NF	-
P. maculiventris (Say) spined soldier bug	-	-	-	-	-	-	-	MB	ON	QC	NB	PE	NS	NF	-
P. placidus Uhler	-	-	-	-	BC	AB	SK	MB	ON	QC	-	-	-	-	-
P. serieventris Uhler	-	-	NT	-	BC	AB	SK	MB	ON	QC	NB	PE	NS	NF	-

Genus RHACOGNATHUS Fieber

R. americanus Stål	-	-	NT	-	-	AB	-	MB	ON	-	-	-	-	-	-

Genus ZICRONA Amyot & Serville

Z. caerulea (Linnaeus)	AK	-	-	-	BC	AB	SK	MB	ON	QC	NB	-	-	-	-

Subfamily PENTATOMINAE

Keys and diagnoses of tribes: Rolston & McDonald 1979 [Western Hemisphere].

Tribe Hyalini

Genus BROCHYMENA Amyot & Serville

Key to species: Larivière 1992.

B. affinis Van Duzee	-	-	-	-	BC	-	-	-	-	-	-	-	-	-	-
B. quadripustulata (Fabricius) fourhumped stink bug; punaise bossue des bois	-	-	-	-	BC	AB	SK	MB	ON	QC	NB	-	NS	-	-

Genus PARABROCHYMENA Larivière

Key to species: Larivière 1994b.

P. arborea (Say)	-	-	-	-	-	-	-	-	ON	QC	-	-	-	-	-

Tribe Pentatomini

Key to genera: Rolston *et al.*1980, Rolston & McDonald 1981, 1984 [Western Hemisphere].

Genus ACROSTERNUM Fieber

Key to species: Rolston 1983 [subgenus Chinavia].

(Subgenus CHINAVIA Orian)

A. hilare (Say)	-	-	-	-	BC	-	-	-	ON	QC	-	-	-	-	-	
A. pensylvanicum (Gmelin)	-	-	-	-	-	-	-	-	ON	QC	-	-	-	-	-	

Genus AELIA Fabricius

A. americana Dallas	AK	YK	NT	-	BC	AB	SK	MB	ON	-	-	-	-	-	-	

Genus BANASA Stål

Key to species: Thomas & Yonke 1981 [Nearctic].

B. calva (Say)	-	-	-	-	BC?	-	-	-	ON	QC	-	-	-	-	-	
B. dimidiata (Say) banasa stink bug; punaise du cormier	-	-	-	-	BC	AB	SK	MB	ON	QC	NB	PE	NS	-	-	
B. sordida (Uhler)	-	-	-	-	BC	-	-	-	ON	-	-	-	-	-	-	

Genus CHLOROCHROA Stål

Key to species: Scudder & Thomas 1987 [Canada].

(Subgenus CHLOROCHROA Stål)

C. congrua Uhler	-	-	-	-	-	AB	-	-	-	-	-	-	-	-	-	
C. granulosa (Uhler)	AK	YK	NT	-	BC	AB	SK	MB	-	-	-	-	-	-	-	
C. ligata (Say)	-	-	NT	-	BC	AB	SK	MB	-	-	-	-	-	-	-	
C. opuntiae Esselbaugh	-	-	-	-	-	AB	-	-	-	-	-	-	-	-	-	[r]
C. persimilis Horváth	-	-	-	-	-	AB?	-	MB	ON	QC	NB	PE	NS	NF	-	
C. rossiana Buxton & Thomas	-	-	-	-	BC	-	-	-	-	-	-	-	-	-	-	
C. uhleri (Stål)	-	-	-	-	BC	AB	SK	MB	-	-	-	-	-	-	-	[A]

(Subgenus RHYTIDOLOMIA Stål)

C. belfragii (Stål)	-	-	-	-	-	-	-	MB	-	-	-	-	-	-	-	[r]
C. faceta (Say)	-	-	-	-	-	AB	SK	-	-	-	-	-	-	-	-	[r]
C. viridicata (Walker)	-	-	-	-	-	AB	-	-	-	-	-	-	-	-	-	[r]

Genus CODOPHILA Mulsant & Rey

Key to species: Thomas 1974.

(Subgenus ANTHEMINIA Mulsant & Rey)

C. remota (Horváth)	-	YK	NT	-	BC	AB	SK	MB	-	-	-	-	-	-	-	

Genus COENUS Dallas

Key to species: Rider 1995.

C. delius (Say)	-	-	-	-	BC	AB	SK	MB	ON	QC	NB	PE	-	-	-	

Genus COSMOPEPLA Stål

Key to species: McDonald 1986.

C. bimaculata (Thomas)	-	-	NT	-	BC	AB	SK	MB	ON	QC	NB	PE	NS	NF	-	[A]
C. conspicillaris (Dallas)	-	-	-	-	BC	-	-	-	-	-	-	-	-	-	-	
C. intergressus (Uhler)	-	-	-	-	BC	-	-	-	-	-	-	-	-	-	-	

Genus DENDROCORIS Bergroth

Key to species: Nelson 1955.

D. pini Montandon	-	-	-	-	BC	-	-	-	-	-	-	-	-	-	-	[r]

* introduced; [] indoors; { } extirpated; ? uncertain record; @ unestablished adventive; $ port-of-entry interception

Genus EUSCHISTUS Dallas

(Subgenus EUSCHISTUS Dallas)

E. conspersus Uhler	-	-	-	-	BC	-	-	-	-	-	-	-	-	-	-	
E. ictericus (Linnaeus)	-	-	-	-	-	-	-	-	ON	QC	-	-	-	-	-	
E. politus Uhler	-	-	-	-	-	-	-	-	ON	QC	-	-	-	-	-	
E. servus euschistoides (Vollenhoven)	-	-	-	-	BC	AB	SK	MB	ON	QC	-	-	NS	-	-	[A]
E. tristigmus luridus Dallas dusky stink bug; punaise à trois taches	-	-	-	-	BC	AB	SK	MB	ON	QC	NB	PE	NS	NF	-	
E. variolarius (Palisot) onespotted stink bug; punaise variée	-	-	-	-	BC	AB	-	-	ON	QC	-	-	-	-	-	[A]

Genus HOLCOSTETHUS Fieber

Key to species: McDonald 1975, 1982.

H. abbreviatus Uhler	-	-	-	-	BC	AB	SK	MB	ON	QC	-	-	-	-	-
H. fulvipes (Ruckes)	-	-	-	-	-	-	-	MB	ON	QC	-	-	-	-	-
H. limbolarius (Stål)	-	-	-	-	BC	AB	SK	MB	ON	QC	-	PE	NS	-	-
H. piceus (Dallas)	-	-	NT	-	BC	AB	SK	MB	ON	QC	-	-	-	-	-
H. tristis (Van Duzee)	-	-	-	-	BC	AB	-	-	-	-	-	-	-	-	-

Genus HYMENARCYS Amyot & Serville

Key to species: Rolston 1973b.

H. aequalis (Say)	-	-	-	-	-	-	SK	-	-	QC	-	-	-	-	-
H. nervosa (Say)	-	-	-	-	-	-	-	-	-	QC	-	-	-	-	-

Genus MENECLES Stål

Key to species: Rolston 1973a.

M. insertus (Say)	-	-	-	-	-	-	-	-	ON	QC	-	-	NS	-	-

Genus MORMIDEA Amyot & Serville

Key to species: Rolston 1978.

(Subgenus MELANOCHILA Stål)

M. lugens (Fabricius)	-	-	-	-	-	-	-	MB	ON	QC	NB	PE	NS	-	-

Genus MURGANTIA Stål

M. histrionica (Hahn) harlequin bug; punaise arlequine	-	-	-	-	-	-	-	-	ON	-	-	-	-	-	-

Genus NEOTTIGLOSSA Kirby

(Subgenus NEOTTIGLOSSA Kirby)

N. trilineata (Kirby)	AK	YK	NT	-	BC	AB	SK	MB	ON	QC	-	-	NS	NF	-
N. undata (Say)	AK	YK	NT	-	BC	AB	SK	MB	ON	QC	NB	PE	NS	NF	-

(Subgenus TEXAS Kirkaldy)

N. tumidifrons Downes	-	-	-	-	BC	-	-	-	-	-	-	-	-	-	-

Genus NEZARA Amyot & Serville

N. viridula (Linnaeus)	-	-	-	-	-	-	-	-	@	$	-	-	-	-	-

Genus PRIONOSOMA Uhler

P. podopioides Uhler	-	-	-	-	BC	-	-	-	-	-	-	-	-	-	-

Genus TEPA Rolston & McDonald

Key to species: Rider 1986.

T. rugulosa (Say)	-	-	-	-	BC	AB	SK	-	-	-	-	-	-	-	-

A agricultural, F forestry, G greenhouse, O ornamental pest; r rare; b biocontrol agent

Genus **THYANTA** Stål

Key to species: Rider & Chapin 1992.

T. accerra McAtee	-	-	-	-	-	-	-	MB	ON	QC	-	-	-	-	-	
T. custator (Fabricius)	-	-	-	-	-	-	-	-	ON	-	-	-	-	-	-	
T. pallidovirens (Stål)	-	-	-	-	BC	-	-	-	-	-	-	-	-	-	-	

Genus **TRICHOPEPLA** Stål

Key to species: McDonald 1976.

T. atricornis Stål	AK	-	-	-	BC	AB	SK	MB	-	-	-	-	-	-	-	
T. aurora Van Duzee	-	-	-	-	BC	-	-	-	-	-	-	-	-	-	-	
T. dubia (Dallas)	-	-	-	-	BC	AB	-	-	-	-	-	-	-	-	-	
T. semivittata (Say)	-	-	-	-	-	-	-	-	ON	QC	-	-	-	-	-	

Tribe Sciocorini

Genus **SCIOCORIS** Fallén

S. microphthalmus Flor	-	YK	NT	-	BC	AB	SK	MB	ON	QC	-	-	-	-	-	

Subfamily PODOPINAE

Genus **AMAUROCHROUS** Stål

Key to species: Barber & Sailor 1953.

A. brevitylus Barber & Sailer	-	-	-	-	-	-	-	-	-	QC	-	-	-	-	-	
A. cinctipes (Say)	-	-	-	-	-	-	-	-	ON	QC	-	-	-	-	-	

Family SCUTELLERIDAE

Key to taxa: Lattin 1964 [North America].

Scutellerids, shield bugs. Phytophagous stink bugs. Dull brownish, with scutellum greatly enlarged, U-shaped, and usually entirely covering abdomen; antennae 5-segmented; membrane of forewing with many veins; prosternum with sulcus; tarsi with 3 segments.

Subfamily EURYGASTRINAE

Genus **EUPTYCHODERA** Bergroth

E. corrugata (Van Duzee)	-	-	-	-	-	AB	-	-	-	-	-	-	-	-	-	[r]

Genus **EURYGASTER** Laporte

E. alternata (Say)	-	-	NT	-	BC	AB	SK	MB	ON	QC	NB	-	NS	-	-	
E. amerinda Bliven	-	-	NT	-	BC	AB	SK	MB	ON	QC	-	-	-	-	-	

Genus **FOKKERIA** Schouteden

F. producta (Van Duzee)	-	-	-	-	-	AB	-	-	-	-	-	-	-	-	-	[r]

Genus **PHIMODERA** Germar

P. binotata (Say)	-	YK	-	-	BC	AB	SK	MB	-	-	-	-	-	-	-	

Genus **VANDUZEEINA** Schouteden

Key to species: Usinger 1930.

V. balli (Van Duzee)	-	-	-	-	BC	AB	-	-	-	-	-	-	-	-	-	
V. borealis Van Duzee	-	YK	-	-	BC	AB	-	MB	ON	-	-	-	-	-	-	

* introduced; [] indoors; { } extirpated; ? uncertain record; @ unestablished adventive; $ port-of-entry interception

Subfamily PACHYCORINAE

Genus ACANTHOLOMIDEA Sailer

A. denticulata (Stål)	-	-	-	-	-	-	-	-	ON	-	-	-	-	-	-	[r]
A. porosus (Germar)	-	-	-	-	BC	-	-	-	-	-	-	-	-	-	-	[r]

Genus HOMAEMUS Dallas

H. aeneifrons aeneifrons (Say)	-	-	-	-	-	-	-	MB	ON	QC	NB	PE	NS	NF	-	
H. aeneifrons consors Uhler	AK	YK	NT	-	BC	AB	SK	-	-	-	-	-	-	-	-	
H. bijugis Uhler	-	-	-	-	BC	AB	SK	MB	-	-	-	-	-	-	-	
H. parvulus (Germar)	-	-	-	-	BC	-	-	-	-	-	-	-	-	-	-	[r]

Genus TETYRA Fabricius

T. bipunctata (Herrich-Schaeffer) shieldbacked pine seed bug	-	-	-	-	-	-	-	-	ON	QC	-	-	-	-	-

Family THYREOCORIDAE

Key to genera and species: McAtee & Malloch 1933.

Negro bugs. Phytophagous. Black, oval and usually shiny; scutellum large, U-shaped, covering most of forewings and abdomen; claval commissure absent; tibiae with two rows of thick black spines; fore tibia not expanded and modified for digging.

Genus CORIMELAENA White

(Subgenus CORIMELAENA White)

C. alpina (McAtee & Malloch)	-	-	-	-	-	-	-	-	-	QC	NB	-	-	-	-
C. lateralis (Fabricius)	-	-	-	-	-	-	-	-	ON	QC	-	-	-	-	-
C. nigra Dallas	-	YK	-	-	BC	AB	SK	MB	ON	QC	NB	-	NS	-	-
C. obscura McPherson & Sailer	-	-	-	-	-	-	-	-	ON	-	-	-	-	-	-
C. pulicaria (Germar)	-	-	-	-	BC	AB	SK	MB	ON	QC	-	-	NS	-	-

(Subgenus PARAPORA McAtee & Malloch)

C. extensa Uhler	-	-	-	-	BC	-	-	-	-	-	-	-	-	-	-
C. incognita (McAtee & Malloch)	-	-	-	-	BC	-	-	-	-	-	-	-	-	-	-

Genus GALGUPHA Amyot & Serville

(Subgenus GALGUPHA Amyot & Serville)

G. aterrima Malloch	-	-	-	-	-	-	-	-	ON	QC	NB	-	NS	-	-
G. atra Amyot & Serville	-	-	-	-	-	-	SK	MB	ON	QC	-	-	NS	NF	-
G. ovalis Hussey	-	-	-	-	BC	AB	-	-	-	-	-	-	-	-	-

(Subgenus NOTHOCORIS McAtee & Malloch)

G. nitiduloides nitiduloides (Wolff)	-	-	-	-	BC	AB	SK	MB	ON	QC	-	-	NS	-	-

References

Allen, R.C. 1969. A revision of the genus *Leptoglossus* Guerin (Hemiptera: Coreidae). Entomologica Americana (New Series) 45:35-140.

Alleyne, E.H. & Morrison, F.O. 1977. Some Canadian poplar aphid galls. The Canadian Entomologist 109:321-328

Andersen, N.M. 1982. The Semiaquatic Bugs (Hemiptera, Gerromorpha): Phylogeny, Adaptations, Biogeography, and Classification. Entomonograph 13. Scandinavian Science Press, Klampenborg, Denmark, 455 pp.

Andersen, N.M. 1990. Phylogeny and taxonomy of water striders, genus *Aquarius* Schellenberg (Insecta, Hemiptera, Gerridae), with a new species from Australia. Steenstrupia 16:37-81.

Andersen, N.M. & Spence, J.R. 1992. Classification and phylogeny of the Holarctic water strider genus *Limnoporus* Stål (Hemiptera, Gerridae). Canadian Journal of Zoology 70:753-785.

Anderson, L.D. 1932. A monograph on the genus *Metrobates* (Hemiptera, Gerridae). University Kansas Science Bulletin 20:297-311.

Annand, P.N. 1928. A contribution toward a monograph of the Adelginae (Phylloxeridae) of North America. Stanford University Publications, University Series, Biological Sciences 6(1). 146 pp.

Archibald, K.D. 1958. The Forest Aphidae of Nova Scotia. Proceedings of the Nova Scotia Institute of Science 24:1-254

Ashlock, P.D. 1958. A revision of the genus *Malezonotus* (Hemiptera-Heteroptera: Lygaeidae). Annals of the Entomological Society of America 51:199-208.

Ashlock, P.D. 1967. A generic classification of the Orsillinae of the World (Hemiptera-Heteroptera: Lygaeidae). University of California Publications in Entomology 48:1-82.

Ashlock, P.D. 1975. Towards a classification of North American Lygaeinae (Hemiptera-Heteroptera: Lygaeidae). Journal of the Kansas Entomological Society 48:27-32.

Ashlock, P.D. 1979. A new *Eremocoris* from California with a key to North American genera of Drymini (Hemiptera: Heteroptera: Lygaeidae). Pan-Pacific Entomologist 55:149-154.

Ashlock, P.D. & Slater, J.A. 1982. A review of the genera of Western Hemisphere Ozophorini with two new genera from Central America (Hemiptera-Heteroptera: Lygaeidae). Journal of the Kansas Entomological Society 55:737-750.

Asquith, A. 1991. A taxonomic revision of the genus *Lopidea* (Heteroptera: Miridae: Orthotylinae) in America north of Mexico. Theses Zoologicae 16:1-281 Koeltz Science Books, Koenigstein.

Bacon, J.A. 1956. A taxonomic study of the genus *Rhagovelia* (Hemiptera, Veliidae) of the Western Hemisphere. University Kansas Science Bulletin 38:695-913.

Bailey, N.S. 1951. The Tingoidea of New England and their biology. Entomologica Americana (New Series) 31:1-140.

Barber, H.G. 1917. Synoptic keys to the Lygaeidae (Hemiptera) of the United States. Part 1. Psyche 24:128-135.

Barber, H.G. 1918. The genus *Plinthisus* Latr. (Lygaeidae-Hemiptera) in the United States. Proceedings of the Entomological Society of Washington 20:108-111.

Barber, H.G. 1924. A new species of *Pselliopus* (Hemiptera: Reduviidae). Proceedings of the Entomological Society of Washington 26:211-213.

Barber, H.G. 1930. Essay on the subfamily Stenopodinae of the New World. Entomologica Americana (New Series) 10:149-238.

Barber, H.G. 1938. A review of the genus *Crophius* Stål, with descriptions of three new species (Hemiptera-Heteroptera: Lygaeidae). Journal of the New York Entomological Society 46:313-319.

Barber, H.G. 1947. Revision of the genus *Nysius* in the United States and Canada (Hemiptera-Heteroptera: Lygaeidae). Journal of the Washington Academy of Sciences 37:354-366.

Barber, H.G. 1952a. The genus *Antillocoris* Kirk. in the United States (Hemiptera-Lygaeidae) Bulletin of the Brooklyn Entomological Society 47:85-87.

Barber, H.G. 1952b. The genus *Pachybrachius* in the United States and Canada with the description of two new species (Hemiptera: Lygaeidae). Journal of the New York Entomological Society 60:211-220.

Barber, H.G. 1953. A revision of the genus *Kleidocerys* Stephens in the United States (Hemiptera: Lygaeidae). Proceedings of the Entomological Society of Washington 55:273-283.

Barber, H.G. & Sailer, R.I. 1953. A revision of the turtle bugs of North America (Hemiptera: Pentatomidae). Proceedings of the Entomological Society of Washington 43:150-162.

Béique, R. & Robert, A. 1963. Les Lygéides de la Province de Québec (Hétéroptères). Annales de la Société Entomologique du Québec 8:70-96.

Béique, R. & Robert, A. 1964. Les Lygéides de la Province de Québec (Hétéroptères) (2e partie). Annales de la Société Entomologique du Québec 9:72-104.

Beirne, B.P. 1956. Leafhoppers (Homoptera: Cicadellidae) of Canada and Alaska. The Canadian Entomologist 88, supplement 2. 180 pp.

Beirne, B.P. 1961. The Cicadas (Homoptera: Cicadidae) and treehoppers (Homoptera: Membracidae) of Canada. Canada Department of Agriculture, mimeograph.

Belton, E.M. & Eidt, D.C. 1999. Common names of insects in Canada [revised]. Entomological Society of Canada, Ottawa. MS-DOS version, http://www.biology.ualberta.ca/esc.hp/homepage.htm

Ben-Dov, Y. 1993. A systematic catalogue of the soft scale insects of the world (Homoptera: Coccoidea: Coccidae) with data on geographical distribution, host plants, biology and economic importance. Flora and Fauna Handbook 9. Sandhill Crane Press. Gainesville, Florida. xxviii+536 pp.

Ben-Dov, Y. 1994. A systematic catalogue of the mealybugs of the world (Insecta: Homoptera: Coccoidea: Pseudococcidae and Putoidae) with data on geographical distribution, host plants, biology and economic importance. Intercept Limited, Andover, UK. 686 pp.

Ben-Dov, Y., Miller, D.R. & Gibson, G. 1996 et seq. ScaleNet. http://www.sel.barc.usda.gov/scalenet/scalenet.htm

Bissel, T. 1978. Aphids on Juglandaceae in North America. Maryland Agriculture Experiment Station MP911. 78 pp.

Blackman, R.L. 1977. The existance of two species of *Euceraphis* (Homoptera: Aphididae) on birch in Western Europe, and a key to European and North American species of the genus. Systematic Entomology 2:1-8

Blackman, R.L. 1987. Morphological discrimination of a tobacco-feeding form from *Myzus persicae* Sulzer (Hemiptera: Aphididae), and a key to New World *Myzus (Nectarosiphon)* species. Bulletin of entomological Research 77:713-730

Blackman, R.L. & Eastop, V.F. 1984. Aphids on the world's crops. Wiley. Chichester, UK. 466 pp.

Blackman, R.L. & Eastop, V.F. 1994. Aphids on the world's trees, an identification and information guide. CAB International. Wallingford, UK. viii+987 pp. +16 plates.

Blatchley, W.S. 1926. Heteroptera or True Bugs of Eastern North America, with especial reference to the Faunas of Indiana and Florida. Nature Publishing Company, Indiana, 1116 pp.

Börner, C. 1952. Europae centralis Aphides, die Blattläuse Mitteleuropas [IX. Schriftverzeichnis]. Mitteilungen der Thüringischen Botanischen Gesellschaft 3(2):372-453

Brailovsky, H. 1978a. Un nuevo género una nueva especie de Lygaeinae (Hemiptera-Heteroptera-Lygaeidae) del Peru. Anales de Instituto de Biologia Universidad Nacional Autónoma México 49 (Serie Zoologia 1):115-122.

Brailovsky, H. 1978b. Estudio del género *Lygaeus* Fabricius 1794, del nuevo mundo, con descripción de cinco nuevae especies (Hemiptera-Heteroptera-Lygaeidae-Lygaeinae). Anales de Instituto de Biologia Universidad Nacional Autónoma México 49 (Serie Zoologia 1):123-166.

Brailovsky, H. 1978c. Contribución al estudio de los Hemiptera-Heteroptera de México: XIV. Una nueva especie de *Neosuris* Barber (Lygaeidae: Udeocorini). Anales de Instituto de Biologia Universidad Nacional Autónoma México 49 (Serie Zoologia 1):167-170.

Brailovsky, H. 1985. Revision del genero *Anasa* Amyot-Serville (Hemiptera-Heteroptera-Coreidae-Coreinae-Coreini). Monografias del Instituto de Biologia del Unversidad Nacional Autónoma de Mexico 2:1-266.

Brooks, A.R. & Kelton, L.A. 1967. Aquatic and semiaquatic Heteroptera of Alberta, Saskatchewan, and Manitoba (Hemiptera). Memoirs of the Entomological Society of Canada 51:1-92.

Brown, P.A. 1989. Keys to the alate *Aphis* (Homoptera) of northern Europe. Occasional Papers in Systematic Entomology 5:1-29

Bullington, S.W. & Kosztarab, M. 1985. Revision of the family Kermesidae (Homoptera) in the Nearctic Region based on adult and third instar females. Studies on the morphology and systematics of scale Insects 12. Virginia Polytechnic Institute and State University Agricultural Experiment Station Bulletin 85-11:i-xii+1-118

Bullington, S.W., Kosztarab, M. & Jiang Guang-Zao 1989. Adult males of the genus *Chionaspis* (Homoptera: Coccoidea: Diaspididae) in North America. Studies on the morphology and systematics of scale Insects 15. Virginia Polytechnic Institute and State University Agriculture Experiment Station Bulletin 88-2:127-198

Burckhardt, D. 1987. Jumping plant lice (Homoptera: Psylloidea) of the temperate neotropical region. Part 1: Psyllidae (subfamilies Aphalarinae, Rhinocolinae and Aphalaroidinae). Zoological Journal of the Linnean Society 89:299-392

Burnham, J. B. 1938. A contribution to a list of the Aphididae of the Maritime Provinces of Canada. The Canadian Entomologist 70:180-188

Carter, C.I. 1971. Conifer Woolly Aphids (Adelgidae) in Britain. Forestry Commision Bulletin 42. 51 pp.

Carvalho, J.C.M. 1955. Keys to the genera of Miridae of the World (Hemiptera). Boletim do Museu Paraense Emilio Goeldi 11:1-151.

Carvalho, J.C.M. & Fontes, A.V. 1981. Mirideos neotropicais CCXXV: Revisäo do genero *Collaria* Provancher no continente Americano (Hemiptera). Experientiae 27:11-46.

Carvalho, J.C.M. & Southwood, T.R.E. 1955. Revisão do complexo *Cyrtorhinus* Fieber-*Mecomma* Fieber (Hemiptera-Heteroptera, Miridae). Boletim do Museu Paraense Emilio Goeldi 11:7-72.

Cassis, G. 1984. A systematic study of the subfamily Dicyphinae (Heteroptera: Miridae) Ph.D. thesis, Oregon State University, 389 pp.

Caudell, A.N. 1901. The genus *Sinea* of Amyot & Serville. Journal of the New York Entomological Society 9:1-11.

Chagnon, G. & Fournier, O. 1948. Contribution à l'étude des Hémiptères aquatiques du Québec. Contributions de l'Institute biologique de l'Université de Montréal 21:66 pp.

Chakrabarti, S & Medda, P.K. 1985. A review of the genus *Ceruraphis* (Homoptera: Aphididae). Oriental Insects 19:125-132

Cheng, L. & Fernando, C.H. 1970. The water-striders of Ontario (Heteroptera: Gerridae). Life Sciences Miscellaneous Publications of the Royal Ontario Museum 1970:1-23.

Chopra, N.P. 1967. The higher classification of the family Rhopalidae (Hemiptera). Transactions of the Royal Entomological Society of London 119:363-399.

Chopra, N.P. 1968. A revision of the genus *Arhyssus* Stål. Annals of the Entomological Society of America 61:629-655.

Cook, E.F. 1984a. *Aphis* (Homoptera: Aphididae) recorded from Compositae in North America, with a key to the species east of the Rocky Mountains and comments on synonymy and redescriptions of some little known forms. Annals of the Entomological Society of America 77:442-449

Cook, E.F. 1984b. *Glabromyzus* and *Utamphorophora* (Homoptera: Aphididae) species of North America. Annals of the Entomological Society of America 77:705-711

Corpuz-Raros, L.A. & Cook, E.F. 1974. A revision of North American *Capitophorus* Van der Goot and *Pleotrichophorus* Börner. Smithsonian Contributions in Zoology 156:1-143

Costa Lima, A., Hathaway, C.R. & Seabra, C.C.A. 1948. Sôbre algumas espécies de Apiomerinae representadas nas nossas coleçoes (Hemiptera: Reduviidae: Apiomerinae). Memórias do Instituto Oswaldo Cruz 45:761-772.

Crawford, D.L. 1914. A monograph of the jumping plant-lice or Psyllidae of the New World. United States National Museum Smithsonian Institution Bulletin 85. x+185 pp.

Dolling, W.R. & Yonke, T.R. 1976. The genus *Coriomeris* in North America. Annals of the Entomological Society of America 69:1147-1152.

Downes, W. 1927. A preliminary list of the Heteroptera and Homoptera of British Columbia. Proceedings of the Entomological Society of British Columbia 23:1-22.

Downes, W. 1935. Additions to the list of B.C. Hemiptera. Proceedings of the Entomological Society of British Columbia 31(1934):46-48.

Dozier, H.L. 1926. The Fulgoridae or plant-hoppers of Mississippi, including those of possible occurrence. Mississippi Agricultural Experiment Station Technical Bulletin 14.

Drake, C.J. 1917. Key to the Nearctic species of *Gargaphia* with the description of a new species (Hem., Heter.). Entomological News 28:227-228.

Drake, C.J. & Chapman, H.C. 1953. Preliminary report on the Pleidae (Hemiptera) of the Americas. Proceedings of the Biological Society of Washington 66:53-59.

Drake, C.J. & Davis, N.T. 1958. The morphology and systematics of the Piesmatidae (Hemiptera), with keys to world genera and American species. Annals of the Entomological Society of America 51:567-581.

Drake, C.J. & Davis, N.T. 1960. The morphology, phylogeny, and higher classification of the family Tingidae, including description of a new genus and species of the subfamily Vianaidinae (Hemiptera: Heteroptera). Entomologica Americana (New Series) 243:1-100.

Drake, C.J. & Harris, H.M. 1934. The Gerrinae of the Western Hemisphere (Hemiptera). Annals of the Carnegie Museum 23:179-240.

Drake, C.J. & Lattin, J.D. 1963. American species of the lacebug genus *Acalypta* (Hemiptera: Tingidae). Proceedings of the United States National Museum 115:331-345.

Drake, C.J. & Ruhoff, F.A. 1965. Lacebugs of the World, a catalog (Hemiptera: Tingidae). Bulletin of the United States National Museum 243:1-634.

Eastop, V.F. 1961. A key for the determination of *Schizaphis* Börner. Entomologist 94:241-246

Eastop, V.F. 1971. Keys for the identification of *Acrythosiphon* (Hemiptera: Aphididae). Bulletin of the British Museum of natural History (Entomology) 26(1):1-115

Eastop, V.F. & D. Hille Ris Lambers. 1976. Survey of the World's Aphids. Dr. W. Junk b.v. The Hague. 573 pp.

Elemjanov, A.F. 1979. Problema razgranichenija semejstv Fulgoridae i Dictyophoridae (Homoptera, Auchenorrhyncha) [The problem of family distinction between the Fulgoridae and the Dictyophoridae (Homoptera, Auchenorrhyncha)]. Trudy Zoologicheskogo Instituta A.N. C.C.C.P. 82:3-22

Essig, E.O. & Abernathy, F. 1952. The aphid genus *Periphyllus*. A systematic, biological, and ecological study. University of California Press, Berkley and Los Angeles. 166 pp

Eyles, A.C. & Carvalho, J.C.M. 1975. Revision of the genus *Dolichomiris*, with a revised key to the genera of Stenodemini (Heteroptera: Miridae). Journal of Natural History 9:257-269.

Fennah, R.G. 1954. The higher classification of the family Issidae (Homoptera: Fulgoroidea) with descriptions of new species. Transactions of the Royal Entomological Society of London 105:455-474

Ferris, G.F. 1937. Atlas of the scale insects of North America. Series 1. Stanford University Press.

Ferris, G.F. 1938. Atlas of the scale insects of North America. Series II. Stanford University Press.

Ferris, G.F. 1941. Atlas of the scale insects of North America. Series III. Stanford University Press.

Ferris, G.F. 1942. Atlas of the scale insects of North America. Series IV. Stanford University Press.

Ferris, G.F. 1950. Atlas of the scale insects of North America. Vol. V. Stanford University Press.

Ferris, G.F. 1953. Atlas of the scale insects of North America. Vol. VI. Stanford University Press.

Ferris, G.F. 1955. Atlas of the scale insects of North America. Vol. VII. Stanford University Press.

Foottit, R.G. & Maw, E. 1997. Aphids of the Yukon. pp. 388-404 *in* Danks, H.V. and Downes, J.A. (Eds.) Insects of the Yukon. Biological Survey of Canada (Terrestrial Arthropods), Ottawa. 1034 pp.

Foottit, R.G. & Richards, W.R. 1993. The Insects and Arachnids of Canada. Part 22. The Genera of the Aphids of Canada (Aphidodea: Aphidoidea and Phylloxeroidea). Agriculture Canada Research Branch. Ottawa. Publication 1885. 766 pp.

Forbes, A.R. & Chan, C.-K. 1989a. Aphids of British Columbia. Agriculture Canada Research Branch Technical Bulletin 1989-1E. 260 pp.

Forbes, A.R. & Chan, C.-K. 1989b. The aphids (Homoptera: Aphididae) of British Columbia. 21. Further additions. Journal of the Entomological Society of British Columbia 86:82-88

Forbes, A.R. & Chan, C.-K. 1991. The aphids (Homoptera: Aphididae) of British Columbia. 20. Further additions. Journal of the Entomological Society of British Columbia 88:7-14

Forbes, A.R. & Chan, C.-K. 1993. The aphids (Homoptera: Aphididae) of British Columbia. 21. Further additions. Journal of the Entomological Society of British Columbia 90:72-82

Fracker, S.B. 1912. A systematic outline of the Reduviidae of North America. Proceedings of the Iowa Academy of Science, 19: 217-252.

Fracker, S.B. 1918. The Alydinae of the United States. Annals of the Entomological Society of America 11:255-280.

Froeschner, R.C. 1960. Cydnidae of the Western Hemisphere. Proceeding of the United States National Museum 111:337-680.

Froeschner, R.C. 1963. The genus *Ceraleptus* Costa in the Western Hemisphere (Hemiptera: Coreidae). Journal of the Kansas Entomological Society 36:109-113.

Gibson, E.H. 1918. The genus *Corythucha* Stål (Tingidae; Heteroptera). Transactions of the American Entomological Society 44:69-104.

Göllner-Scheiding, U. 1975. Revision der Galtung *Stictopleurus* Stål, 1872 (Heteroptera, Rhopalidae) Deutsches Entomologisches Zeitung. 22:1-60.

Göllner-Scheiding, U. 1976. Revision der Gattung *Liorhyssus* Stål, 1870 (Heteroptera, Rhopalidae). Deutsche Entomologische Zeitschrift 23: 181-206.

Göllner-Scheiding, U. 1978. Revision der Gattung *Harmostes* Burm., 1835 (Heteroptera, Rhopalidae) und einige Bermerkungen zu den Rhopalinae. Mitteilungen aus dem Zoologische Museum in Berlin 54:257-311.

Göllner-Scheiding, U. 1983. General-Katalog der Familie Rhopalidae (Heteroptera). Mitteilungen aus dem Zoologische Museum in Berlin. 59:37-189.

Gualtieri, L. & McCleod, D.G.R. 1994. Atlas of aphids trapped in agricultural crops. Agriculture and Agri-food Canada Research Branch Publication 1901/E. 66 pp. / Atlas des pucerons piégés dans les champs agricoles. Publication 1901/F. 66 pp.

Guldemond, J. A. 1991. Biosystematic and morphometric study of the *Cryptomyzus galeopsidis/alboapicalis* complex (Homoptera, Aphididae), with a key to and notes on the *Crypotomyzus* species of Europe. Netherlands Journal of Zoology 41:1-31

Halbert, S.E. 1991. A new species of *Anoecia* (Homoptera: Aphididae) on rhizomes of *Equisetum laevigatum*. Proceedings of the Entomological Society of Washington 93:760-766

Hamid, A. 1975. A systematic revision of the Cyminae (Heteroptera: Lygaeidae) of the World with a description of the morphology, biology, phylogeny and zoogeography. Occasional Publications of the Entomological Society of Nigeria 14:1-179.

Hamilton, K.G.A. 1982. The spittlebugs of Canada (Homoptera: Cercopidae). The Insects and arachnids of Canada, 10.

Hamilton, K.G.A. 1983. Introduced and native leafhoppers common to the old and new worlds (Rhynchota: Homoptera: Cicadellidae). The Canadian Entomologist 115: 473-511.

Hamilton, K.G.A. 1985. Leafhoppers of ornamental and fruit trees in Canada. Agriculture Canada publication 1779.

Hamilton, S.W. 1983. *Neortholomus*, a new genus of Orsillini (Hemiptera-Heteroptera: Lygaeidae: Orsillinae). University of Kansas Science Bulletin 52:197-234.

Harper, A.M. 1959. Gall aphids on poplar in Alberta I. Descriptions of galls and distributions of aphids. The Canadian Entomologist 91:489-495

Harper, A.M. & Bradley, G.A. 1987. Aphids of Alberta. Agriculture Canada Research Branch Technical Bulletin 1987-7E. 24 pp.

Harrington, B.J. 1980. A generic level revision and cladistic analysis of the Myodochini of the World (Hemiptera: Lygaeidae: Rhyparochrominae). Bulletin of the American Museum of Natural History 167:45-116.

Harris, H.M. 1928. A monographic study of the hemipterous family Nabidae as it occurs in North America. Entomologica Americana 9:1-97.

Hart, E.R. 1986. Genus *Zelus* Fabricius in the United States, Canada, and Northern Mexico (Hemiptera: Reduviidae). Annals of the Entomological Society of America 79:535-548.

Heie, O.E. 1979. Revision of the aphid genus *Nasonovia* Mordwilko, including *Kakimia* Hottes and Frison, with keys and descriptions of the species of the world (Homoptera: Aphididae). Entomologica scandinavica, Supplementum 9:1-105

Heie, O.E. 1980. The Aphidoidea (Hemiptera) of Fennoscandia and Denmark. 1. General Part. The families Mindaridae, Hormaphididae, Thelaxidae, Anoeciidae, and Pemphigidae. Fauna entomologica scandinavica 9. 1-236

Heie, O.E. 1982. The Aphidoidea (Hemiptera) of Fennoscandia and Denmark. 2. The family Drepanosiphidae. Fauna entomologica scandinavica 11:1-176

Heie, O.E. 1986. The Aphidoidea (Hemiptera) of Fennoscandia and Denmark. 3. Family Aphididae: subfamily Pterocommatinae: tribe Aphidini of subfamily Aphidinae. Fauna entomologica scandinavica 17:1-314

Heie, O.E. 1992. The Aphidoidea (Hemiptera) of Fennoscandia and Denmark. 4. Family Aphididae: Part 1 of tribe Macrosiphini of subfamily Aphidinae. Fauna entomologica scandinavica 25:1-188

Heie, O.E. 1994. The Aphidoidea (Hemiptera) of Fennoscandia and Denmark. 5. Family Aphididae: Part 2 of tribe Macrosiphini of subfamily Aphidinae. Fauna entomologica scandinavica 28:1-242

Heie, O.E. 1995. The Aphidoidea (Hemiptera) of Fennoscandia and Denmark. 6. Family Aphididae: Part 3 of tribe Macrosiphini of subfamily Aphidinae, and family Lachnidae. Fauna entomologica scandinavica 31:1-217

Henry, T.J. 1976. Review of the genus *Reuteria* Puton 1875, with descriptions of two new species (Hemiptera: Miridae). Entomological News 87:61-74.

Henry, T.J. 1979a. Review of the *Ceratocapsus lutescens* group, with descriptions of seven new species from the eastern United States (Hemiptera: Miridae) Proceedings of the Entomological Society of Washington 81:401-423.

Henry, T.J. 1979b. Review of the New World species of *Myiomma* with descriptions of eight new species (Hemiptera: Miridae: Isometopinae). Proceedings of the Entomological Society of Washington 81:552-569.

Henry, T.J. 1982a. Genus *Parthenicus* in the eastern United States, with descriptions of new species (Hemiptera: Miridae). Florida Entomologist 65:354-366.

Henry, T.J. 1982b. The onion plant bug genus *Labopidicola* (Hemiptera: Miridae): Economic implications, taxonomic review, and description of a new species. Proceedings of the Entomological Society of Washington 84:1-15.

Henry, T.J. 1983. The garden fleahopper genus *Halticus* (Hemiptera: Miridae): resurrection of an old name and key to species of the Western Hemisphere. Proceedings of the Entomological Society of Washington 85:607-611.

Henry, T.J. 1991. *Melanotrichus whiteheadi*, a new crucifer-feeding plant bug from the southeastern United States, with new records for the genus and a key to the species of eastern North America (Heteroptera: Miridae: Orthotylinae). Proceedings of the Entomological Society of Washington 93:449-456.

Henry, T.J. 1994. Revision of the myremocomorphic plant bug genus *Schaffneria* Knight (Heteroptera: Miridae: Orthotylinae). Proceedings of the Entomological Society of Washington 96:701-712.

Henry, T.J. 1997a. Phylogenetic analysis of family groups within the Infraorder Pentatomomorpha (Hemiptera: Heteroptera), with emphasis on the Lygaeoidea. Annals of the Entomological Society of America 90:275-301.

Henry, T.J. 1997b. Cladistic analysis and revision of the stilt bug genera of the world (Heteroptera: Berytidae). Contributions to the American Entomological Institute 30(1):1-100.

Henry, T.J. & Froeschner, R.C. 1988. Catalog of the Heteroptera, or True Bugs, of Canada and the Continental United States. E.J. Brill. Leiden, New York, København, Köln, xx+958 pp.

Henry, T.J. & Herring J.L. 1979. Review of the genus *Corticoris* with descriptions of two new species from Mexico (Hemiptera: Miridae: Isometopinae). Proceedings of the Entomological Society of Washington 81:82-96.

Henry, T.J. & Kim, K.C. 1984. Genus *Neurocolpus* Reuter (Heteroptera: Miridae): taxonomy, economic implications, hosts, and phylogenetic review. Transactions of the American Entomological Society 110:1-75.

Henry, T.J. & Wheeler Jr., A.G. 1986. *Melanorhopala froeschneri* (Heteroptera: Tingidae): a new lace bug from eastern United States, with notes on host plant and habits, description of fifth instar, and key to species of the genus. Journal of the New York Entomological Society 94:235-244.

Herring, J.L. 1966. The genus *Orius* of the Western Hemisphere (Hemiptera: Anthocoridae). Annals of the Entomological Society of America 59:1094-1109.

Herring, J.L. 1980. A review of the cactus bugs of the genus *Chelinidea* with the description of a new species (Hemiptera: Coreidae). Proceedings of the Entomological Society of Washington 82:237-251.

Hille Ris Lambers, D. 1966. Notes on California aphids with descriptions of new genera and new species (Homoptera: Aphididae). Hilgardia 37(15):569-623.

Hille Ris Lambers, D. 1970. A study of *Tetraneura* Hartig, 1841 (Homoptera, Aphididae), with descriptions of a new subgenus and new species. Bolletino zoologica agraria e bachicoltura (ser. II) 9:21-101

Hille Ris Lambers, D. & Hottes, F.C. 1962. Notes on aphids from Alaska. Entomologische Berichten 22:112-120

Hodgson, C. J. 1994. The Scale Insect Family Coccidae: An identification manual to genera. CAB International. Wallingford. vi+639 pp.

Hodkinson, I.D. 1988. The Nearctic Psylloidea (Insecta: Homoptera): an annotated check list. Journal of Natural History 22:1179-1243

Hoebeke, E.R. & Wheeler Jr., A.G. 1982. *Rhopalus* (*Brachycarenus*) *tigrinus*, recently established in North America, with a key to the genera and species of Rhopalidae in eastern North America (Hemiptera: Heteroptera). Proceedings of the Entomological Society of Washington 84:213-224.

Hoy, J.M. 1963. A catalogue of the Eriococcideae (Homoptera: Coccoidea) of the World. New Zealand Department of Scientific and Industrial Research Bulletin 150. 260 pp.

Hungerford, H.B. 1933. The genus *Notonecta* of the World. University of Kansas Science Bulletin 21:5-195.

Hungerford, H.B. 1948. The Corixidae of the Western Hemisphere (Hemiptera). University of Kansas Science Bulletin 32:5-827.

Hungerford, H.B. 1954. The genus *Rheumatobates* Bergroth (Hemiptera-Gerridae). University of Kansas Science Bulletin 36:529-588.

Hungerford, H.B. & Evans, N.E. 1934. The Hydrometridae of the Hungarian National Museum and other studies in the family (Hemiptera). Annales Musei Nationalis Hungarici 28:31-112.

Hussey, R.E. 1953. Concerning some North American Coreidae (Hemiptera). Bulletin of the Brooklyn Entomological Society 48:29-34.

Ilharco, F.A. & van Harten, A. 1987. Systematics pp. 51-77 **in** Minks, A.K. & Harrwwijn, P. [editors]. World Crop Pests, volume 2A. Aphids. Their biology, natural enemies and control. Elsevier. Amsterdam. xx+450 pp

Jansson, A. 1972. Systematic notes and new synonymy in the genus *Cenocorixa* (Hemiptera: Corixidae). The Canadian Entomologist 104:449-459.

Jensen, A.S. 1998. Redescription of *Macrosiphum impatientis* (Williams), another rose aphid from eastern North America, with key to related species (Hemiptera: Aphididae). Proceedings of the Entomological Society of Washington. 100: 32-42.

Jensen, A.S. & Stoetzel, M. 1999. An examination of the North American aphid species currently placed in *Ovatus* van der Goot (Hemiptera: Aphididae) with the description of a new genus. Proceedings of the Entomological Society of Washington 101: 39-56

Journet, A.R.P & Vickery, V.R. 1979. Studies on Nearctic *Craspedolepta* Enderlein 1921 (Homoptera: Psylloidea): taxonomic revision. Lyman Entomological Museum and Research Laboratory Memoir No. 7. 164 pp.

Kelton, L.A. 1961. Synopsis of the Nearctic species of *Stenodema* Laporte, and description of a new species from Western Canada (Hemiptera: Miridae). The Canadian Entomologist 93:450-455.

Kelton, L.A. 1963. Synopsis of the genus *Orius* Wolff in America north of Mexico (Heteroptera: Anthocoridae). The Canadian Entomologist 95:631-636.

Kelton, L.A. 1964. Revision of the genus *Reuteroscopus* Kirkaldy 1905 with descriptions of eleven new species (Hemiptera: Miridae). The Canadian Entomologist 96:1421-1433.

Kelton, L.A. 1965a. *Diaphnidia* Uhler and *Diaphnocoris* Kelton in North America (Hemiptera: Miridae). The Canadian Entomologist 97:1025-1030.

Kelton, L.A. 1965b. *Chlamydatus* Curtis in North America (Hemiptera: Miridae). The Canadian Entomologist 97:1132-1140.

Kelton, L.A. 1966a. Synopsis of the genus *Tetraphleps* Fieber in North America (Hemiptera: Anthocoridae). The Canadian Entomologists 98:199-204.

Kelton, L.A. 1966b. Review of the species of *Teratocoris* Fieber, with description of a new species from the Nearctic region (Hemiptera: Miridae). The Canadian Entomologist 98:1265-1271.

Kelton, L.A. 1967. Synopsis of the genus *Lyctocoris* in North America and description of a new species from Quebec (Heteroptera: Anthocoridae). The Canadian Entomologist 99:807-814.

Kelton, L.A. 1968. Revision of the North American species of *Slaterocoris* (Heteroptera: Miridae). The Canadian Entomologists 100:1121-1137.

Kelton, L.A. 1971a. Revision of the species of *Trigonotylus* in North America (Heteroptera: Miridae). The Canadian Entomologist 103:685-705.

Kelton, L.A. 1971b. Review of *Lygocoris* species found in Canada and Alaska (Heteroptera: Miridae) Memoirs of the Entomological Society of Canada 83:87 pp.

Kelton, L.A. 1972a. Four new species of *Bolteria*, with a key to North American species and a note on the species found in Canada (Heteroptera: Miridae). The Canadian Entomologist 104:627-640.

Kelton, L.A. 1972b. Species of *Dichrooscytus* found in Canada, with descriptions of four new species (Heteroptera: Miridae). The Canadian Entomologist 104:1033-1049.

Kelton, L.A. 1973. A new species of *Semium* (Heteroptera: Miridae) from Mexico with new records on distribution for two other species. The Canadian Entomologist 105:1583-1584.

Kelton, L.A. 1976. The genus *Elatophilus* Reuter in North America with descriptions of two new species (Heteroptera: Anthocoridae). The Canadian Entomologist 108:631-634.

Kelton, L.A. 1977. Species of the genus *Pinalitus* Kelton found in North America (Heteroptera: Miridae). The Canadian Entomologist 109:1549-1554.

Kelton, L.A. 1978. The Anthocoridae of Canada and Alaska. Heteroptera: Anthocoridae. The Insects and Arachnids of Canada. Part 4. Agriculture Canada Research Branch Publication 1639:101 pp.

Kelton, L.A. 1979a. *Labopidea* Uhler in North America, with descriptions of a new species and a new genus (Heteroptera: Miridae). The Canadian Entomologist 111:753-758.

Kelton, L.A. 1979b. A new genus *Brooksetta*, near *Ilnacora* Reuter, with new synonymy and new combinations for 15 species currently placed in *Melanotrichus* Reuter (Heteroptera: Miridae). The Canadian Entomologist 111:949-954.

Kelton, L.A. 1980. The plant bugs of the Prairie Provinces of Canada. The Insects and Arachnids of Canada. Part 8. Agriculture Canada Research Branch Publication 1703:408 pp.

Kelton, L.A. 1985. Species of the genus *Fulvius* Stål found in Canada (Heteroptera: Miridae: Cylapinae). The Canadian Entomologist 117:1071-1073.

Kelton, L.A. & Herring, J.L. 1978. Two new species of *Neoborella* Knight (Heteroptera: Miridae) found on dwarf mistletoe, *Arceuthobium* spp. The Canadian Entomologist 110:779-780.

Kelton, L.A. & Knight, H.H. 1959. A new species of *Paradacerla* from Mexico, and synopsis of the genus in North America (Hemiptera: Miridae). The Canadian Entomologist 91:122-126.

Kelton, L.A. & Knight, H.H. 1962. *Mecomma* Fieber in North America (Hemiptera: Miridae). The Canadian Entomologist 94:1296-1302.

Kelton, L.A. & Knight, H.H. 1970. Revision of the genus *Platylygus*, with descriptions of 26 new species (Hemiptera: Miridae). The Canadian Entomologist 102:1429-1460.

Kiritshenko, A.N. & Scudder, G.G.E. 1973. Some new genera and species of Rhyparochrominae (Hemiptera: Lygaeidae) from the Soviet Union, with a key to the genera of Gonianotini. Journal of Natural History 7:133-151.

Kittle, P.D. 1977. A revision of the genus *Trepobates* Uhler (Hemiptera: Gerridae). Ph.D. thesis, University of Arkansas.

Knight, H.H. 1921. Monograph of the North American species of *Deraeocoris* (Heteroptera: Miridae). University of Minnesota Agricultural Experimental Station Technical Bulletin 1:75-210.

Knight, H.H. 1926. A key to the North American species of *Macrolophus* with descriptions of two new species (Hemiptera: Miridae). Entomological News 37:313-316.

Knight, H.H. 1928. Key to the species of *Hadronema* Uhler with descriptions of five new species (Hemiptera: Miridae). The Canadian Entomologist 60:177-182.

Knight, H.H. 1930. A new key to *Paracalocoris* with descriptions of eight new species (Heteroptera, Miridae). Annals of the Entomological Society of America 23:810-827.

Knight, H.H. 1941. The plant bugs, or Miridae, of Illinois. Illinois Natural History Survey Bulletin 22:1-234.

Knight, H.H. 1963. Review of the genus *Ilnacora* Reuter with descriptions of ten new species (Heteroptera, Miridae). Iowa State Journal of Science 38:161-178.

Knight, H.H. 1965. A new key to species of *Reuteroscopus* Kirk. with descriptions of new species (Heteroptera, Miridae). Iowa State Journal of Science 40:101-120.

Knight, H.H. 1968a. Review of genus *Orectoderus* Uhler with a key to the species (Heteroptera, Miridae). Iowa State Journal of Science 42:311-318.

Knight, H.H. 1968b. Taxonomic review: Miridae of the Nevada test site and the western United States. Brigham Young University Science Bulletin (Biology Series) 9(3):1-282.

Knight, H.H. 1970. Review of the genus *Slaterocoris* Wagner, with a key and description of new species (Heteroptera, Miridae). Iowa State Journal of Science 45:233-267.

Knight, H.H. & Schaffner, J.C. 1976. New and old species of the genus *Ilnacora* Reuter (Heteroptera, Miridae). Iowa State Journal of Research 50:399-407.

Kopp, D.D. & T.R. Yonke. 1973a. The treehoppers of Missouri: part 1. Subfamilies Centrotinae, Hoplophorioninae, and Membracinae (Homoptera: Membracidae). Journal of the Kansas Entomological Society 46: 42-64.

Kopp, D.D. & T.R. Yonke. 1973b. The treehoppers of Missouri: part 2. Subfamily Smiliinae; tribes Acutalini, Cerisini, and Polyglyptini (Homoptera: Membracidae). Journal of the Kansas Entomological Society 46: 233-276.

Kopp, D.D. & T.R. Yonke. 1973c. The treehoppers of Missouri: part 3. Subfamily Smiliinae; tribe Smiliini (Homoptera: Membracidae). Journal of the Kansas Entomological Society 46: 375-421

Kopp, D.D. & T.R. Yonke. 1974. The treehoppers of Missouri: part 4. Subfamily Smiliinae; tribe Telamonini (Homoptera: Membracidae). Journal of the Kansas Entomological Society 47: 80-130.

Kormilev, N.A. 1954. Notas sobre Coreidae Neotropicales II. (Hemiptera) Merocorinae de la Argentina y paises limitrofes. Revista Ecuatoriana de Entomologia y Parasitologia 2:153-186.

Kormilev, N.A. 1962. Revision of Phymatinae (Hemiptera, Phymatidae). Philippine Journal of Science 89:287-486.

Kormilev, N.A. & Froeschner, R.C. 1987. Flat bugs of the world. A synonymic list (Heteroptera: Aradidae). Entomography 5:1-246.

Kosztarab, M. 1996. Scale insects of northeastern North America: Identification, biology and distribution. Virginia Museum of Natural History Special Publication 3. viii+650 pp.

Kosztarab, M. & Kosztarab M.P. 1988. A selected bibliography of the Coccoidea. Third supplement. Virginia Polytechnic Institute and State University, Agricultural Experiment Station Bulletin #88-1. 252 pp.

Kozár, F., Humble, L.M., Foottit, R.G. & Otvos I.S. 1989. New and little known scale insects (Homoptera: Coccoidea) from British Columbia. Journal of the Entomological Society of British Columbia 86:70-77

Kritsky, G. 1978. The North American and Caribbean species of *Systelloderus* (Hemiptera: Enicocephalidae). Entomological News 89:65-73.

La Rivers, I. 1948. A new species of *Pelocoris* from Nevada, with notes on the genus in the United States (Hemiptera: Naucoridae). Annals of the Entomological Society of America 41:371-376.

Larivière, M.-C. 1992. Description of *Parabrochymena*, new genus, and redefinition and review of *Brochymena* Amyot and Audinet-Serville (Hemiptera: Pentatomidae) with considerations of natural history, chronological affinities, and evolutionary relationships. Memoirs of the Entomological Society of Canada 163:75 pp.

Larivière, M.-C. 1994a. Biodiversity of *Nabicula* Kirby species (Hemiptera: Nabidae) in Canada: faunistic review, bioecology, biogeography. The Canadian Entomologist 126:327-378.

Larivière, M.-C. 1994b. *Parabrochymena* Larivière (Hemiptera: Pentatomidae): systematics, natural history, chronological affinities, and evolutionary relationships, with a biogeographic analysis of *Parabrochymena* and *Brochymena* Amyot and Audinet-Serville. The Canadian Entomologist 126:1193-1250.

Larochelle, A. 1983. Les Punaises Terrestres (Hétéroptères:Géocorises) du Québec. Fabreries, Supplement 3:1-513.

Lattin, J.D. 1964. The Scutellerinae of America north of Mexico (Hemiptera: Heteroptera: Pentatomidae). Ph.D. thesis, University of California, Berkeley, 350 pp.

Lattin, J.D. & Stanton, N.L. 1993. A review of the genus *Melanocoris* Champion with remarks on distribution and host tree associations (Hemiptera: Heteroptera: Anthocoridae). Journal of the New York Entomological Society 101:95-107.

Lauck, D.R. & Menke, A.S. 1961. The higher classification of the Belostomatidae. Annals of the Entomological Society of America 54:644-657.

Lent, H. & Wygodzinsky, P. 1947. Contribuiçao ao conhecimento do 'Reduviinae' Americanos; (Reduviidae, Hemiptera). Revista Brasileira de Biologia 7:341-368.

Leonard, D.E. 1968. A revision of the genus *Blissus* (Heteroptera: Lygaeidae) in eastern North America. Annals of the Entomological Society of America 61:239-250.

Lindberg, H. 1958. Hemiptera Heteroptera from Newfoundland, collected by the Swedish-Finnish Expedition of 1949 and 1951. Acta Zoologica Fennica 96:1-25.

Liu Tong-Xian, Kosztarab, M. & Rhoades, M. 1989. Biosystematics of the adult females of the genus *Chionaspis* (Homoptera: Coccoidea: Diaspididae) of North America, with emphasis on polymorphism. Studies on the morphology and systematics of scale Insects 15. Virginia Polytechnic Institute and State University Agriculture Experiment Station Bulletin 88-2:1-126

MacGillivray, M.E. 1952. Some aphids of the genus *Aphis* from the Maritime Provinces of Canada. The Canadian Entomologist 84:74-76

MacGillivray, M.E. 1955. Further aphids of the Maritime Provinces (Homoptera: Aphidae). The Canadian Entomologist 87:331-337

MacGillivray, M.E. 1958. A study of the genus *Masonaphis* Hille Ris Lambers, 1939 (Homotera: Aphididae). Temminckia 10:1-131

Maldonado-Capriles, J. 1990. Systematic Catalogue of the Reduviidae of the World (Insecta: Heteroptera). Caribbean Journal of Science, (special edition) 694 pp.

Martin, J.H. 1987. An identification guide to common whitefly pest species of the world (Homoptera, Aleyrodidae). Tropical Pest Management 33:298-322

Matsuda, R. 1977. The Aradidae of Canada. Hemiptera: Aradidae. The Insects and Arachnids of Canada. Part 3. Research Branch Canada Department of Agriculture Publication 1634:116 pp.

McAtee, W.L. 1914. Key to the Nearctic genera and species of Geocorinae (Heteroptera: Lygaeidae) Proceedings of the Biological Society of Washington 27:125-136.

McAtee, W.L. & Malloch, J.R. 1925. Revision of bugs of the family Cryptostemmatidae in the collection of the United States National Museum. Proceedings of the United States National Museum 67(13):1-42.

McAtee, W.L. & Malloch, J.R. 1933. Revision of the subfamily Thyreocorinae of the Pentatomidae (Hemiptera-Heteroptera). Annals of the Carnegie Museum 21:191-411.

McDonald, F.J.D. 1975. Revision of the genus *Holcostethus* in North America (Hemiptera:Pentatomidae). Journal of the New York Entomological Society 82(1974):245-258.

McDonald, F.J.D. 1976. Revision of the genus *Trichopepla* (Hemiptera: Pentatomidae) in North America. Journal of the New York Entomological Society 84:9-22

McDonald, F.J.D. 1982. Description of the male genitalia of *Holcostethus hirtus* (Van Duzee) with a revised key to North American species (Hemiptera: Pentatomidae). Journal of the New York Entomological Society 90:5-7.

McDonald, F.J.D. 1986. Revision of *Cosmopepla* Stål (Hemiptera: Pentatomidae). Journal of the New York Entomological Society 94:1-15.

McKinnon, C.N. & Polhemus, J.T. 1986. Notes on the genus *Ioscytus* with the description of a new species and key to species (Hemiptera; Heteroptera: Saldidae). Journal of the New York Entomological Society 94:431-441.

McPherson, J.E. 1982. The Pentatomoidea (Hemiptera) of northeastern North America, with emphasis on the fauna of Illinois. Southern Illinois University Press, Carbondale and Edwardsville, 240 pp.

McPherson, J.E., Taylor, S.J. & Keffer, S.L. 1992. Evaluation of characters to distinguish *Fitchia aptera* and *F. spinosula* (Heteroptera: Reduviidae). Florida Entomologist 75:222-230.

Menke, A.S. 1958. A synopsis of the genus *Belostoma* Latreille, of America north of Mexico, with the description of a new species (Hemiptera: Belostomatidae). Bulletin of the Southern California Academy of Sciences 57:154-174.

Menke, A.S. 1963. A review of the genus *Lethocerus* in North and Central America, including the West Indies (Hemiptera: Belostomatidae). Annals of the Entomological Society of America 56:261-267.

Metcalf, Z.P. 1923. The Fulgoroidea of eastern North America. Journal of the Elisha Mitchell Society 38:139-230 + 70 plates.

Miller, D.R. 1992. Systematic analysis of *Acanthococcus* species (Homoptera: Coccoidea: Eriococcidae) in the western United States. Transactions of the American Entomological Society 118:1-106

Miller, D.R. & McKenzie, H.L. 1970. Reivew of the mealybug genus *Heterococcus* (Homoptera: Coccoidea: Pseudococcidae) with a description of a new species. Annals of the Entomological Society of America 63:438-453

Miller, D.R. & Miller, G.L. 1993. Eriococcidae of the eastern United States (Homoptera). Contributions of the American Entomological Institute 27(1). 91 pp.

Moore, G.A. 1950. Check-list of Hemiptera of the Province of Quebec. Contributions de l'Institute biologique de l'Université de Montréal 26:49 pp.

Moore, G.A. 1955. A new species of *Agnocoris* from Illinois, and a synopsis of the genus in North America. Proceedings of the Entomological Society of Washington. 57:175-180.

Moore, G.A. 1956. *Agnocoris rubicundus* in North America (Hemiptera, Miridae). Journal of the Kansas Entomological Society 29:37-39.

Morrison, H. & Morrison, E.R. 1965. A selected bibliography of the Coccoidea. First supplement. USDA-ARS Miscellaneous Publication #987. 44 pp.

Morrison, H. & Renk, A.V. 1957. A selected bibliography of the Coccoidea. USDA-ARS Miscellaneous Publication #734. 222 pp.

Moulds, M.S. 1990. Australian cidadas. New South Wales University, Kensington. 217 pp.

Mound, L.A. & Halsey, S.H. 1978. Whitefly of the world. A systematic catalogue of the Aleyrodidae (Homoptera) with host plant and natural enemy data. British Museum (Natural History) and John Wiley & Sons. Chichester. 340 pp.

Nelson, G.H. 1955. A revision of the genus *Dendrocoris* and its generic relationships (Hemiptera, Pentatomidae). Proceedings of the Entomological Society of Washington 57:49-66.

Nielson, M.W. 1968. The leafhopper vectors of phytopathogenic viruses (Homoptera, Cicadellidae): taxonomy, bioloigy, and virus transmission. USDA-Agricultural Research Service Technical Bulletin 1382.

Oman, P.W. 1949. The Nearctic leafhoppers. A generic classification and check list. Memoirs of the Entomological Society of Washington. 3. 253 pp.

Ossiannilsson, F. 1981. The Auchenorrhyncha (Homoptera) of Fennoscandia and Denmark Part 2; the families Cicadidae, Cercopidae, Membracidae, and Cicadellidae (excl. Deltocephalinae). Fauna entomologica scandinavica 7 (2).

Ossiannilsson, F. 1983. The Auchenorrhyncha (Homoptera) of Fennoscandia and Denmark Part 3; the family Cicadellidae: Deltocephalinae, catalogue, literature and index. Fauna entomologica scandinavica 7 (3).

Osuna, E. 1984. Monografia de la tribu Anisoscelidini (Hemiptera; Heteroptera, Coreidae). 1. Revisión Genérica. Boletín de Entomología Venezolana 3:77-148.

O'Donnell, J.E. 1986. Systematics of Western Hemisphere Lygaeini (Insecta: Hemiptera: Lygaeidae). Ph.D. thesis, University of Connecticut, 253 pp.

O'Shea, R. 1980. A generic revision of the Acanthocerini (Hemiptera: Coreidae: Coreinae). Studies on Neotropical Fauna and Environment 15:57-80.

Packauskas, R. J. 1994. Key to the subfamilies and tribes of the New World Coreidae (Hemiptera), with a checklist of published keys to genera and species. Proceedings of the Entomological Society of Washington 96:44-53.

Palmer, M.A. 1952. Aphids of the Rocky Mountain Region. Thomas Say Foundation Vol. 5. A.B. Hirschfeld Press. Denver, CO. 452 pp.

Parshley, H.M. 1923. Records of Nova Scotian Hemiptera-Heteroptera. Proceedings of the Acadian Entomological Society 1922:102-108.

Pelletier, Y. & Boiteau, G. 1986. Aphids of the Maritime Provinces of Canada: The Dr. Marjorie Ellen MacGillivray collection. Agriculture Canada Research Branch Technical Bulletin 1986-11E

Pepper, J.O. & Tissot, A.N. 1973. Pine-feeding species of *Cinara* in the Eastern United States (Homoptera: Aphididae). Florida Agricultural Experiment Stations Monograph Series 3. 159 pp. + 6 pl.

Perez Guerra, G. & Kosztarab, M. 1992. Biosystematics of the family Dactylopiidae (Homoptera) with empahasis on the life cycle of *Dactylopius coccus* Costa. Studies on the morphology and systematics of scale Insects 16. Virginia Polytechnic Institute and State University Agricultural Experiment Station Bulletin 92-11. vi+90 pp.

Picchi, V.D. 1977. A systematic review of the genus *Aneurus* of North and Middle America and the West Indies (Hemiptera: Aradidae). Quaestiones Entomological 13:255-308.

Polhemus, J.T. 1985. Shore bugs (Heteroptera, Hemiptera; Saldidae). A world overview and taxonomy of Middle American forms. The Different Drummer, Englewood, Co. 252 pp.

Poppius, B. 1909. Beiträge zur Kenntuis der Anthocoriden. Acta Societatis Scientiarum Fennicae 37(9):1-43.

Quednau, F.W. 1966. A list of aphids from Quebec with descriptions of two new species (Homoptera: Apidoidea). The Canadian Entomologist 98:415-430

Quednau, F.W. & Remaudière, G. 1994. Le genre *Myzocallis* Passerini, 1860: Classification mondiale des sous-genres et nouvelles espèces Paléarctiques (Homoptera: Aphididae). The Canadian Entomologist 126:303-326

Razafimahatratra, V. 1981. 1981. A revision of the genus *Deraeocoris* Kirschbaum (Heteroptera: Miridae) from western America north of Mexico. Ph.D. thesis, Oregon State University, 235 pp.

Readio, J. & Sweet, M.H. 1982. A review of the Geocorinae of the United States east of the 100th meridian (Hemiptera: Lygaeidae). Miscellaneous Publications of the Entomological Society of America 12(2):1-91.

Remaudière, G. 1983. Pucerons nouveaux et peu connus de Mexique. 5e note: un *Utamphorophora* nouveau (Hom. Aphididae). Annales de la Société entomologique de France (N.S.) 19: 227-233

Remaudière, G. & Muños Viveros, A.L. 1985. Pucerons nouveaux et peu connus de Mexique. 6e note: biologie et taxonomie du genre *Muscaphis* et description de *M. mexicana* n.sp. (Hom. Aphididae). Annales de la Société entomologique de France (N.S.) 21:433-447

Remaudière, G. & Muños Viveros, A.L. 1993. Révision du genre *Carolinaia* et description de nouveaux taxa (Homoptera, Aphididae). Insecta Mundi 6:43-58

Remaudière, G. & Remaudière, M. 1997. Catalogue des Aphididae du monde (Homoptera-Aphidoidea). INRA Éditions, Versaille. 478 pp.

Remaudière, G. & Seco Ferández, M.V. 1990. Claves de Pulgones alados de la región mediterránea. Universidad de Léon. Volumen 1, Introduccion y claves. 110 pp.; Volumen 2, Illustarciones y lista alfabetica de los taxones. 205 pp.

Richards, W.R. 1960. Synopsis of the genus *Rhopalosiphum* in Canada (Homoptera: Aphididae). The Canadian Entomologist 92:1-51

Richards, W.R. 1961. North American *Börnerina* Bramstedt and *Betulaphis* Glendenning (Homoptera: Aphididae). The Canadian Entomologist 93:622-625

Richards, W.R. 1963a. Two new Dactynotine aphids (Homoptera). The Canadian Entomologist 95: 254-257.

Richards, W.R. 1963b. The Myzaphidines of Canada (Homoptera: Aphididae). The Canadian Entomologist 95:680-704

Richards, W.R. 1967a. A review of *Plocamaphis* Oestlund with descriptions of three new species (Homoptera: Aphididae). The Canadian Entomologist 98:835-851

Richards, W.R. 1967b. The *Pterocomma* of Canada and Greenland with notes on the phyletic position of the Pterocommatini (Homoptera: Aphididae). The Canadian Entomologist 99:1015-1040

Richards, W.R. 1968. A revision of the world fauna of *Tuberculatus*, with descriptions of two new species from China (Homoptera: Aphididae). The Canadian Entomologist 100:561-596

Richards, W.R. 1969. *Cepegillettea viridis* a new aphid from Ontario, with a review of the genus (Homoptera: Aphididae). The Canadian Entomologist 101:1121-1162

Richards, W.R. 1971. A synopsis of the world fauna of the Saltusaphidinae, or sedge aphids (Homoptera: Aphididae). Memoirs of the Entomological Society of Canada 80:1-97

Richards, W.R. 1972a. The Chaitophorinae of Canada (Homoptera: Aphididae). Memoirs of the Entomological Society of Canada 87:1-109

Richards, W.R. 1972b. A review of the *Solidago*-inhabiting aphids in Canada with descriptions of three new species (Homoptera: Aphididae). The Canadian Entomologist 104:823-824

Richards, W.R. 1972c. Three new species of *Aulacorthum* from British Columbia, with a key to the Canadian species (Homoptera: Aphididae). The Canadian Entomologist 104:1017-1023

Richards, W.R. 1972d. *Acyrthosiphon pedicularis*, a new aphid from the District of Keewatin, with a key to the Canadian species (Homoptera: Aphididae). The Canadian Entomologist 104:1149-1152

Richards, W.R. 1973. A review of *Phyllaphis*, with a description of a new species from Ontario (Homoptera: Aphididae). The Canadian Entomologist 105:1573-1581

Rider, D.A. 1986. A new species and new synonymy in the genus *Tepa* Rolston and McDonald (Hemiptera: Pentatomidae). Journal of the New York Entomological Society 94:552-558.

Rider, D.A. 1995. Review of the genus *Coenus* Dallas, with the description of *C. explanatus*, new species (Heteroptera: Pentatomidae). Journal of the New York Entomological Society 103:39-47.

Rider, D.A. & Chapin, J.B. 1992. Revision of the genus *Thyanta* Stål, 1862 (Heteroptera: Pentatomidae). II. North America, Central America, and the West Indies. Journal of the New York Entomological Society 100: 42-98.

Robinson, A.G. 1964. *Asiphonaphis* Wilson and Davis in North America (Homoptera: Aphidiae). The Canadian Entomologist 98:1252-1259

Robinson, A.G. 1979. Annotated List of Aphids (Homoptera: Aphididae) of Northwest Canada, Yukon and Alaska. The Manitoba Entomologist 13:23-29

Robinson, A.G. 1980. A key to *Macrosiphum (Sitobion)* spp. (Homoptera: Aphididae) on ferns in North America, with descriptions of four new species. The Canadian Entomologist 98:1252-1259

Robinson, A.G. 1984. A key to the genus *Nearctaphis* (Homoptera: Aphididae) in North America. The Canadian Entomologist 116:289-292

Robinson, A.G. 1985. Annotated list of *Uroleucon (Uroleucon, Uromelan, Satula)* (Homoptera: Aphidiae) of America north of Mexico, with keys and descriptions of new species. The Canadian Entomologist 117:1029-1054

Robinson, A.G. 1986. Annotated list of *Uroleucon* (Lambersius) (Homoptera: Aphidiae) of America north of Mexico, with a key and descriptions of new species. The Canadian Entomologist 118:559-576

Robinson, A.G. 1987. Annotated list of *Macrosiphoniella* (Homoptera: Aphididae) of America north of Mexico, with a key and description of a new species. The Canadian Entomologist 119:913-919

Robinson, A.G. & Chen, Ya-Hwai 1969. Species of *Aphis* L. on *Cornus* species in North America. The Canadian Entomologist 101:521-527

Robinson, A.G. & Halbert, S.E. 1989. Review of *Obtusicauda* (Homoptera: Aphididae) of America north of Mexico, with a key to species. The Canadian Entomologist 121:525-531

Robinson, A.G. & Lamb, R.J. 1991. Aphids of Manitoba. Agriculture Canada Research Branch Technical Bulletin 1991-7E. 29 pp.

Robinson, A.G. & Rojanavongse, V. 1976. A key to the nearctic species of *Aphis* L. (Homoptera: Aphididae) on *Ribes* spp., with descriptions of two new species from Manitoba. The Canadian Entomologist 108:159-164

Rojanavongse, V. & Robinson, A.G. 1976. Nearctic species of *Aphis* L. (Homoptera: Aphididae) on Umbelliferae, with a key to those found in Manitoba. The Canadian Entomologist 108:57-60

Rojanavongse, V. & Robinson, A.G. 1977. Species of *Aphis* (Homoptera: Aphididae) in Manitoba, with a key, and descriptions of new species. The Canadian Entomologist 109:649-661

Rolston, L.H. 1973a. The genus *Menecles* Stål (Hemiptera: Pentatomidae). Journal of the New York Entomological Society 80:234-237.

Rolston, L.H. 1973b. A review of *Hymenarcys* (Hemiptera: Pentatomidae). Journal of the New York Entomological Society 81:111-117.

Rolston, L.H. 1978. A revision of the genus *Mormidea* (Hemiptera: Pentatomidae). Journal of the New York Entomological Society 86:161-219.

Rolston, L.H. 1983. A revision of the genus *Acrosternum* Fieber, subgenus *Chinavia* Orian, in the Western Hemisphere (Hemiptera: Pentatomidae). Journal of the New York Entomological Society 91:97-176.

Rolston, L.H. & McDonald, F.J.D. 1979. Keys to the diagnosis for the families of Western Hemisphere Pentatomoidea, subfamilies of Pentatomidae, and tribes of Pentatominae (Hemiptera). Journal of the New York Entomological Society 87:189-207.

Rolston, L.H. & McDonald, F.J.D. 1981. Conspectus of Pentatomini genera of the Western Hemisphere. Part 2 (Hemiptera: Pentatomidae). Journal of the New York Entomological Society 88:257-272.

Rolston, L.H. & McDonald, F.J.D. 1984. A conspectus of Pentatomini of the Western Hemisphere. Part 3. (Hemiptera: Pentatomidae). Journal of the New York Entomological Society 92:69-86.

Rolston, L.H., McDonald, F.J.D. & Thomas, D.B., Jr. 1980. A conspectus of Pentatomini genera of the Western Hemisphere. Part 1. (Hemiptera: Pentatomidae). Journal of the New York Entomological Society 88:120-132.

Russel, L.M., Kosztarab, M. & Kosztarab M.P. 1974. A selected bibliography of the Coccoidea. Second supplement. USDA-ARS Miscellaneous Publication #1281. 122 pp.

Sailer, R.I. 1948. The genus *Trichocorixa* (Corixidae, Hemiptera), pp. 289-407 (in) Hungerford, H.B. The Corixidae of the Western Hemisphere (Hemiptera). University of Kansas Science Bulletin 32:1-827.

Schaefer, C.W. (editor) 1996. Studies on Hemipteran phylogeny. Thomas Say Publications in Entomology: Proceedings. Entomological Society of America. Lanham MD, USA. 244 pp.

Schell, D.V. 1943. The Ochteridae (Hemiptera) of the Western Hemisphere. Journal of the Kansas Entomological Society 16:29-47.

Schouteden, H. 1905. Heteroptera Fam. Pentatomidae Subfam. Graphosomatinae. Genera Insectorum Fascicule 30:1-46.

Schouteden, H. 1907. Heteroptera Fam. Pentatomidae Subfam. Asopinae (Amyoteinae). Genera Insectorum Fascicule 52:1-82.

Schuh, R.T. 1995. Plant bugs of the World (Insecta: Heteroptera: Miridae). Systematic Catalog, Distributions, Host Lists, and Bibliography. New York Entomological Society. 1329 pp.

Schuh, R.T., Galil, B. & Polhemus, J.T. 1987. Catalog and bibliography of Leptopodomorpha (Heteroptera). Bulletin of the American Museum of Natural History 185(3):243-406.

Schuh, R.T. & Schwartz, M.D. 1988. Revision of the New World Pilophorini (Heteroptera: Miridae: Phylinae). Bulletin of the American Museum of Natural History 187(2):101-201.

Schuh, R.T. & Slater, J.A. 1995. True Bugs of the World (Hemiptera: Heteroptera). Classification and Natural History. Cornell University Press, Ithaca and London. 336 pp.

Schwartz, M.D. 1984. A revision of the black grass bug genus *Irbisia* Reuter (Heteroptera: Miridae). Journal of the New York Entomological Society 92:193-306.

Schwartz, M.D. 1989. A review of the late season Stendomini of southwest North America, and a description of the new genus *Caracoris* from Brazil (Heteroptera: Miridae). American Museum Novitates 2955: 36 pp.

Schwartz, M.D. & Foottit, R.G. 1998. Revision of the Nearctic species of the genus *Lygus* Hahn, with a review of the Palaearctic species (Heteroptera: Miridae). Memoirs on Entomology, International. 10:i-iv, +428 pp.

Schwartz, M.D. & Schuh, R.T. 1999. New genera and species of conifer-inhabiting phyline plant bugs from North America (Heteroptera: Miridae). Journal of the New York Entomological Society 107(2-3):204-237

Schwartz, M.D., Scudder, G.G.E. & Henry, T.F. 1991. The first Nearctic records of two Holarctic species of *Polymerus* Hahn, with remarks on a monophyletic species-group (Heteroptera: Miridae: Mirinae). The Canadian Entomologist 123:721-743.

Scudder, G.G.E. 1961. Some Heteroptera new to British Columbia. Proceedings of the Entomological Society of British Columbia. 58:26-29.

Scudder, G.G.E. 1962a. The World Rhyparochrominae (Hemiptera: Lygaeidae). 1. New synonyms and generic changes. The Canadian Entomologist 94:764-773.

Scudder, G.G.E. 1962b. New Heterogastrinae (Hemiptera) with a key to the genera of the world. Opuscula Entomologica 27:117-127.

Scudder, G.G.E. 1962c. The Ischnorhynchinae of the world (Hemiptera: Lygaeidae). Transactions of the Royal Entomological Society of London 114:163-194.

Scudder, G.G.E. 1971. The Gerridae (Hemiptera) of British Columbia. Journal of the Entomological Society of British Columbia 68:3-10.

Scudder, G.G.E. 1977. An annotated checklist of the aquatic and semiaquatic Hemiptera (Insecta) of British Columbia. Syesis 10:31-38.

Scudder, G.G.E. 1979. Hemiptera. *in* Danks, H. V., editor. Canada and its insect fauna. Memoirs of the Entomological Society of Canada 108:329-348

Scudder, G.G.E. 1981. Two new species of Lygaeinae (Hemiptera: Lygaeidae) from Canada. The Canadian Entomologist 113:747-753.

Scudder, G.G.E. 1984. Two new genera of Rhyparochromini (Hemiptera: Lygaeidae) from North America. The Canadian Entomologist 116:1293-1300.

Scudder, G.G.E. 1986. Additional Heteroptera new to British Columbia. Journal of the Entomological Society of British Columbia. 83:63-65.

Scudder, G.G.E. 1991. The stilt bugs (Heteroptera: Berytidae) of Canada. The Canadian Entomologist 123:425-438.

Scudder, G.G.E. 1997. True Bugs (Heteroptera) of the Yukon. pp. 241-336 (in) Danks, H.V. and Downes, J.A. (Eds.) Insects of the Yukon. Biological Survey of Canada (Terrestrial Arthropods), Ottawa.

Scudder, G.G.E. 1999. The genus *Peritrechus* Fieber (Hemiptera: Rhyparochromidae) in North America. Journal of the New York Entomological Society 107(2-3):268-276.

Scudder, G.G.E. & Jamieson, G.S. 1972. The immature stages of *Gerris* (Hemiptera) in British Columbia. Journal of the Entomological Society of British Columbia 69:72-79.

Scudder, G.G.E. & Thomas Jr., D.B. 1987. The green stink bug genus *Chlorochroa* Stål (Hemiptera: Pentatomidae) in Canada. The Canadian Entomologist 119:83-93.

Sharma, M.L. 1969. Bibliographie des Aphidoidea. Bibliography of Aphidoidea. Librairie de la Cité universitaire. Sherbrooke, Québec. 293 pp.

Sharma, M.L. 1971. Bibliographie des Aphidoidea / Bibliography of Aphidoidea, tome II. Université de Sherbrooke. Sherbrooke, Québec. 221 pp.

Sharma, M.L. 1972. Bibliographie des Aphidoidea / Bibliography of Aphidoidea, tome III. Université de Sherbrooke. Sherbrooke, Québec. 207 pp.

Sharma, M.L. 1977. Bibliographie des Aphidoidea / Bibliography of Aphidoidea, tome IV. Université de Sherbrooke. Sherbrooke, Québec. 214 pp.

Sites, R.W. & Polhemus, J.T. 1994. Nepidae (Hemiptera) of the United States and Canada. Annals of the Entomological Society of America 87:27-42.

Slater, J.A. 1954. Notes on the genus *Labops* Burmeister in North America, with the description of three new species (Hemiptera: Miridae). Bulletin of the Brooklyn Entomological Society 49:57-65, 89-94.

Slater, J.A. 1955. A revision of the subfamily Pachygronthinae of the world (Hemiptera: Lygaeidae). Philippine Journal of Science 84:1-160.

Slater, J.A. 1964. A Catalogue of the Lygaeidae of the World. University of Connecticut, Storrs. 2 vols. 1668 pp.

Slater, J.A. 1979. The systematics, phylogeny, and zoogeography of the Blissinae of the world (Hemiptera, Lygaeidae). Bulletin of the American Museum of Natural History 165:1-180.

Slater, J.A. 1980. Systematic relationships of the Antillocorini of the Western Hemisphere (Hemiptera: Lygaeidae). Systematic Entomology 5:199-226.

Slater, J.A. 1988. The Ozophorini of the western United States and Baja California (Hemiptera: Lygaeidae). Journal of the New York Entomological Society 96:91-109.

Slater, J.A. & Baranowski, R.M. 1978. How to know the true bugs (Hemiptera-Heteroptera). Wm. C. Brown Co., Dubuque, Iowa, 256 pp.

Slater, J.A. & O'Donnell, J.E. 1995. A Catalogue of the Lygaeidae of the World (1960-1994). New York Entomological Society, 410 pp.

Slater, J.A., Sweet, M.H. & Brailovsky, H. 1993. Two new species of *Slaterobius* Harrington with comments on the ecology and distribution of the genus (Hemiptera: Lygaeidae). Proceedings of the Entomological Society of Washington 95:590-602.

Smith, C.F. 1972. Bibliography of the Aphididae of the World. North Carolina Agricultural Experiment Station. Tech. Bul. 216. 717 pp.

Smith, C.F. 1974. Keys to and descriptions of the genera of Pemphigini in North America (Homoptera: Aphididae: Pemphiginae). Bulletin of the North Carolina Agricultural Experiment Station 226:1-17

Smith, C.F. 1985. Pemphiginae in North America. pp. 277-302 *in* Proceedings of the International Aphidological Symposium at Jablonna, 1981. Polsk. Akad. Nauk, Warsaw. 510 pp.

Smith, C.F. 1992. A key to many of the common alate aphids of North Carolina (Aphididae: Homoptera). North Carolina Agricultural Research Service, North Carolina State University, Technical Bulletin 299. 92 pp.

Smith, C.F. & Dillery, D.G. 1968. The genus *Drepanaphis* Del Guercio (Homoptera: Aphididae). Annals of the Entomological Society of America 61:185-204

Smith, C.F. & Knowlton, G.F. 1977. The genus *Rhopalosiphoninus* Baker (Homoptera: Aphididae) in North America. U.S. Department of Agriculture Co-operative Plant Pest Reporter 2:75-80

Smith, C.F. & Parron, C.S. 1978. An Annotated List of Aphididae (Homoptera) of North America. North Carolina Agricultural Experimental Station Technical Bulletin #255. 428 pp.

Smith, C.F. & Robinson, A.G. 1975. The genus *Myzodium* with the description of *M. knowltoni*, new species (Homoptera: Aphididae). Proceedings of the Entomological Society of Washington 77:481-486

Smith, C.L. 1980. A taxonomic revision of the genus *Microvelia* Westwood (Heteroptera: Veliidae) of North America including Mexico. Ph.D. thesis, University of Georgia, 372 pp.

Smith, C.L. & Polhemus, J.T. 1978. The Veliidae (Heteroptera) of America north of Mexico-keys and check list. Proceedings of the Entomological Society of Washington 80:56-68.

Sorensen, J.T. 1994. A revision of the aphid genus *Essigella* (Homoptera: Aphididae: Lachninae): its ecological associations with, and evolution on, Pinaceae hosts. Pan-Pacific Entomologist 70(1):1-102

Sorensen, J.T., Campbell, B.C., Gill, R.J. & Steffen-Campbell, J.D. 1995. Non-monophyly of Auchenorryncha ("Homoptera"), based upon 18s rDNA phylogeny: Eco-evolutionary and cladistic implications within pre-Heteroptodea Hemiptera (s.l.) and a proposal for new monophyletic suborders. Pan-Pacific Entomologist 7:31-60

Southwood, T.R.E. & Leston, D. 1959. Land and Water Bugs of the British Isles. Frederick Warne and Co. Ltd., London and New York, 436 pp.

Spence, J.R. & Scudder, G.G.E. 1978. Larval taxonomy and distribution of *Gerris pingreensis* and *G. incognitus* (Hemiptera: Gerridae). Journal of the Entomological Society of British Columbia 75:41-45.

Stonedahl, G.M. 1988. Revision of the mirine genus *Phytocoris* Fallén (Heteroptera: Miridae) for western North America. Bulletin of the American Museum of Natural History 188(1):1-257.

Stonedahl, G.M. 1990. Revision and cladistic analysis of the Holarctic genus *Atractotomus* Fieber (Heteroptera: Miridae: Phylinae). Bulletin of the American Museum of Natural History 198:1-88.

Stonedahl, G.M. & Schwartz, M.D. 1986. Revision of the plant bug genus *Pseudopsallus* Van Duzee (Heteroptera: Miridae). American Museum Novitates 2842:1-58.

Strickland, E.H. 1953. An annotated list of the Hemiptera (s.l.) of Alberta. The Canadian Entomologist. 85:193-214.

Stroyan, H.L.G. 1969. Notes on some species of *Cavariella* Del Guercio, 1911 (Homoptera: Aphdoidea). Proceedings of the Royal Entomological Society of London (B) 38:7-19 + 2 pl.

Stroyan, H.L.G. 1977. Handbooks for the Identification of British Insects. Vol. II Part 4(a). Homoptera, Aphidoidea (part), Chaitophoridae and Callaphididae. Royal Entomological Society of London, London, 130 pp.

Stroyan, H.L.G. 1984. Handbooks for the Identification of British Insects. Vol. II Part 6. Aphids - Pterocommatinae and Aphidinae (Aphidini) Homoptera, Aphididae. Royal Entomological Society of London, London, 232 pp.

Stål, C. 1868. Bidrag till Hemipternas Systematik. Öfversigt af Kongl. Vetenskaps-Akademiens Förhandlinger, Stockholm 24:491-560.

Sweet, M.H. 1967. The tribal classification of the Rhyparochrominae (Heteroptera: Lygaeidae). Annals of the Entomological Society of American 60:208-226.

Sweet, M.H. 1977. Elevation of the seed bug *Eremocoris borealis* (Dallas) from synonymy with *Eremocoris ferus* (Say) (Hemiptera: Lygaeidae). Entomological News 88:169-176.

Sweet, M.H. 1986. *Ligyrocoris barberi* (Heteroptera: Lygaeidae) a new seedbug from the southeastern United States with a discussion of its ecology, life cycle, and reproductive isolation. Journal of the New York Entomological Society 94:281-290.

Sweet, M.H. 1991. A new genus and new species of Rhyparochrominae (Hemiptera: Lygaeidae) from western North America. Journal of the New York Entomological Society 99:471-477.

Szelegiewicz, H. 1969. Mszyce - Szkodniki roslin wektory choróob wirusowych i producenci spadzi - Bibliographia. Panstwowoe Wydawnictwo Rolnicze Lesne. Warszawa. 250 pp.

Szerlip, S.L. 1980. Biosystematic revision of the genus *Apiomeris* (Hemiptera: Reduviidae) in North and Central America. Ph.D. thesis, University of California, Berkley, CA.

Tishechkin, D. Yu. 1998. Acoustic signals of Issidae (Homoptera, Cicadinea, Fulgoroidea) compared with signals of some other Fulgoroidea with notes on taxonomic status of the subfamily Caliscelinae. Zoologicheskij Zhurnal 77:1257-1265

Thomas, D.B. 1974. The genus *Codophila* Mulsant in North America (Hemiptera: Pentatomidae). Pan-Pacific Entomologist 50:441-442.

Thomas, D.B. 1991. The Acanthosomatidae (Heteroptera) of North America. Pan-Pacific Entomologist 67:159-170.

Thomas, D.B. 1992. Taxonomic synopsis of the Asopini Pentatomidae (Heteroptera) of the Western Hemisphere. Thomas Say Foundation 16:1-156.

Thomas, D.B. & Yonke, T.R. 1981. A review of the Nearctic species of the genus *Banasa* Stål (Hemiptera: Pentatomidae). Journal of the Kansas Entomological Society 54:233-248.

Todd, E.L. 1955. A taxonomic revision of the family Gelastocoridae (Hemiptera). University of Kansas Science Bulletin 37:277-475.

Torre-Bueno, J.R. de la, 1946. A synopsis of the Hemiptera-Heteroptera of America north of Mexico. Part III. Family XI-Lygaeidae. Entomologica Americana (New Series) 26:1-141.

Truxal, F.S. 1953. A revision of the genus *Buenoa* (Hemiptera, Notonectidae). University of Kansas Science Bulletin 35:1351-1523.

Tuthill, L.D. 1943. The Psyllids of America North of Mexico (Psyllidae: Homoptera). Iowa State College Journal of Science 17:443-660

Usinger, R.L. 1930. Two new species of *Vanduzeeina* from California (Scutelleridae, Hemipt.). Pan Pacific Entomologist 6:131-133.

Usinger, R.L. 1936. Studies in the American Aradidae with descriptions of new species (Hemiptera). Annals of the Entomological Society of America 29:490-516.

Usinger, R.L. 1938. Review of the genus *Gastrodes* (Lygaeidae, Hemiptera). Proceedings of the California Academy of Sciences (4) 23:289-301.

Usinger, R.L. 1944. A review classification of the Reduvioidea with a new subfamily from South America (Hemiptera). Annals of the Entomological Society of America 36:602-618.

Usinger, R.L. 1966. Monograph of Cimidicae (Hemiptera-Heteroptera). Thomas Say Foundation 7:1-585.

Usinger, R.L. & Matsuda, R. 1959. Classification of the Aradidae. British Museum, London, 410 pp.

Van Duzee, E.P. 1916. Review of the genus *Macrotylus* Fieb. (Hemiptera). Pomona Journal of Entomology and Zoology 8:5-11.

Vinokurov, N.N. 1977. The systematics and intraspecific variability of capsid bugs of the genus *Capsus* F. (Heteroptera: Miridae). Entomologischeskoye Obozreniye 56:103-115 (English translation in Entomological Review 56:76-85).

Vinokurov, N.N. 1982. Capsid-bugs of the genus *Leptopterna* (Heteroptera: Miridae) in the fauna of the USSR and adjacent territories. Akad. Nauk-USSR 105 (1981):93-115.

Voegtlin, D.J. 1984. Notes on *Hyadaphis foeniculi* and redescription of *Hyadaphis tataricae* (Homoptera: Aphididae). The Great Lakes Entomologist 17:55-67

Voegtlin, D.J. 1995. Notes on the *Mindarus* spp. (Homoptera: Aphididae) of North America with descriptions of two new species. Proceedings of the Entomological Society of Washington 97: 178-196

Voegtlin, D.J. & Bridges, C.A. 1988. Catalog of the *Cinara* species of North America (Homoptera: Aphididae). Special Publications, Illinois Natural History Survey 8:1-55

Walley, G.S. 1929. Two new species of *Eremocoris* with notes and a key to the species of *Peritrechus* (Hemip., Lygaeidae). The Canadian Entomologist 61:41-44.

White, I.M. & Hodkinson, I.D. 1985. Nymphal taxonomy and systematics of the Psylloidea (Homoptera). Bulletin of the British Museum (Natural History) 50(2):153-301

Wygodzinsky, P. & Schmidt, K. 1991. Revision of the New World Enicocephalomorpha (Heteroptera). Bulletin of the American Museum of Natural History 200: 265 pp.

Wygodzinsky, P. & Usinger, R.L. 1964. The genus *Reduvius* Fabricius in western North America (Reduviidae, Hemiptera, Insecta). American Museum Novitates 2175:1-15.

Wygodzinsky, P. 1947. Sôbre duas novas espécies de "Emesinae" do Brasil, com notas sôbre *Stenolemoides arizonensis* (Banks) (Reduviidae, Hemiptera). Revistas Brasileira de Biologia 6: 509-519.

Wygodzinsky, P. 1966. A Monograph of the Emesinae (Reduviidae, Hemiptera) Bulletin of the American Museum of Natural History 133:1-614.

Zheng, L.-Y. & Slater, J.A. 1984. A revision of the lygaeid genus *Pseudopachybrachius* (Hemiptera). Systematic Entomology 9:95-115.

Index to Generic and Subgeneric Names

A

D

E

Index to Specific and Subspecific Names

C

D

H

M

N

P

Q

R

S

T

W

X

Y

Z